Carbon Nanotubes for I

Aida Todri-Sanial • Jean Dijon • Antonio Maffucci

Editors

Carbon Nanotubes for Interconnects

Process, Design and Applications

 Springer

Editors
Aida Todri-Sanial
CNRS-LIRMM/University of Montpellier
Montpellier, France

Jean Dijon
CEA LITEN
Grenoble, France

Antonio Maffucci
University of Cassino and Southern Lazio
Cassino, Italy

ISBN 978-3-319-80642-6   ISBN 978-3-319-29746-0   (eBook)
DOI 10.1007/978-3-319-29746-0

This Springer imprint is published by Springer Nature
The registered company is Springer International Publishing AG Switzerland

To our families!

Preface

Our modern society has gained enormously from novel miniaturized microelectronic products with enhanced functionality at ever-decreasing cost. However, as size goes down, interconnects become major bottlenecks irrespective of the application domain. Current electrical copper (Cu) interconnects will approach their physical limits and may no longer be able to keep pace with a processor's data throughput. Aggressive scaling has aggravated Cu resistivity increase due to electron scattering, and even more severely, it introduced electromigration issues. Mass transport along interfaces and grain boundaries in Cu interconnects is one of the most imminent issues to be addressed for future technology nodes.

Interconnect innovation with novel material such as carbon nanotubes has been the focus of extensive research with the goal of propagating terabits/second at femtoJoule per bit. Carbon nanotubes have sparked a lot of interest because of their desirable properties such as large electron mean free path, mechanical strength, high thermal conductivity, and large current carrying capacity. The advent of carbon nanotube as new material for back-end of line interconnects is a direct result of active research in academia, research laboratories, and industry over the past decade. Today, carbon nanotube integration takes many forms depending on the applications. As of the time of writing, there are already some prototypes driven to characterize the performance, alignment density, ampacity, and energy efficiency of carbon nanotubes.

As a direct outcome of many years of active research, there are many important documentations on carbon nanotubes. However, a book dedicated to carbon nanotube interconnects covering aspects from process, design, and application is lacking. The idea of a book on carbon nanotube interconnects dates back to more than 2 years ago. While, our initial idea was to co-author a book, we soon realized that such endeavor would be challenging given the required multidisciplinary knowledge from material science, nanosciences, physics, modeling, and simulation to circuit analysis. We revisited our plan and decided to edit a book instead with contributions from experts in academia, research laboratories, and industry.

After careful planning, we identified and invited chapter contributions from an impressive lineup of highly qualified scientists. It took a full 1 year for planning, writing, and editing.

This book aims to give an in-depth look into the process, growth, characterization, modeling, and simulation of carbon nanotube interconnects as local and global on-chip interconnects for 2D, 3D, and monolithic 3D integration. The book is organized into two parts. The first part provides an in-depth overview of the process and growth of carbon nanotubes and their electrical and thermal properties. In the second part, each chapter is dedicated to investigate carbon nanotube interconnects for different applications such as signaling interconnects, power delivery network interconnects, through-silicon-via material, and monolithic 3D integration interconnect material. This book is particularly beneficial to researchers, students, and engineers that are already working or beginning to work on carbon nanotube interconnects.

This book would have not been possible without a team of highly qualified and dedicated people. We are particularly grateful to Charles Glaser for initiating this undertaking and for his encouragements. Jessica Lauffer worked along side with us and provided us with the necessary editorial support. Aida Todri-Sanial is grateful to the continued support of her work on carbon nanotubes from the French National Center for Scientific Research (CNRS) and European Commission on H2020 CONNECT project. Jean Dijon is grateful to the continued support from his work on carbon nanotube by the "Commissariat à l'énergie atomique et aux énergies alternatives" (CEA) and by the European Commission which is funding carbon interconnect projects since 2005.

Last but not least, we are extremely thankful to authors who accepted our invitation and contributed chapters to this book. We hope that the readers will find this book useful in their pursuits of carbon nanotube interconnects.

Aida Todri-Sanial dedicates this book to Tom, Arnaud, Pellumbesha, Pandeli, Parid, Arnisa, and David. Jean Dijon dedicates this book to Marie-Lise, Pierre-Luc, Laury-Anne, and his grand-children. Antonio Maffucci dedicates this book to Michela, Lucrezia, Bea, and Minù.

Montpellier, France   Aida Todri-Sanial
Grenoble, France   Jean Dijon
Cassino, Italy   Antonio Maffucci
December 2015

Contents

Contributors

Dominique Baillargeat XLIM UMR CNRS 7252, Université de Limoges/CNRS, Limoges, France

Y.M. Banadaki Division of Electrical and Computer Engineering, Louisiana State University, Baton Rouge, LA, USA

College of Engineering, Southern University, Baton Rouge, LA 70813, USA

Ahmet Ceyhan Intel Corporation, Hillsboro, OR, USA

Wenchao Chen College of Information Science and Electronic Engineering, Zhejiang University, Hangzhou, China

Jean Dijon CEA-LITEN/DTNM, Commissariat à l'énergie atomique et aux énergies alternatives, Grenoble Cedex, France

Ryoichi Ishihara Department of Microelectronics, Delft University of Technology, Delft, The Netherlands

Franz Kreupl TUM, Munich, Germany

X.H. Liu School of Physics, Liaoning University, Liaoning, China

Antonio Maffucci Department of Electrical and Information Engineering, University of Cassino and Southern Lazio, Cassino, Italy

INFN—LNF, Frascati, Italy

Sergey A. Maksimenko Institute for Nuclear Problem, Belarus State University, Minsk, Belarus

Giovanni Miano Department of Electrical Engineering and Information Technology, University of Naples Federico II, Naples, Italy

Subhasish Mitra Stanford University, Stanford, CA, USA

Azad Naeemi Georgia Institute of Technology, Atlanta, GA, USA

Richard N. Salaway Department of Mechanical and Aerospace Engineering, University of Virginia, Charlottesville, VA, USA

Max Marcel Shulaker Stanford University, Stanford, CA, USA

Gregory Ya. Slepyan School of Electrical Engineering, Tel Aviv University, Tel-Aviv, Israel

A. Srivastava Division of Electrical and computer Engineering, Louisiana State University, Baton Rouge, LA, USA

E.B.K. Tay CINTRA CNRS/NTU/THALES, UMI 3288, Research Techno Plaza, Singapore, Singapore

NOVITAS, School of EEE, Nanyang Technological University, Singapore, Singapore

Aida Todri-Sanial CNRS-LIRMM/University of Montpellier, Montpellier, France

Alexey N. Volkov Department of Mechanical Engineering, University of Alabama, Tuscaloosa, AL, USA

Sten Vollebregt Department of Microelectronics, Delft University of Technology, Delft, The Netherlands

Hai Wei Stanford University, Stanford, CA, USA

Bernard K. Wittmaack Department of Materials Science and Engineering, University of Virginia, Charlottesville,VA, USA

H.-S. Philip Wong Stanford University, Stanford, CA, USA

Wen-Yan Yin College of Information Science and Electronic Engineering, Zhejiang University, Hangzhou, China

Wen-Sheng Zhao Key Lab of RF Circuits and Systems of Ministry of Education, Microelectronic CAD Center, Hangzhou Dianzi University, Hangzhou, China

Leonid V. Zhigilei Department of Materials Science and Engineering, University of Virginia, Charlottesville, VA, USA

Part I
Process and Design

Chapter 1
Overview of the Interconnect Problem

Ahmet Ceyhan and Azad Naeemi

1.1 Overview

The exponential growth of the electronics industry has been guided by continued dimensional scaling of silicon–based CMOS technology for over four decades. Numerous companies have pursued smaller and faster transistors for years because miniaturization of transistors has enabled significant improvements in the transistor performance and power; higher transistor density for improved functionality, complexity, and performance of microchips; and reduction in the cost for a single transistor. At the center of these advancements has been Moore's law, which, combined with Dennard's guidelines for classical scaling introduced in 1974 [1], has determined the industry target to double the number of transistors on a microchip approximately every 18–24 months.

In recent years, the semiconductor industry has needed many innovative material and device–structure solutions to overcome significant threats to continued dimensional scaling. During the last decade, limitations to scaling started with the challenges in reducing the thickness of the gate dielectric material. This challenge stemmed from fundamental quantum laws that governed quantum-mechanical tunneling of electrons from the gate to the channel [2, 3]. Even though the gate dielectric scaling stopped for a few technology generations in effort to keep gate leakage current under control, transistor performance improvement was still maintained thanks to the introduction of the revolutionary strained–silicon technology [4, 5]. The gate dielectric scaling problem was eventually resolved by

A. Ceyhan
Intel Corporation, Hillsboro, OR, USA
e-mail: ahmet.ceyhan@intel.com

A. Naeemi (✉)
Georgia Institute of Technology, Atlanta, GA, USA
e-mail: azad@gatech.edu

© Springer International Publishing Switzerland 2017
A. Todri-Sanial et al. (eds.), *Carbon Nanotubes for Interconnects*,
DOI 10.1007/978-3-319-29746-0_1

replacing SiO_2 with a high-κ dielectric material [6], which allowed increasing the physical thickness of the gate oxide to reduce the probability of electron tunneling, while providing a thinner electrical equivalent for better electrostatic control of the channel, and improved transistor performance. Also, the polycrystalline silicon gate was replaced with a metal gate because the poly gate was not compatible with the high-κ material [7]. Finally, the 22-nm technology generation announced the revolutionary departure from planar CMOS by introducing fully depleted tri-gate transistors [8], which utilize the vertical dimension to extend the electrostatic control of the gate to three sides of a fin for improved performance at a smaller supply voltage and reduced short channel effects.

Besides smaller and faster transistors, the semiconductor industry requires fast and dense interconnects to manufacture high-performance microchips. Interconnect performance, however, degrades with dimensional scaling. Resistance increases as the dimensions get smaller and the total capacitance increases due to the high density of interconnects. Therefore, the number of metal layers has gradually increased over the years [9], providing the possibility to route fine-pitch interconnects for high density at some metal layers, and wider and thicker interconnects for improved delay at other metal layers. In the last decade, Aluminum (Al) has been replaced by Copper (Cu) to improve the resistance–capacitance (RC) delay of interconnects because Cu offers increased conductivity compared to Al [10], and has a higher resistance to electromigration [9]. Furthermore, in effort to reduce the capacitance associated with interconnects, which directly determines both the interconnect RC delay and the interconnect dynamic power dissipation, progressively lower-κ dielectric materials have been introduced in many generations of technology [9]. These new materials, new processes, and the increase in the number of metal layers have enabled interconnect scaling for various technology generations. The 22-nm technology node comprises 9 Cu layers with an ultra-low-κ dielectric material [8].

All of these innovative solutions in the last decade have come to reality as a result of enormous investments in research and development. Even though utilizing the vertical dimension in both the device and the chip levels is expected to govern the technological advancements in the near future, the semiconductor industry is expected to continue facing major challenges to continue scaling during the next decade. One of the two major challenges is to extend the use of 193-nm immersion lithography tools to ultra-scaled technology nodes through optimized multiple-patterning and computational-lithography techniques, until extreme ultraviolet (EUV) light lithography, which makes use of light at a wavelength of 13 nm, is ready. This chapter focuses on another major challenge, namely interconnects, which still constitute significant limitations to the performance of microchips despite the aforementioned innovations.

The research pipeline of the semiconductor industry involves increasingly radical potential solutions to carry technology advancement through dimensional scaling to beyond conventional CMOS. Many companies encourage and conduct research on emerging device and interconnect technologies, such as carbon-based devices [11, 12] and interconnects [13, 14], nano-electromechanical systems (NEMS) [15], optical or photonic interconnects [16, 17], and even non-charge-based systems [18], to extend Moore's law to beyond-2020 technology generations. However, any device technology that offers advantages in performance, power dissipation, or ease in dimensional scaling will suffer from the same interconnect challenges that the semiconductor industry faces today. Therefore, the interconnect considerations for future technologies require a comprehensive evaluation. This evaluation starts with the limitations of the existing technology as explained in this chapter.

1.2 The Interconnect Structure Design Challenge

The essential goal for interconnects is to provide communication between two points on a microchip. However, the metrics that govern the intended performance of interconnects depend strongly on their functionality in the chip. Interconnects can carry power/ground signals to the logic gates, data signals between logic gates, or clock signals to the sequential elements in the design. This functionality–performance relationship directly determines the optimal physical parameters associated with the design of the interconnect structure.

Power/ground interconnects are very prone to electromigration [19] because they carry large direct currents from the power source to the transistors. The degree to which electromigration can be destructive to the power wires is a strong function of the current density. Therefore, power wires must be designed such that they can carry the maximum current density in the circuit. The main performance metric for power/ground interconnects is their signal reliability. They must be designed to provide a very low resistance path from the power source to the transistors to keep the drop in the supply voltage (IR drop) at a minimum such that the transistors can operate at the nominal voltage value with a high performance. In addition, the simultaneous switching noise (SSN) must be minimized because it can often lead to the degradation of the integrity of the power signal by causing distortions, which may cause unstable logic gate operation.

Unlike power/ground interconnects, signal interconnects are bidirectional and they carry small alternating currents, which means that they are not prone to electromigration. The main performance metrics for signal interconnects are delay, energy-per-bit, and crosstalk. At the current technology generations and for short signal interconnects, their latency is mostly determined by their capacitance. However, for long distances, the resistance of the interconnect has a significant contribution to the total latency. Similar to the SSN problem, crosstalk noise due to the coupling capacitance between signal interconnects needs to be minimized as it may cause unexpected behavior by slowing down/speeding up the connection or may distort the data signal to the extent that the implemented function is altered. Furthermore, as there are a large number of signal interconnects in a chip, the area they occupy is a significant parameter to optimize.

Interconnects that carry clock signals are more prone to electromigration than signal interconnects due to the large alternating currents that they carry across the chip. The most important metric for clock design is skew and the uncertainty of interconnect delay may have a large contribution to the skew value. Therefore, reducing uncertainty is a significant performance metric for interconnects on the clock network. The clock interconnect network needs to be designed to have a low resistance and capacitance to reduce the total latency and minimize the total power dissipation as clock signals are unique in the sense that their activity factor is 1.

The challenge is to design an interconnect structure such that all of these different requirements are co-optimized since all of these wires are routed together. This is a very strong constraint since it requires comprehensive optimization of the interconnect technology that accounts for the diversity of interconnects in terms of length, functionality, and the type and size of their drivers and receivers. Furthermore, optimizing many of the parameters that are mentioned in this section not only requires challenging innovations in materials, architecture, and circuit design at ultra-scaled technology nodes, but also some of these solutions introduce new trade-offs between each other.

All of the interconnect performance metrics, such as delay, power dissipation, clock uncertainty, crosstalk, bandwidth, power supply reliability, and even area, depend on per unit length resistance, capacitance, and inductance of the interconnect structure. In the next section, the trend in the intrinsic interconnect parameters is discussed in detail.

1.3 Intrinsic Interconnect Parameters

In the current technology, interconnect networks are implemented in copper (Cu) with an inter-layer dielectric material that has a low effective dielectric constant value to reduce the total effective capacitance. Copper/low-κ interconnects are implemented using a damascene process that comprises four major steps. First, the dielectric material is patterned and the trench locations are defined using photomasks and lithography techniques. The next step deposits a barrier material, typically consisting of Tantalum Nitride (TaN) in conjunction with Tantalum (Ta), over the dielectric that covers the trench walls and the bottom of the trench and acts to prevent diffusion of Cu atoms into the dielectric [20]. The remaining volume in the trench is filled with Cu using electrochemical deposition. Last step of the damascene process is to remove excess Cu using chemical-mechanical polishing (CMP).

1.3.1 Interconnect Resistance

The resistivity of the Cu wire is determined by the geometrical dimensions of the trench and the amount of valuable real-estate that is lost to the barrier material. As the dimensions of the Cu wires reduce, size effects such as electron scatterings at wire surfaces and grain boundaries, and line edge roughness (LER) cause the effective resistivity to increase rapidly [21]. The effective resistivity in the presence of LER and grain and sidewall boundary scatterings is given by Eq. (1.1) as [22],

$$\rho_{eff} = \frac{\rho_0}{\sqrt{1-(u/W)^2}} \left(GB_{scat} + 0.45\,(1-p)\,\frac{\lambda}{W}\left(W/T + \left(1-(u/W)^2\right)\right)\right).$$

(1.1)

In this equation, ρ_0 is the resistivity of the bulk material, u the LER, λ the mean free path (MFP) of electrons, W the width of the interconnect, and T the thickness of the interconnect. The specularity parameter, p, is a fit parameter and represents the percentage of electrons that scatter specularly at the sidewalls of the interconnect. Its value can vary between 0 and 1, where a specularity of 1 means a specular reflection at the surface of the conductor for all electrons. This means that the collisions are elastic and the momentum of electrons are conserved in the direction that they are traveling. At the other extreme, $p=0$, all electrons that collide at the surface of the conductor undergo diffusive scattering. The impact of the quantum-mechanical barrier that conducting electrons encounter at the grain boundaries is captured in the GB_{scat} term, which is given as

$$GB_{scat} = \frac{1}{3}\left[\frac{1}{3} - \frac{\alpha}{2} + \alpha^2 - \alpha^3 \ln\left(1 + \frac{1}{\alpha}\right)\right]^{-1},$$

(1.2)

$$\alpha = \frac{\lambda}{d}\frac{R}{1-R}.$$

(1.3)

The average distance that an electron travels between grain boundaries is denoted by d. The reflectivity parameter, R, represents the fraction of electrons that are scattered by the aforementioned quantum-mechanical barrier. It assumes values between 1 and 0 representing complete scattering and complete transmission at the grain boundaries, respectively.

There are many experimental studies in literature that focus on the resistivity of Cu wires at the sub-100-nm geometrical dimensions reporting various values of specularity and reflectivity parameters, which cover a range of Cu resistivity values. Figure 1.1 illustrates the Cu resistivity normalized to the bulk value of 1.8 μΩ cm calculated considering the International Technology Roadmap for Semiconductors (ITRS) projections [23] for the minimum interconnect width, barrier material thickness, and aspect ratio. The distance between the grain boundaries is assumed to be equal to the width of the interconnect and the LER is assumed to be 40 % of the nominal interconnect width. Three different pairs of p and R parameter values are illustrated. The fourth line corresponds to a hypothetical scenario that represents a single-crystal Cu structure. This structure can be manufactured using a subtractive process [24], where, unlike the current dual-damascene process, the grain size is not limited by the height or the width of a trench. The goal is to effectively eliminate the

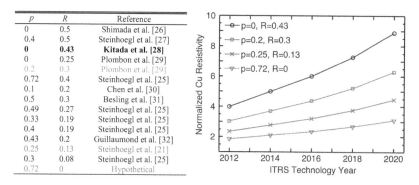

p	R	Reference
0	0.5	Shimada et al. [26]
0.4	0.5	Steinhoegl et al. [27]
0	**0.43**	**Kitada et al. [28]**
0	0.25	Plombon et al. [29]
0.2	0.3	Plombon et al. [29]
0.72	0.4	Steinhoegl et al. [25]
0.1	0.2	Chen et al. [30]
0.5	0.3	Besling et al. [31]
0.49	0.27	Steinhoegl et al. [25]
0.33	0.19	Steinhoegl et al. [25]
0.4	0.19	Steinhoegl et al. [25]
0.43	0.2	Guillaumond et al. [32]
0.25	0.13	Steinhoegl et al. [21]
0.3	0.08	Steinhoegl et al. [25]
0.72	0	Hypothetical

Fig. 1.1 Minimum size Cu wire resistivity normalized to the bulk Cu resistivity, taken as 1.8 μΩ cm. Interconnect parameters such as barrier thickness, aspect ratio, and minimum width are taken from ITRS projections. Mean free path of electrons is taken as 40 nm

impact of grain boundary scatterings on the overall resistivity increase by creating a grain size that is much larger than the MFP of electrons. The specularity parameter for this scenario is assumed to be an optimistic value of 0.72 [25].

In 2020, assuming single-crystal Cu, a typical change in the specularity parameter, $p = 0.4 \rightarrow 0$, would induce only $\sim 1.5 \times$ increase in the resistivity of minimum size wires, which have a width of 12 nm. Assuming fully specular sidewall scatterings, however, the change $R = 0.1 \rightarrow 0.7$ induces $\sim 8 \times$ increase in resistivity. Improving the LER from 40 % to 20 % of the linewidth reduces the resistivity by less than 15 %. Hence, changes in grain boundary scatterings have a greater impact in determining the Cu resistivity over surface scatterings and LER.

1.3.2 Interconnect Capacitance

The per unit length capacitance associated with a wire has three main components, which are the line-to-line capacitance between neighboring wires on the same metal layer, line-to-ground capacitance between any layer and the substrate, and the crossover capacitance between any layer and its neighboring upper and lower orthogonal layers. Ignoring the capacitance to the substrate, the inter-layer and intra-layer components of the interconnect capacitance are illustrated in Fig. 1.2 assuming that metal levels M1 and M3 are grounded and are denoted by C_g and C_m, respectively. The parameters W, T, H, S, and AR represent the interconnect width, height, inter-layer dielectric height, spacing between neighboring wires on the same layer, and the aspect ratio of the wires, respectively.

The neighboring wires on the same metal layer and the different metal layers are separated from each other by dielectric materials whose effective permittivity values directly determine the magnitude of the total interconnect capacitance. The effective

Fig. 1.2 The conventional Cu/low-κ interconnect configuration illustrates the interconnect capacitance components assuming grounded upper and lower layer plates

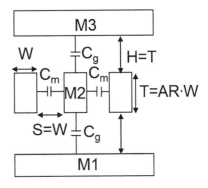

per unit length capacitances of this structure can be calculated by using field-solvers or approximated by compact models [33] as

$$c_m = \epsilon \frac{T}{S} \left(1 - 1.897 e^{\frac{-H}{0.31S} - \frac{-T}{2.474S}} + 1.302 e^{\frac{-H}{0.082S}} - 0.1292 e^{\frac{-T}{1.326S}} \right)$$
$$+ 1.722 \left(1 - 0.6548 e^{\frac{-W}{0.3477H}} \right) e^{\frac{-S}{0.651H}}, \tag{1.4}$$

$$c_g = \epsilon \frac{W}{H} + 1.086 \left(1 + 0.685 e^{\frac{-T}{1.343S}} - 0.9964 e^{\frac{-S}{1.421H}} \right) \left(\frac{S}{S+2H} \right)^{0.0476} \left(\frac{T}{H} \right)^{0.337}. \tag{1.5}$$

The actual interconnect capacitance depends on the switching patterns of the wire and its neighbors. When an interconnect switches, its neighbors may stay quiet, switch in the same direction, or switch in the opposite direction giving rise to a total per unit length interconnect capacitance value of $c = 2c_g + 2c_m$, $c = 2c_g$, and $c = 2c_g + 4c_m$, respectively. The average interconnect capacitance ($c = 2c_g + 2c_m$) is considered for the comparisons that are presented in the following sections.

As the minimum dimensions shrink, the per unit length interconnect capacitance stays almost constant assuming that the aspect ratio and the dielectric material are the same. However, as more and more devices are crammed up per unit area on the chip, the number of interconnects, hence the aggregate interconnect capacitance, increases. As a consequence, the total dynamic power dissipation associated with interconnects increases. Interconnect dynamic power contributes to more than 50 % of the total dynamic power dissipation of the chip and it is a major performance limiter at future technology nodes [34]. Furthermore, since the per unit length resistance increases rapidly at ultra-scaled technology generations, reducing the per unit length capacitance can help significantly in reducing line delay. Lower-κ dielectric materials can reduce this capacitance and reduce latency, power dissipation, and crosstalk issues on the chip. However, there are reliability concerns related to using low-κ dielectric materials, which are discussed later.

1.4   Impact on Interconnect Metrics

This section focuses on the impact of the trend in the aforementioned interconnect parameters on the principal interconnect performance metrics. Arguably the most important performance metrics for interconnects are RC delay and energy-per-bit. The intrinsic latency of an RC-limited interconnect is proportional to

$$\tau_{RC} \sim \rho \in \frac{L^2}{HT},\tag{1.6}$$

where ρ, ϵ, L, T, and H are the interconnect resistivity, dielectric permittivity, interconnect length, thickness, and inter-layer dielectric height, respectively. His- torically, the delay of short local and intermediate interconnects has been much smaller compared to the delay of switches. The length of the short local and longer intermediate level interconnects is scaled with technology scaling. Therefore, the L^2/HT term in Eq. (1.6) stayed constant. Due to the large widths of the wires, the resistivity has remained close to the bulk resistivity of the conducting material. All in all, the intrinsic RC delay of these interconnects remained constant with technology scaling resulting in an increasing delay trend compared to gate delays, but has been much smaller in magnitude than the gate delays. The length of long global interconnects, however, did not scale with technology scaling since they ran across the chip. The chip area, hence the length of the longest global interconnect, usually increased at each new technology node. Therefore, the delay of global interconnects increased with technology scaling making global interconnects the more serious interconnect problem.

The interconnect delay can be reduced by: (1) reducing metal resistivity using new materials, (2) scaling insulator permittivity (ϵ), (3) reducing the interconnect length (L) using novel architectures, and (4) reverse scaling metal height (H) and insulator thickness (T). A variety of solutions have materialized in order to mitigate the global interconnect problem over the years. Some of these include: switching to the Cu/low-κ interconnect technology and increasing the aspect ratio of wires to introduce a lower $\rho\epsilon$ product, using many core architectures and switching to three- dimensional (3-D) integration to reduce the maximum global interconnect length, and reverse scaling.

At the current technology generations, the delay of short local and intermediate interconnects can no longer be determined by just the output resistance of transistors and interconnect capacitance. The resistivity term starts to dominate the intrinsic RC delay equation. Figure 1.3 illustrates that the intrinsic interconnect latency and the energy-delay product (EDP) of a 10-gate-pitch-long interconnect quickly become comparable to or worse than those of Si-CMOS switches. The intrinsic interconnect latency is calculated by $0.4r_{int}c_{int}L^2$ and the energy-per-bit is calculated by $0.5c_{int}LV^2$, where r_{int} and c_{int} are the per unit length resistance and capacitance values calculated by the expressions described in the previous sections using ITRS projected interconnect parameters, L is the interconnect length in nanometers, and

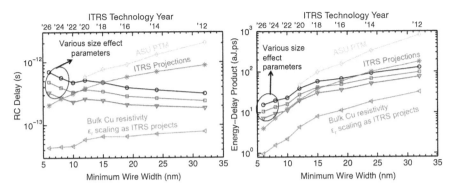

Fig. 1.3 Intrinsic latency (*left*) and EDP (*right*) comparison between a 10-gate-pitch-long interconnect at the minimum pitch metal level assuming various size effect parameters and an inverter cell with 5× the strength of a minimum size inverter cell is plotted at multiple ITRS technology years

V is the supply voltage. Logic transistor density projections from ITRS are used to calculate the gate pitch length for each technology node. The gate pitch length is calculated by $\sqrt{A_{die}/N_{gates}}$, which can be rewritten as $\sqrt{4/D_t}$, where D_t is the logic transistor density, assuming that all gates are 2-input NANDs with four transistors.

The gate delay and EDP projections for devices are calculated from ITRS as well as using the Arizona State University (ASU) predictive technology models (PTMs) for FinFET devices [35], which are based on the industry standard BSIM-CMG model. The PTMs are created for the 20-, 16-, 14-, 10-, and 7-nm technology nodes corresponding to the even years between 2012 and 2020, respectively. The nominal supply voltage values that are used in calibrating the PTMs are unchanged. Based on these models, the characterization of the 5× inverter cell is performed assuming that the cell is loaded by a fanout of three similar cells and running transient simulations on HSPICE. All related parameters are tabulated in Table 1.1 for the 2026 ITRS technology year projections.

To compensate for the increasing RC delay trend, one approach can be to increase the aspect ratio to reduce the per unit length resistance of the conducting material as plotted in the upper left corner of Fig. 1.4 for minimum-width wires in the ITRS technology year 2020. However, increasing the aspect ratio results in having a larger per unit length capacitance as plotted in the upper right corner of Fig. 1.4. As a consequence, the intrinsic RC delay of the interconnect experiences the law of diminishing returns with increasing aspect ratio within the window of interest, which is taken to be between one and five in this example.

To estimate the impact of the change in the aspect ratio on the circuit delay, it is assumed that a 5× minimum size inverter gate drives a load of three similar gates through an interconnect of varied length that is modeled by a distributed RC network as illustrated in Fig. 1.5. The delay of this circuit can be calculated by

$$\tau = 0.7R_{dr}\left(C_{out} + C_{load}\right) + 0.7R_{dr}c_{int}L + 0.7r_{int}LC_{load} + 0.4r_{int}c_{int}L^2, \quad (1.7)$$

Table 1.1 ITRS projected interconnect and device parameter values used to evaluate the *RC* delay and EDP at the 2026 technology year

Parameter	Description	Value
W	Width of the interconnect taken as half metal 1 (M1) pitch	6 nm
V	Supply voltage	0.57 V
$I_{d,sat}$	Saturation current density for a minimum sized nFET	2308 μA/μm
$C_{g,nonideal}$	Nonideal gate capacitance for a minimum sized nFET	0.418 aF/m
ϵ_r	Relative permittivity constant of the dielectric material	1.87
t_b	Thickness of the conformal barrier material	0.5 nm
AR	Aspect ratio (thickness/width) of the interconnect	2.2
D_t	Logic transistor density	32,179 Mtransistors/cm^2
GP	Gate pitch	111.49 nm

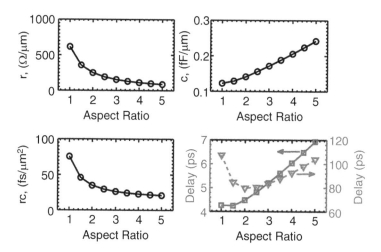

Fig. 1.4 Resistance and capacitance per unit length, intrinsic interconnect *rc* delay per unit length squared, and the total circuit delay assuming short (3 μm, ~10 gate pitches, *solid line*) and longer (45 μm, ~150 gate pitches, *dashed line*) interconnects, respectively, are plotted versus aspect ratio at the 7-nm technology node. Size effect parameters are taken as $p = 0$ and $R = 0.43$

where R_{dr}, C_{out}, C_{load}, c_{int}, r_{int}, and L are the resistance and output capacitance of the driver, load capacitance, interconnect capacitance, and resistance per unit length, and length of the interconnect, respectively. The output capacitance of the driver is assumed to be equal to its input capacitance. As the plot on the bottom right corner of Fig. 1.4 illustrates, there can be an optimal aspect ratio for a given interconnect length. For short signal interconnects that are only about 10-gate-pitch long, the

Fig. 1.5 The schematic and equivalent circuit model of a simple circuit comprising a driver loaded by an interconnect and a receiver

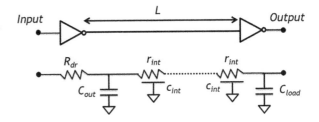

aspect ratio that offers the smallest delay is as low as 1.5, whereas for 150-gate-pitch long interconnects, it is around 2.

Impact of aspect ratio on delay is determined by the relative changes of per unit length values of resistance and capacitance of the wire, and its length. For short lengths, the resistance of the driver dominates the total wire resistance; and increasing capacitance per unit length as a result of increasing the aspect ratio always hurts short wires in terms of delay. For sufficiently long interconnects, for which the total wire resistance is larger than the driver resistance, increasing the aspect ratio can help until the increase in capacitance per unit length becomes dominant over the gain in resistance per unit length. In Fig. 1.4, only minimum-width interconnects, which are routed in the lower metal levels, are considered assuming pessimistic size effect parameters. In an actual interconnect stack, wire widths vary, hence there may be different optimal aspect ratio values for different metal levels. This adds to the complexity of designing an optimal interconnect stack.

1.5 Impact at the Circuit Level

The early electronic equipments comprised only a few dozen components, which could be interconnected using hand-soldering techniques [36]. As the electronic systems became more complex, however, this manufacturing procedure of manually assembling circuits with discrete components quickly became costly, bulky, and unreliable. The exponential increase of the number of interconnections with the increasing number of circuit components was a major limiting factor for making more complex electronic systems. Therefore, the first interconnect problem was directly related to the complexity of routing the large number of interconnects in a system.

As an attempt to simplify this manufacturing process from the interconnect perspective, Danko and Abrahamson announced the Auto-Sembly process in 1949

[37], which would later evolve into the standard printed circuit board fabrication process. The worldwide pursuit of a method to reduce the cost, improve the performance, and reduce the size and weight of electronic equipments gave its fruits in late 1950s with the announcement of the integrated circuit (IC) [36]. Texas Instrument's Jack Kilby came up with a method to integrate a transistor, a capacitor, and a resistor on the same semiconductor material and connected them with soldered wires. Fairchild Semiconductor's Robert Noyce independently formed two transistors with three diffusion regions on a common substrate, using one of the transistors as a pair of diodes; the junctions as capacitors; metal leads over an oxide layer as resistors where required; and planar interconnections [36]. The substantial cost reduction in producing electronic equipment enabled by the transition from interconnecting discrete transistors to integrated circuits has led to tremendous research and development to achieve integration on increasingly larger scales [37]. As mentioned before, the substantial increase in the number of transistors on a chip has given rise to a modified version of the early interconnect problem. To route the tremendous number of wires on a microchip in the same footprint, progressively larger number of metal layers have been implemented.

This multilevel interconnect structure is not only a solution to the routability problem due to the large number of connections, but also a necessity for better performance. Short interconnects that carry signals between transistors that are relatively close to each other, within a certain functional block, are routed at local interconnect levels with fine pitches for high density. As interconnects get longer, they are made wider and thicker to reduce the associated resistance per unit length; hence, delay as explained by Eq. (1.6) that was introduced in the earlier sections.

Considering the complexity of today's modern electronic chips, it is important to build on the material- and circuit-level trend analysis regarding the performance of the interconnect technology presented earlier and evaluate the system-level performance. There can be two complementary approaches to attack this problem. The system behavior can be simulated using compact mathematical models from the material level all the way to the architectural level such that the impact of various technology parameters on the overall performance parameter that is being optimized can be studied and multiple design options can be explored quickly to narrow down the design space for the designer. This approach requires a priori knowledge of the wiring requirements of the system. Alternatively, the real system can be designed for multiple scenarios using the design flow for taping out an actual chip to monitor the changes in the performance parameter and make decisions depending on the outcome. This approach requires a full process design kit and a hardware description language (HDL) description of a circuit. The advantage of the former approach is that it can save significant design time and allow for investigating a broader range of design options in a limited amount of time whereas the advantage of the latter approach lies in providing the real chip information, which increases its accuracy.

1.5.1 Physical Design of Circuit Blocks at Future Technology Generations

In this section, HDL descriptions are used to synthesize, place, route and optimize multiple circuit blocks at future technology generations to illustrate the impact of interconnects on power, performance, and area (PPA) optimizations of future electronic chips. This is a very comprehensive methodology that follows the same design flow used for real chip tape-out and encompasses the diversity of interconnects in terms of length, functionality, and the type and size of drivers and receivers. However, it requires that a full standard cell and interconnect library be built. Conclusions are drawn from timing-closed GDSII-level layouts with detailed routing.

This section focuses on designing simple circuits assuming a predetermined number of metal levels and evaluates the impact of the interconnect performance degradation on the circuit performance in terms of maximum clock frequency, total area, and power dissipation. The predictive libraries that are used in this analysis are created based on the Nangate 45 nm open cell library [38]. First, physical parameters to define device and interconnect layers for each technology node are determined based on the scaling trends projected by the ITRS. These definitions are used to create a library exchange format file (.lef), which has the layout information for the standard cells in each technology generation, and an interconnect technology file (.ict), which has the interconnect structure information. The assumption is that the same set of functions exist in the standard cell library in each technology node. This assumption ignores the possibility that new functions may need to added to the library for PPA optimization for a fair analysis of the interconnect impact. The .lef and .ict files are used to generate a simple capacitance table file (.capTbl) and an elaborate technology file with parasitics information. The design flow uses the simple capacitance table for estimating the performance of the unrouted design and determines the placement of the cells based on these estimates. The parasitic extraction engine uses the more elaborate technology file after detailed routing of the design for better accuracy. Formerly introduced PTMs are used to model transistors at each technology generation and modified *RC*-extracted SPICE netlists for standard cells at each technology node are used to perform library characterization and generate predictive timing and power libraries (.lib and .db).

These predictive libraries are used to synthesize the HDL definitions of multiple circuits with different characteristics in Synopsys Design Compiler [39]. Placement, routing, and optimizations are performed using Cadence Encounter [40]. Timing and power analysis is performed using Synopsys Primetime [41].

Since circuit blocks, instead of a full chip, are designed, the number of metal levels required to route each design is determined based on the worst case scenario simulations for interconnect considerations. As described in the rest of this section, the number of metal levels for the Nangate library is ten, but a maximum of six metal levels have been enough to route all the designs that are investigated.

1.5.1.1 Interconnect and Standard Cell Definitions

The interconnect structure and the layer dimensions are derived from the existing Nangate 45 nm library assuming a scaling factor of roughly 0.7× at each new technology node. The interconnect width, thickness, and the resistivity values normalized to the bulk Cu resistivity for each layer are tabulated in Table 1.2. The resistivity values are calculated using the same set of size effect parameters and the mathematical equations that are outlined in the earlier sections. There are five resistivity scenarios tabulated in order of decreasing resistivity values.

As described earlier, the traditional Cu interconnect manufacturing process utilizes a multilayer barrier layer deposition step before Cu electrochemical deposition to reduce the probability of the metal diffusing into or interacting with the surrounding dielectric material or device regions. This multilayer is required not only for reliability issues, but also because Cu does not have good adhesion to the TaN-based barrier/liner layer, which is resolved by the growth of a metallic Ta layer on top of the TaN barrier. This adhesion layer can allow the Cu seed to nucleate more smoothly. Each of the barrier, adhesion, and seed layers has a finite thickness and occupies useful volume within the trench where the Cu metallization should occupy. Considering that most materials have much higher resistivity than Cu, the barrier material will increase the effective resistance of the interconnect structure. However, at ultra-scaled interconnect dimensions, it is not this high resistivity that is the critical contributor to performance degradation, but the total thickness of the multilayer. This is because the higher the resistance of the barrier material, the more current will flow through the Cu filled fraction of the trench. Due to the increasing Cu resistivity at smaller dimensions as previously explained, this Cu filled fraction must be as large as possible. Four of the five resistivity scenarios, namely ρ_2, ρ_3, ρ_4, and ρ_5, assume ITRS projected values for the barrier thickness. The fifth scenario assumes thicknesses of 3.5, 3, and 2.5 nm at all metal levels of the 22-, 11-, and 7-nm technology generations, respectively. These numbers are projected to estimate the resistivity increase through a slower barrier thickness scaling path than the ITRS projections provided that the Cu ratio for the local metal levels is larger than or equal to 50 %.

To save time in creating standard cell libraries at each new technology node, the predictive standard cell libraries are obtained by scaling the existing 45-nm library data where applicable. For instance, the library exchange format (.lef) file for the original 45 nm library is modified using the dimensional scaling factors to generate .lef files for the predictive technology libraries. In addition, timing/power characterization of the cells is performed based on the *RC*-extracted SPICE netlist files from the Nangate 45 nm library after modifying the files with the appropriate transistor models for each technology and scaling the cell internal parasitic resistance and capacitance values. The scaling factors for the parasitics are calculated considering that the shape of the cells, and the length and width of the internal interconnects are changed by the dimensional scaling factor. The per unit length capacitances within the cell are assumed to remain the same resulting in a total capacitance value that is scaled by the dimensional scaling factor. Therefore,

Table 1.2 Width, thickness, and effective Cu resistivity values normalized to the bulk resistivity of 1.8 μΩ cm for interconnect layers at the 45-, 22-, 11-, and 7-nm technology nodes based on the Nangate open cell library interconnect stack and assuming a dimensional scaling factor of 0.7× at each new technology generation

	Width (nm)	Thickness (nm)	ρ_1 $p=0, R=0.43$	ρ_2 $p=0, R=0.43$	ρ_3 $p=0.2, R=0.3$	ρ_4 $p=0.25, R=0.13$	ρ_5
M1	70	130	–	2.81	2.3	1.94	1.68
	35	65	5.1	4.49	3.44	2.7	2.14
	17.4	32.5	12.98	7.75	5.63	4.13	3.01
	10.8	20.2	29.47	13.29	9.36	6.58	4.52
M2:M3	70	140	–	2.79	2.29	1.93	1.67
	35	70	5.05	4.44	3.41	2.67	2.12
	17.4	35	12.8	7.67	5.56	4.07	2.96
	10.8	21.8	28.97	13.13	9.24	2.35	4.44
M4:M6	140	280	–	1.84	1.62	1.46	1.35
	70	140	2.67	2.53	2.09	1.77	1.54
	35	70	4.73	3.85	2.98	2.35	1.89
	21.8	43.6	7.75	5.75	4.26	3.2	2.41

Row labels for technology nodes (left column): 45 nm, 22 nm, 11 nm, 7 nm (repeated for each of M1, M2:M3, M4:M6).

Table 1.3 Cell delays at various interconnect scenarios calculated at a medium input slew/output load case. Input slew = 18.75 ps (14.06 ps for DFF), output load = 0.64/0.88/1.76/3.2 fF at 45/22/11/7-nm technology nodes, respectively

Cell type	Technology (nm)	Cell delay (ps)				
		ρ_1	ρ_2	ρ_3	ρ_4	ρ_5
INV	45	–	43.55	43.56	43.62	43.61
	22	20.29	20.28	20.25	20.25	20.24
	11	12.54	11.62	10.84	10.78	10.62
	7	13.04	11.6	10.79	9.78	9.06
NAND2	45	–	49.05	49.05	49.17	49.17
	22	24.33	24.32	24.32	24.3	24.29
	11	14.57	14.17	13.74	13.6	13.53
	7	15.55	13.42	12.71	12.39	11.49
DFF	45	–	122.9	122.82	122.76	122.77
	22	48.37	48.26	48.08	47.94	47.55
	11	23.4	22.76	22.62	22.24	21.83
	7	24.34	20.17	19.66	18.94	18.04

the internal cell capacitances in the 32-nm node become roughly 0.7× the original values in the 45-nm node. The cell internal resistance values are more complicated to calculate because their value is a function of the size effect parameters, the cross-sectional dimensions, and barrier material thickness.

Instead of characterizing each cell in the library, the modified *RC*-extracted SPICE netlists for minimum size INV, NAND2, and DFF cells in the new libraries are characterized. The delay characterization results for these cells are tabulated in Table 1.3. The characterization data for the Nangate 45 nm library counterparts of the cells in each technology node are compared with the results obtained for the three standard cells under consideration and scaling factors are calculated for each of their performance parameters. These scaling factor values for the three standard cells are averaged out to calculate final scaling factors for each performance parameter and these final scale factors are used to modify the original 45 nm library Liberty file (.lib).

As demonstrated in Table 1.3, the cell delay highly depends on the interconnect scenario at sub-11-nm technology nodes. Considering a minimum size inverter and comparing the most optimistic and the most pessimistic scenarios for the interconnect resistivity, the cell delay increases by 18.1 and 44 % at the 11- and 7-nm technology nodes, respectively. This moderate change is due to the interconnects within the cell, which are short.

1.5.1.2 Experiment Setup and Results

Having demonstrated the cell-level impact of the interconnect performance degradation with dimensional scaling, this section focuses on full-chip layout experiment results for three different categories of circuit blocks. These three different categories of circuits are represented by an encryption circuit (AES), a low-density parity check (LDPC) circuit, and a Fast Fourier Transform (FFT) circuit. LDPC represents a wire-dominated group of circuits with a very high routing demand. FFT represents circuits with a highly regular layout. Most cells in the FFT circuits that communicate with each other are clustered together and there are a small number of connections between these smaller clusters. The third group of circuits whose regularity lies somewhere between the former two groups are represented by the AES circuit, which is a random logic circuit with a fair amount of routing demand. There are small clusters of cells within the AES circuit similar to FFT, but the communication between these smaller clusters is much higher compared to FFT. Since the routing requirements of these three types of circuits are vastly different, so are the interconnect impact on their performances.

The experiment results in this section are for a setup with the target maximum utilization taken as 85 %. For highly wire-dominated designs, this utilization number is reduced to provide enough routing tracks. For instance, due to the high wiring demand of the LDPC circuit, the initial utilization is lowered to 25 %, which drastically increases the total footprint. Similarly, the number of metal levels for each design is determined based on the wiring demand of the circuit. For fair comparison and evaluation of the interconnect impact, the minimum number of metal levels that ensures routability in the worst case scenario for each circuit is used for each scenario of that circuit. The minimum numbers of metal layers that are required to route FFT, AES and LDPC circuits are four, five and six, respectively. Table 1.4 demonstrates the physical design quality for each technology node and interconnect scenario.

1.5.1.3 Critical Path Delay

For all the designs that are reported in Table 1.4, the minimum clock period value decreases if size effects can be mitigated from scenario ρ_1 towards ρ_5. The relative change in the critical path delay increases as technology scales, especially at the 11-nm technology node and beyond. For the AES circuit, the difference in the circuit speed can be as high as 52 and 98 % at the 11- and 7-nm technology nodes, respectively. These values are 90 and 143 % for LDPC, and 71 and 104 % for the FFT circuit. Therefore, irrespective of the circuit size and type, there is a drastic reduction in circuit speed due to interconnect performance degradation as dimensional scaling continues. Furthermore, the improvement in the intrinsic device speed at each new technology node translates into smaller and smaller returns in the circuit speed due to the impact of the wires. In fact, in all of the circuits that are presented, the circuit speed degrades beyond the 11-nm technology node for severe

Table 1.4 Placement and routing results for all designs for the AES, FFT, and LDPC circuits at multiple technology generations and considering various interconnect scenarios

Circuit	Tech. (nm)	Design	Period (ps)	Iso-performance results							
				Target (ps)	Cell Count	Buffer Count	WNS (ps)	Total Power (mW)	Net Power (mW)	Cell Power (mW)	Leakage Power (mW)
AES	45	ρ_2	714	714	17,559	5121	0	18.35	9.9	8.03	0.422
		ρ_5	710	714	16,907	4818	0	18.05	9.802	7.849	0.403
	22	ρ_1	236	236	20,050	6538	0	20.4	9.703	10.18	0.517
		ρ_2	226	236	19,818	6379	+2	20.31	9.656	10.14	0.515
		ρ_5	216	236	19,818	6354	+7	20.15	9.51	10.13	0.511
	11	ρ_1	164	164	17,651	5725	+3	10.29	4.6	5.394	0.291
		ρ_2	134	164	17,257	5547	+17	9.696	4.507	4.899	0.29
		ρ_3	126	164	17,695	5518	+34	9.634	4.517	4.832	0.285
		ρ_4	118	164	17,381	5219	+36	9.49	4.473	4.737	0.28
		ρ_5	108	164	17,411	5091	+41	9.396	4.46	4.671	0.265
	7	ρ_1	202	202	17,647	5769	+1	6.094	2.114	3.763	0.217
		ρ_2	148	202	15,908	4582	+29	5.362	1.875	3.334	0.153
		ρ_3	120	202	15,604	4537	+24	5.161	1.866	3.145	0.15
		ρ_4	110	202	14,425	3855	+44	5.086	2.04	2.902	0.144
		ρ_5	102	202	12,382	2665	+40	4.457	1.801	2.531	0.125
LDPC	45	ρ_2	1260	1260	78,047	28,442	0	178	124.7	51	2.222
		ρ_5	1100	1260	75,051	26,793	0	167.5	117.7	47.82	2.044
	22	ρ_1	620	620	60,495	22,092	0	88.136	57.65	28.85	1.636
		ρ_2	590	620	59,844	18,658	0	86.097	57.25	27.25	1.597
		ρ_5	500	620	57,129	16,601	+2	81.76	53.82	26.54	1.405
	11	ρ_1	570	570	45,583	8711	0	30.28	19.81	9.67	0.798
		ρ_2	390	570	43,333	6987	+1	28.05	18.59	8.782	0.677
		ρ_5	300	570	40,975	5007	+1	26.48	17.68	8.227	0.576
	7	ρ_1	680	680	50,735	13,744	0	19.19	10.04	8.39	0.752
		ρ_2	470	680	45,111	8699	0	16.96	9.79	6.597	0.567
		ρ_5	280	680	39,106	5178	+2	14.45	7.91	6.1	0.438
FFT	11	ρ_1	480	480	231,865	18,754	+2	154.847	61.98	88.14	4.727
		ρ_2	350	480	230,716	17,716	+7	153.783	61.32	87.76	4.703
		ρ_5	280	480	230,608	17,502	+21	150.999	59.53	86.8	4.669
	7	ρ_1	590	590	236,174	22,881	+2	102.3	33.02	65.41	3.871
		ρ_2	370	590	233,350	20,498	+10	100.19	32.01	64.33	3.849
		ρ_5	240	590	231,457	18,473	+16	98.42	31.5	63.09	3.609

size effect scenarios. Therefore, it is not enough to improve the device intrinsic properties beyond the 11-nm technology node to improve the circuit speed. It is critical to mitigate size effects and find solutions to manufacture thin barrier/liner regions. For instance, the speed of the AES circuit will degrade by 10 % from the 11- to the 7-nm technology node if the interconnect size effects are as severe as ρ_2

for both technology nodes. By mitigating size effects towards ρ_4 during the shift to the 7-nm technology node, there is still a potential that this circuit speed can be improved by 18 % instead.

1.5.1.4 Power Dissipation

To investigate the impact on power dissipation, iso-performance simulations are run for each design at the frequency that each circuit can support for all the experimental setups, which corresponds to the minimum clock period value that is estimated for the simulations in the worst interconnect scenario, namely ρ_1. The switching activity is taken as 0.2 for primary inputs and 0.1 for sequential cell outputs for the calculation of the total power dissipation. The total power dissipation is calculated as the sum of (1) the net switching power, which is the power dissipated in charging the interconnect capacitance and cell pin input capacitances, (2) the cell internal power, which is the power dissipated within each cell including the short circuit power, and (3) the cell leakage power. The percentage contributions of each of these components to the total power dissipation, hence the impact of interconnect performance on total power dissipation, may change depending on the type of the circuit. As demonstrated in Table 1.4, this impact increases with technology scaling for all of the three circuits that are investigated here.

For the AES circuit, the percentage increase in total power is 9.51 and 36.73 % at the 11- and 7-nm technology nodes, respectively, when comparing the results for the most pessimistic and the most optimistic interconnect scenarios. The cell internal power is the main component of power that experiences the largest change as the interconnect scenarios are swept. This is due to both the increase in the number of buffers in the system and the upsizing of some of the gates on the critical paths to meet timing constraints. The number of buffers increases by 12.45 and 116.5 % at the 11- and 7-nm technology nodes, respectively. The net switching power is also affected by these changes through the insertion of extra input pin capacitance, but the overall impact is not as pronounced as for the cell internal power since the fraction due to the interconnect capacitance changes only slightly. The extra buffers and larger gates directly affect the change in the total cell leakage power as well, but the leakage power is a small component of the total power in this analysis.

The percentage increase in total power when comparing the scenarios with ρ_1 and ρ_5 results for the LDPC is 14.35 and 32.8 % at the 11- and 7-nm technology nodes, respectively, similar to the AES circuit results. The significant difference in the impact of interconnects on the percentage change for the critical path delay between the AES and LDPC circuits does not reflect to the power dissipation results in the same way due to the difference in the circuit type. The high wiring demand for the LDPC results in a much more pronounced interconnect capacitance impact on the total power dissipation. This total interconnect capacitance is much larger compared to the total input pin capacitance, which shadows the power dissipation increase due to the larger change in the buffer count for the LDPC than the AES.

The percentage increase in the number of buffers is 73.97 and 165.4 % at the 11- and 7-nm technology nodes, respectively.

Results for the FFT circuit indicate that the significant change in the critical path delay is not translated to the results for the power dissipation. Comparing the two extreme cases, the percentage increase in total power is only 2.55 and 3.94 % at the 11- and 7-nm technology nodes, respectively. Since most of the cells that communicate with each other are placed closely by the routing tool to minimize the total wirelength and there are a small number of connections between these clusters, the cells on the critical path are a very small portion of this large circuit. Therefore, even at the 7-nm technology node, the percentage increase in the number of buffers is only 23.9 %.

In short, the impact of the interconnect performance degradation on the power dissipation of circuit blocks is less inclined with intuition compared to the impact on the circuit speed. The dependence of the interconnect impact on the power dissipation of circuit blocks increases the complexity of designing circuits for which power dissipation is a significant metric to optimize at future technology nodes.

1.5.2 Impact of Vias

As technology scales, via resistance has an increasingly significant impact on the circuit speed and needs to be considered in optimizing the BEOL architecture. The analysis so far has concentrated on line resistance and assumed an optimistic value for via resistance. Based on a circuit model considering an inverter driving a similar inverter through a variable-length, horizontal interconnect at the third metal level, it was shown that there is a critical length below which via resistance has a large contribution to the line delay [42]. In this section, via resistance impact is analyzed based on the same physical design methodology that was described so far considering the resistance increase for vias due to both dimensional scaling and possible misalignment to the upper and lower metal levels. The test case is the AES circuit designed with the interconnect scenario ρ_l as defined before.

Synopsys Raphael [43] is used to estimate via resistance at the 7-nm technology node for both the ideal and misaligned via structures. The simulation structure is illustrated in Fig. 1.6. The barrier material resistivity is assumed to be 500 μΩ cm. The horizontal run length, L_z, for the top, M_U, and bottom, M_L, metal levels are assumed to be very small to avoid any impact on the final estimated via resistance value. The misalignment length, L_{mis}, is calculated as a percentage of the ideal via width and is varied from 0 to 50 % of the width value. The vertical length of via, L_{via}, is based on the layer definitions as determined during library construction. V1–V3 resistance values for well-aligned, 10-, 20-, 30-, 40-, and 50 %-misaligned cases are 311.4, 313.9, 333.21, 360.6, 397.9, and 456.2 Ω, respectively, at the 7-nm technology node. Via dimensions and resistance values for all via layers are tabulated in Table 1.5 for three different cases considering optimistic resistance values (CASE A) that was considered in earlier sections, a realistic scenario

Fig. 1.6 The well-aligned and misaligned via structures used to calculate via resistances for various scenarios are illustrated

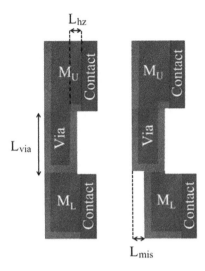

Table 1.5 Via dimensions and the corresponding resistance values as calculated by the aforementioned experimental setup are tabulated

			Resistance (Ω)		
			Optimistic	Ideal	50 % Misaligned
Via levels	Width (nm)	L_{via} (nm)	CASE A	CASE B	CASE C
V1–V3	10.8	18.7	24.08	311.4	456.2
V4–V6	21.8	45.1	14.25	30.46	67.18
V7–V8	62.2	127.6	0.68	2.97	4.1
V9	124.4	311.2	0.41	0.98	1.28

Table 1.6 Placement and routing results for the AES circuit under multiple via resistance scenarios are tabulated

		Iso-performance results					
CASE	Minimum clock period (ps)	Target period (ps)	Cell count	Via count	Buffer count	Wirelength (mm)	Total power (mW)
A	202	230	17,457	124,681	5744	24.39	5.246
B	210	230	17,736	121,695	5801	25.35	5.805
C	230	230	18,011	121,641	6177	26.73	5.964

for well-aligned vias (CASE B), and a 50 %-misaligned via scenario (CASE C). The impact of via resistance on the circuit performance and power at the 7-nm technology node is quantified and tabulated in Table 1.6. If the same netlist is used to recalculate the critical path delay of the circuit, a 18.57 % increase between CASE A and CASE B results is observed. For a better comparison, however, the design automation tools should be given the correct set of via resistance values such that the timing-driven placement, routing, and optimization steps accurately consider the correct via scenario.

When the design is optimized for the correct via resistance values, the impact of via resistance on circuit speed is only 3.96 % between CASE A and CASE B. Therefore, the timing-driven placement and routing tools can compensate for the increasing via resistance if the correct values are provided during the design process. Comparing the results for all cases, there is a monotonic reduction in the number of vias per standard cell and a monotonic increase in the total wirelength. This means that the placement and routing tools work to use a smaller number of vias even though the number of standard cells in the design increases, mainly due to a larger number of buffers, while running longer wires to connect them. Therefore, it can be concluded that the trade-off between using shorter wires to connect two points by changing the metal layer through a via and using a slightly longer wire for the same connection avoiding a via connection shifts towards the latter option as via resistance is increased.

The overall impact of via resistance comparing CASE A and CASE C results for the AES circuit design is to reduce the maximum circuit speed by 13.86 % and to increase the total power dissipation by 13.69 %. Therefore, at future technology nodes, the interconnect challenges are not limited to the horizontal wires, but extended to the vertical via connections as well.

1.6   Impact at the Full-chip Level

The block-level analysis that was described in the previous section is very time consuming and it is hard to extend this approach to a full microprocessor design due to the long simulation time. However, there is still a lot of things that can be estimated regarding the impact of the interconnect metrics at the system level based on stochastic wirelength distribution models [44], which give a priori knowledge of the wiring demand of a design. By combining this knowledge with the appropriate circuit models, the wiring requirements of a design can be estimated and the interconnect structure can be optimized.

Having established the interconnect domination in the PPA optimization for future circuit blocks whose interconnect architecture was derived from an existing 45-nm technology node library in the previous section, this section focuses on optimizing the multilevel interconnect network architecture for the number of metal levels considering ITRS projections for the die area, the number of logic gates, the clock frequency, and all interconnect related parameters such as the aspect ratio, minimum allowed pitch, barrier material thickness, and inter-layer dielectric constant.

1.6.1 System Modeling Based on Wirelength Distribution

The wiring requirements of a system can be estimated based on the well-established Rent's rule, which describes the relationship between the number of signal terminals, T, and the number of gates, N, through a power law given as

$$T = kN^P, \tag{1.8}$$

where k and P are empirical constants describing the average number of signal input and output terminals per gate and the degree of wiring complexity of the system, respectively. Assuming that Rent's rule is valid for any arbitrary closed curve of N gates within a system and describes the input/output (I/O) requirements of this collection, a complete wirelength distribution model has been developed. This distribution can then be used to design a multilevel interconnect architecture and optimize the system behavior with any key constraints of interest including the power dissipation, the number of metal levels, area, and performance.

Logic gates are modeled using the aforementioned PTMs. The interconnect requirements for memories and the logic network are vastly different. The memory structure is regular, more device limited, and needs fewer number of metal levels whereas the logic network is more random, more interconnect limited in both power and performance, and requires more metal levels. Therefore, this section considers only the logic network on the die to reduce the complexity of the analysis. The goal of this section is to illustrate the impact of the trend in interconnect parameters on the design of the multilevel interconnect network and the consequent increase in the required number of metal levels.

The complete wirelength distribution can be described by the interconnect density function that gives the number of interconnects in the system that have a certain length in gate pitches. The minimum interconnect length is assumed to be one-gate-pitch long. The cumulative interconnect density function, which gives the total number of interconnects below a certain length, can be calculated by integrating the interconnect density function between one gate pitch and the length of interest. The interconnect density function for an example 120-million-logic-gate core is illustrated in Fig. 1.7.

The principle question in designing a multilevel interconnect network is to optimize the interconnect widths such that the estimated circuit performance can be maintained while making sure that the limited wiring resources are used wisely to enable routability. While wider interconnects mean better performance, fewer of them can be routed due to the physical dimensions of the die. To address the issue of limited resources, the interconnect density function is used to compare the estimated wiring demand of a certain metal level with what can be supplied by that metal level depending on its pitch. This relationship is given by

$$e_{w,n}A = \chi p_n \sqrt{\frac{A}{N_{gates}}} \int_{L_{n-1}}^{L_n} li(l)dl. \tag{1.9}$$

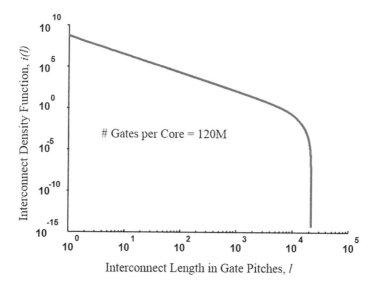

Fig. 1.7 The number of wires for each length in gate pitches is plotted for a 120-million-logic-gate core

The left-hand side of this equation represents the available area for routing wires, where $e_{w,n}$ is the net wiring efficiency of the nth metal level and A is the area of the logic core. The right-hand side represents the area required for routing wires that have lengths between L_{n-1} gate pitches and L_n gate pitches. L_n and p_n are the maximum length normalized to gate pitches and wire pitch of the nth metal level in nanometers, respectively.

From the performance point of view, the interconnect pitch of each metal level is determined assuming that the maximum RC time delays of minimum-pitch short interconnects in the lower levels and custom-pitch longer interconnects in the upper levels are 25 and 90 % of the clock period, respectively [45]. The RC time delay of interconnects without repeaters is calculated by

$$\tau_{RC} = 4.4 \frac{\rho(AR,p)}{p^2 \cdot AR} c_{int} L^2 \frac{A}{N_{gates}}, \tag{1.10}$$

where ρ, c_{int}, and AR are the resistivity, capacitance per unit length, and aspect ratio of the interconnect, respectively. Assuming that repeaters are inserted for optimal EDP in long interconnects, RC time delay can be calculated by [46],

$$\tau_{RC} = \left(\frac{0.7}{\delta} + 0.7\gamma + \frac{0.4}{\gamma} + 0.7\delta\right) \sqrt{\frac{\rho(AR,p)c_{int}R_o C_o}{AR \cdot (p/2)^2}} L \sqrt{\frac{A}{N_{gates}}},$$
$$\gamma = \left(0.73 + 0.07ln\Phi_{gate}\right)^2,$$
$$\delta = \left(0.88 + 0.07ln\Phi_{gate}\right)^2, \tag{1.11}$$
$$\Phi_{gate} = P_{dynamic}/\left(P_{dynamic} + P_{leakage}\right).$$

In this set of equations, R_o, C_o, $P_{dynamic}$, and $P_{leakage}$ are the resistance, capacitance, and dynamic and leakage power dissipations of the minimum size inverter, respectively.

The net wiring efficiency of a certain metal level is calculated considering via blockage due to repeaters inserted in the upper level interconnects and connections to signal wires in the upper layers [47], and the power/ground via blockage [48]

$$e_{w,n} = e_r \cdot \left(1 - e_{pgnd}\right) \cdot \left(1 - e_{via,n}\right),$$

$$e_{via,n} = \sqrt{\frac{2\left(N_{wires\_above,n} + N_{rep\_above,n}\right)\left(p_n + s\lambda\right)^2}{A}}, \tag{1.12}$$

where e_r is the router efficiency typically assumed to be equal to 0.5, e_{pgnd} is the fraction of area used by routing power and ground wires, and $e_{via,n}$ is the via blockage factor associated with the nth level. $N_{wires\_above,n}$ and $N_{rep\_above,n}$ are the number of wires and repeaters above the nth metal level, respectively. λ is the design rule unit equal to half the minimum feature size and s is a via covering factor equal to 3.

The architecture is initially designed without any repeater-inserted levels. Repeaters are then inserted starting from the topmost level, which accommodates the longest interconnect, and continued downwards for each metal level until the die area available for repeater insertion is all used or repeater insertion no longer improves the chip performance. The methodology is calibrated and validated by designing the multilevel interconnect network for a commercial 22-nm technology quad-core microprocessor that contains 1.4 billion transistors on a 160 mm^2 die [49].

1.6.1.1 Resistivity Impact on the Number of Metal Levels

In this section, the impact of the resistivity increase that is studied in earlier sections on the multilevel network design is highlighted. Figure 1.8 illustrates the required number of metal levels at each technology generation between 2012 and 2020. The number of cores in the system is assumed to double every other year. In all technology nodes, the smallest number of metal levels can be achieved if single-crystal Cu interconnects with infinite grain sizes can be grown. This number is taken as the baseline at each technology node in determining the percentage increase of metal levels for other scenarios. The percentage increase in the number of metal levels worsens with technology scaling. Consequently, in 2020, the number of metal levels may increase by as much as 33.8 % due to the resistivity increase of the conducting material requiring three extra metal levels to route all interconnects.

Figure 1.9 plots the delay distribution function for the optimal designs for the two extreme cases of resistivities in the ITRS technology year 2012. ITRS projection for the delay of a minimum size device in this year, τ_0, is used to divide interconnects into two groups based on their delay, τ_{RC}. Groups I and II comprise interconnects

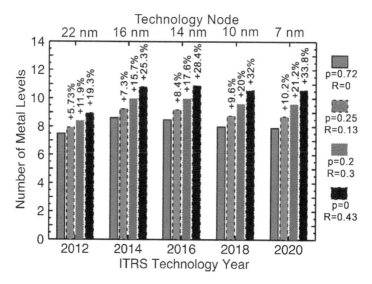

Fig. 1.8 Number of metal levels is plotted versus the technology year considering a range of size effect parameters. Mitigating size effects can reduce the number of metal levels significantly

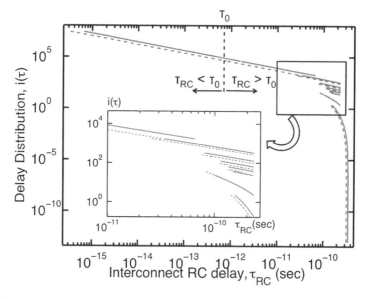

Fig. 1.9 Interconnect delay distribution calculated for the worst case of size effects (*straight line*) and single-crystal Cu assumption (*dashed line*) in 2012

Table 1.7 Average delays for the wires in the multilevel interconnect architecture design

	All (ps)	%	Group I (fs)	%	Group II (ps)	%
			2012, $\tau_0 = 0.57$ ps			
Best	1.41	–	19.82	–	27.08	–
Worst	1.81	28	25.57	29	27.41	1
Toy	2.5	77	25.57	29	37.93	40
			2020, $\tau_0 = 0.19$ ps			
Best	1.01	–	8.36	–	12.87	–
Worst	1.63	61	12.75	52	14.71	14
Toy	2.5	148	12.75	52	22.89	78

with τ_{RC} smaller and larger than τ_0, respectively. Table 1.2 shows the average wire delays in these two groups for both designs in 2012 and 2020.

It is shown in Table 1.7 that the change in average delay due to size effects in Group II is only 1 % when the multilevel interconnect network is optimally designed considering the interconnect resistivity increase. The toy example in Table 1.7 emphasizes the significance of design optimization with the proper interconnect resistivity values by showing that the average delay in Group II would have increased by 40 % if the design for $p = 0.72$ and $R = 0$ was used for $p = 0$ and $R = 0.43$ as well.

Due to the disparity between the impacts of scaling on the delay performance of interconnects and devices, the critical wirelength, L_0, where the interconnect delay becomes as large as τ_0, shortens with each new technology node. Combined with smaller wire widths, some of the interconnects which are longer than L_0 are thin enough to drastically suffer from size effects. Consequently, the average wire delay of Group II increases by 14 % as shown in Table II for 2020, even when the interconnect architecture is optimized based on the proper resistivity increase considerations and three extra metal levels are added.

1.6.1.2 Barrier Thickness Impact on the Number of Metal Levels

This section uses the same methodology to design an optimized multilevel interconnect structure to illustrate the impact of the barrier material thickness on the required number of metals to meet the design constraints. Figure 1.10 shows the impact of the total barrier/liner and adhesion layer thickness scaling on the required number of metal levels in the optimized multilevel interconnect network in 2020, assuming that this bilayer thickness can be extended down to ∼3–4 nm [20]. The percentage values on each bar in Fig. 1.10 represent the increase in the number of metal levels taking the ITRS projections, which is 1.1 nm, as reference for each resistivity scenario as introduced before. If the bilayer is 3.5 nm thick, the increase in the number of metal levels over the reference scenario ranges between 12 and

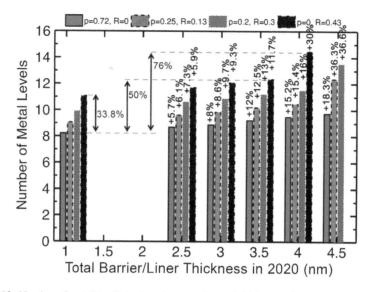

Fig. 1.10 Number of metal levels is plotted versus the total thickness of the barrier/liner layer at the 7-nm technology node for various size effect parameters. The thickness of the bilayer should be scalable to 3.5 nm

13 %. Combined with the impact of size effect parameters, the number of metal levels may increase by as much as 50 %. For certain combinations of size effect parameters and bilayer thicknesses, it is not possible to come up with a design that meets the ITRS clock frequency projections.

1.6.1.3 Interconnect Variability Impact on the Number of Metal Levels

Variations in wire width and spacing affect resistance and capacitance associated with interconnects. The resistance–capacitance product and its percentage variation with width are plotted for minimum size wires at the 7-nm technology node in Fig. 1.11. Assuming a perfect Gaussian distribution for the width of the wire, it is equally likely to get a wire that is 44 % slower or 20 % faster than the nominal delay [50] due to the large change in the resistivity of the material. As the interconnect pitch is increased, variation in RC delay reduces.

This variation in wire delay is taken into account during the design of the interconnect architecture by introducing a variation term in the equations that are described in the earlier sections. This method assumes that the width of the longest wire that is routed in each metal level is different from the nominal width by an amount determined by the variation parameter; and the pitch of each metal level is optimized accordingly. Although each interconnect in a metal level may have a different variation parameter, by considering the worst case for the longest interconnect, reliable operation is ensured for all the other wires. Comparing the

Fig. 1.11 RC delay per unit length squared for various width values considering an interconnect pitch of 24 nm in 2020 and size effect parameters $p = 0$ and $R = 0.43$. The *inset* figure shows the percentage variation in *rc* delay versus the variation in width as a percentage of the nominal width value for various interconnect pitches

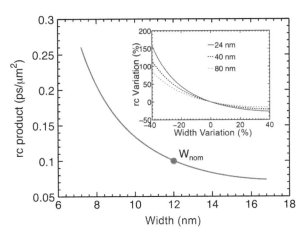

total wire area required for the nominal design and the worst case scenario for a $-20\,\%$ variation in the wire width shows only a 4 % increase due to the variations.

It is important to note here that there is a significant limitation that this methodology introduces and that is the number and pitch of each metal level are a free variable. In an interconnect design where these parameters are constant or in an SoC design where there are many minimum size metal levels, the impact of variations would be more severe as numerous wires would require being routed at metal levels that are wider in pitch causing routability problems. In fact, it has been reported that ignoring wire delay variability can lead to $\sim 12\,\%$ more violating paths in a design [50].

1.7 Reliability Challenges

Previous sections have concentrated on the challenges in designing interconnect structures from the timing and power performance point of view. With aggressive dimensional scaling, the reliability physics of both the interconnect and the dielectric material have become a significant concern for interconnects in future ultra-scaled technology generations. Ensuring this reliability becomes increasingly challenging with smaller dimensions because of the increasing electric fields and the inherent weak mechanical properties of the lower-κ dielectric materials. Degradation in the mechanical, thermal, electrical, and adhesion properties of the lower-κ dielectric materials makes future interconnect structures vulnerable to reliability issues. In this section, electromigration and TDDB issues with Cu interconnect structures are considered.

1.7.1 Cu Electromigration

Electrical current causes the metal atoms in the interconnect to migrate towards the anode. In time, this migration can cause voids in the interconnect, which can increase the total resistance of the wire and significantly impact performance. If the voids in the interconnect grow large enough, the connection may be completely cut off and functionality of the chip may be broken. The severity of electromigration is directly related to the current density. As device and interconnect dimensions shrink and the drive current, hence the switching speed, of the devices increases, interconnects need to be able to carry higher densities of current reliably. However, the electromigration lifetime decreases as wire dimensions decrease because the critical void size that can cause a fail gets smaller. Therefore, it takes a shorter time for the void to become destructive.

It was mentioned in earlier sections that some of the solutions to interconnect design challenges introduce new trade-offs. One of these trade-offs is between the *RC* delay improvement through capacitance per unit length reduction by switching to low-κ dielectric materials and the negative impact of this transition on the electromigration lifetime of future interconnects. Due to the low thermal conductivities of low-κ dielectric materials, there is more joule heating for a given current density, which causes the wires to heat up quickly. This high temperature operation increases the rate of electromigration and reduces the average time to fail.

In the typical Cu damascene process, the weakest point for electromigration-induced failure is the interface between the Cu and the dielectric layer. Introducing a metallic capping layer at this interface can significantly improve the electro-migration lifetime of wires. However, cost efficient optimization of the process to manufacture ultra-thin cap layers with high electromigration blocking ability is a major barrier for optimizing future advanced interconnects. An alternative methodology to improve the electromigration lifetime is to introduce impurities in the Cu interconnect, which can significantly reduce Cu diffusion at the grain boundaries, and Cu-cap layer and other interfaces. This methodology introduces a trade-off between Cu resistivity and electromigration lifetime since the Cu resistivity increases with introducing impurity atoms, which degrades *RC* delay performance.

1.7.2 Time-Dependent Dielectric Breakdown

TDDB is a measure of the dielectric reliability of the interconnect structure. Historically, due to the low electric fields and large dielectric thicknesses, BEOL dielectric reliability has received very little attention. However, rapid shrinking of the interconnect pitch, hence the spacing between neighboring wires, has given rise to a quick increase in the lateral electric field across the dielectric in between. Combined with the introduction of low-κ dielectric materials to reduce crosstalk and

RC delay associated with wires, TDDB performance has become a major metric to consider during the design of advanced interconnect technologies.

TDDB performance is very sensitive to the materials and the processes that are used to manufacture the interconnect structures. Typically, the breakdown strength of the dielectric material is lower if there is a high density of defect sites in the dielectric, CMP-induced damage on the dielectric material, electromigration of Cu atoms into the dielectric, and misalignment between neighboring wires, hence reduced spacing. All of these factors that contribute to lower breakdown strength degrade with dimensional scaling, consequently creating a very challenging dielectric reliability problem.

TDDB performance can be improved by minimizing electric field enhancement. Considering that the minimum feature size on the die will continue to shrink in accordance with Moore's law at future technology generations, the only way to achieve this is by optimizing the manufacturing processes and ensuring that the dimensions are perfectly controlled with minimal roughness on the edges of the interconnect structures and minimal misalignment between the vias and the upper or lower metal layers. Overcoming the inherent critical dimension and overlay variation associated with multiple patterning technologies using 193 nm lithography tools is a challenging task at ultra-scaled dimensions. Another methodology that can help achieve adequate TDDB performance is to minimize Cu diffusion into the dielectrics and the capping layer. As described in the previous sections, this can be ensured by using a conformal barrier material in the trench. Even though a thick barrier material can improve TDDB lifetime, it will degrade interconnect resistance, hence the timing performance of the chip, significantly.

1.8 Conclusion and Outlook

Interconnects are an ever growing challenge to continue improving the performances of electronic chips. There is no value to creating better transistors unless they can be connected to each other efficiently. Therefore, any practical implementation of future integrated circuits will require a comprehensive understanding f the limitations imposed on the chip performance by the degrading performance of interconnects and the key tradeoffs between power-performance-area (PPA) optimizations, manufacturability and reliability. This chapter has aimed to illustrate these major interconnect related limitations to the steady advancement of CMOS technology and highlighted the key tradeoffs from the material level to the system level.

Future interconnect structures will continue to require innovations in material, interconnect structure and architecture. The research from the materials perspective has concentrated on reducing the Cu interconnect resistance through new technologies that mitigate the increase in Cu resistivity by decreasing electron scatterings at wire surfaces and grain boundaries by modifying the liner material and increasing grain sizes, respectively; and finding novel materials with lower resistivities.

Furthermore, as the rest of this book illustrates, replacing Cu interconnects with bundles of densely packed carbon nanotubes (CNT) and multi-layer graphene nanoribbons (GNR) have gathered significant attention as a potential solution to the degrading interconnect performance. CNT interconnects can be aligned both vertically for via applications and horizontally for routing. To replace Cu, low resistance CNT structures with high quality and high density CNTs and good ohmic contacts are needed.

One of the solutions to the interconnect problem from the performance perspective is three-dimensional (3D) integration. Despite not solving all interconnect problems, this technology is anticipated to be pivotal for the future growth of the semiconductor industry. 3D integration can reduce the global interconnect length through stacking of devices on multiple layers an connecting them together thereby reducing both the total resistance and capacitance associated with interconnects. This can result in reduced wire delay and significant reduction in energy per operation. 3D integration can significantly enhance the interconnect resources and can provide improved bandwidth and throughput. These 3D architectures can be manufactured by using various approaches including wire-bonded structures where connections between chips go through the board, microbumps, capacitive or inductive coupling, through-silicon-vias (TSVs), and monolithic 3D integration where each tier of devices is sequentially fabricated on top of another and connected through monolithic inter-tier vias. Some of these approaches have better performance than others.

Among these options, the TSV technology is one of the more promising approaches and it is relatively closer to the market. The performance of this technology highly depends on the dimensions and distribution of TSVs and their parasitics. TSVs have large pitches and associated parasitics, which limit their performance and put critical limits on their drivers. The alternative approach, the monolithic 3D integration, have advantages over the TSV technology. However, this is a longer term solution and still requires a mature process technology. Since each tier of devices is fabricated sequentially, there is no need for die alignment as in the TSV technology and the inter-tier via dimensions are much smaller than TSVs. These small via dimensions can enable a much better integration density and reduce the associated parasitic interconnect components significantly.

References

1. Dennard R, Gaensslen F, Rideout V, Bassous E, LeBlanc A (1974) Design of ion-implanted MOSFETs with very small dimensions. IEEE J Solid-State Circuits 9:256–268
2. Buchanan D (1999) Scaling the gate dielectric: materials, integration and reliability. IBM J Res Dev 43:245–264
3. Yeo Y, Lu Q, Lee W, King T-J, Hu C, Wang X, Ma T (2000) Direct tunneling gate leakage current in transistors with ultrathin silicon nitride gate dielectric. IEEE Electron Device Lett 21:540–542

4. Bai P et al (2004) A 65 nm logic technology featuring 35 nm gate lengths, enhanced channel strain, 8 Cu interconnect layers, low-κ ILD and 0.57 mm^2 SRAM cell. In: IEDM Technical Digest, pp 657–660
5. Antoniadis D, Aberg I, Ni Chleirigh C, Nayfeh O, Khakifirooz A, Hoyt J (2006) Continuous MOSFET performance increase with device scaling: the role of strain and channel material innovations. IBM J Res Dev 50:363–376
6. Chau R, Datta S, Doczy M, Kavalieros J, Metz M (2003) Gate dielectric scaling for high–performance CMOS: from SiO$_2$ to high–κ. In: International Workshop on Gate Insulator, pp 124–126
7. Auth C et al (2008) 45nm high–κ + metal gate strain–enhanced transistors. In: Symposium on VLSI Technology, pp 128–129
8. Auth C et al (2012) A 22nm high–performance and low–power CMOS technology featuring fully–depleted tri–gate transistors, self–aligned contacts and high density MIM capacitors. In: Symposium on VLSI Technology, pp 131–132
9. Bohr M (2009) The new era of scaling in an SoC world. In: IEEE International Solid State Circuits Conference, pp 23–28
10. Edelstein D et al (1997) Full copper wiring in a sub–0.25 mm CMOS ULSI technology. In: IEDM Technical Digest, pp 773–776
11. Lin Y-M et al (2010) 100–GHz transistors from wafer scale epitaxial graphene. Science 327:662
12. Bachtold A, Hadley P, Nakanishi T, Dekker C (2001) Logic circuits with carbon nanotube transistors. Science 294:1317–1320
13. Naeemi A, Meindl J (2007) Design and performance modeling for single-walled carbon nanotubes as local, semiglobal, and global interconnects in gigascale integrated systems. IEEE Trans Electron Devices 54:26–37
14. Naeemi A, Meindl J (2009) Compact physics-based circuit models for graphene nanoribbon interconnects. IEEE Trans Electron Devices 56:1822–1833
15. Nathanael R, Pott V, Kam H, Jeon J, Liu T-J (2009) 4-Terminal relay technology for complementary logic. In: IEDM Technical Digest, pp 1–4
16. Beausoleil R et al (2008) Nanoelectronic and nanophotonic interconnect. Proc IEEE 96:230–246
17. Krishmamoorthy A et al (2009) Computer systems based on silicon photonic interconnects. Proc IEEE 97:1337–1361
18. Behin-Aein B, Datta D, Salahuddin S, Datta S (2010) Proposal for an all-spin logic device with built-in memory. Nat Nanotechnol 5:266–270
19. Gambino J, Lee T, Chen F, Sullivan T (2009) Reliability challenges for advanced copper interconnects: electromigration and time-dependent dielectric breakdown (TDDB). In: IEEE international symposium on the physical and failure analysis of integrated circuits, pp 677–684
20. Kaloyeros A, Eisenbraun ET, Dunn K, Van der Straten O (2011) Zero thickness diffusion barriers and metallization liners for nanoscale device applications. Chem Eng Commun 198:1453–1481
21. Steinhoegl W, Schindler G, Engelhardt M (2005) Unraveling the mysteries behind size effects in metallization systems. Semicond Int 28:34–38
22. Lopez G, Davis J, Meindl J (2009) A new physical model and experimental measurements for copper interconnect resistivity considering size effects and line-edge roughness (LER). In: IEEE international interconnect technology conference, Sapporo, pp 231–234, 1–3 June 2009
23. International Technology Roadmap for Semiconductors (2013) Online http://www.itrs.net
24. Wu F, Levitin G, Hess W (2010) Low-temperature etching of Cu by hydrogen-based plasmas. ACS Appl Mater Interfaces 2:2175–2179
25. Steinhoegl W, Schindler G, Steinlesberger G, Traving M, Engelhardt M (2005) Comprehensive study of the resistivity of copper wires with lateral dimensions of 100 nm and smaller. J Appl Phys 97:023706-1-023706-7
26. Shimada M, Moriyama M, Ito K, Tsukimoto S, Murakami M (2006) Electrical resistivity of polycrystalline Cu interconnects with nanoscale linewidth. J Vac Sci Technol B 24(1):190–194

27. Steinhoegl W, Schindler G, Steinlesberger G, Traving M, Engelhardt M (2004) Impact of line edge roughness on the resistivity of nanometer-scale interconnects. Microelectron Eng 76(1–4):126–130
28. Kitada H et al (2007) The influence of the size effect of copper interconnects on RC delay variability beyond 45 nm technology. In: IEEE IITC, pp 10–12
29. Plombon JJ, Andideh E, Dubin VM, Maiz J (2006) Influence of phonon, geometry, impurity, and grain size on copper line resistivity. Appl Phys Lett 89(11):113124–113124-3
30. Chen H-C, Chen H-W, Jeng S-P, Wu C-MM, Sun JY-C (2006) Resistance increase in metal nano-wires. In: International symposium on VLSI technology, systems and applications, pp 1–2
31. Besling WFA, Broekaart M, Arnal V, Torres J (2004) Line resistance behaviour in narrow lines patterned by a TiN hard mask spacer for 45 nm node interconnects. Microelectron Eng 76(1–4):167–174
32. Guillaumond J et al (2003) Analysis of resistivity in nano-interconnect: full range (4.2–300 K) temperature characterization. In: IEEE IITC, Burlingame, pp 132–134
33. Chern J-H, Huang J, Arledge L, Li P-C, Yang P (1992) Multilevel metal capacitance models for CAD design synthesis systems. IEEE Electron Device Lett 13(1):32–34
34. Magen N, Kolodny A, Weiser U, Shamir N (2004) Interconnect power dissipation in a microprocessor, In: International workshop on system level interconnect prediction, pp 7–13
35. Sinha S, Yeric G, Chandra V, Cline B, Cao Y (2012) Exploring sub-20 nm FinFET design with predictive technology models. In: Design automation conference, San Francisco, pp 283–288
36. Kilby J (1976) Invention of the integrated circuit. IEEE Trans Electron Devices 23:648–654
37. Danko S (1951) New developments in the Auto-Sembly technique of circuit fabrication. In: Proceedings of the national electronics conference, pp 542–550
38. Nangate (2011) Nangate FreePDK45 Open Cell Library. Online http://www.nangate.com
39. Synopsys (2012) Synopsys Design Compiler, version: 2012.06-SP5. Online http://www.synopsys.com
40. Cadence Design Systems (2013) Encounter digital implementation system, version: 2013.1. Online http://www.cadence.com
41. Synopsys (2011) Synopsys PrimeTime, version: 2011.06-SP3-2. Online http://www.synopsys.com
42. Chen J-C, Standaert T, Alptekin E, Spooner T, Paruchuri V (2014) Interconnect performance and scaling strategy at 7 nm node. In: IEEE International interconnect technology conference, pp 93–96
43. Synopsys (2012) Synopsys Raphael, version: 2012.06. Online http://www.synopsys.com
44. Davis J, De V, Meindl J (1998) A stochastic wire-length distribution for gigascale integration (GSI) – part I: derivation and validation. IEEE Trans Electron Devices 45:580–589
45. Sekar D, Naeemi A, Sarvari R, Meindl J (2007) IntSim: A CAD tool for optimization of multilevel interconnect networks. In: ICCAD, San Jose, pp 560–567, 4–8 November 2007
46. Sekar D, Venkatesan R, Bowman K, Joshi A, Davis J, Meindl J (2006) Optimal repeaters for sub-50 nm interconnect networks. In: IEEE international interconnect technology conference, pp 199–201
47. Chen Q, Davis J, Zarkesh-Ha P, Meindl JD (2000) A compact physical via blockage model. IEEE Trans VLSI Syst 8(6):689–692
48. Sarvari R, Naeemi A, Zarkesh-Ha P, Meindl JD (2007) Design and optimization for nanoscale power distribution networks in gigascale systems. In: IEEE IITC, pp 190–192
49. Damaraju S et al (2012) A 22 nm IA multi-CPU and GPU system-on-chip. In: IEEE international solid-state circuits conference digest of technical papers. San Francisco, CA, pp 56–57
50. Nassif SR, Nam G–J, Banerjee S (2013) Wire delay variability in nanoscale technology and its impact on physical design. ISQED, Santa Clara, pp 591–596, 4–6 March 2013

Chapter 2
Overview of Carbon Nanotube Interconnects

A. Srivastava, X. H. Liu, and Y. M. Banadaki

2.1 Introduction

At present, electronic information technology has become an important drive force
that promotes social and economic progress. Integrated circuit (IC) as a core and
foundation of the electronic information technology has a great influence on the
daily life of human being. The semiconductor technology and IC industry have
become an important symbol to embody a country's comprehensive scientific and
technological capability. In order to improve circuit's performance and increase
number of transistors on a chip, microelectronic devices have been continuously
reduced in dimension according to Moore's law [1] and scaling rule [2]. According
to the 2013 International Technology Roadmap for Semiconductors (ITRS 2013),
the feature size of semiconductor devices will reduce to 22 nm in 2016 and 10 nm
in 2025 [3] in very large scale integrated (VLSI) circuits. For the first generation
interconnect material aluminum (Al) [4], an increase in electric resistance and
capacitance due to increasing wire length and decreasing wire interval as dimension
scales down had led to large signal delays [5] and poor tolerance to electromigration

A. Srivastava (✉)
Division of Electrical and Computer Engineering, Louisiana State University,
Baton Rouge, LA 70803, USA
e-mail: eesriv@lsu.edu; ashok@ece.lsu.edu

X.H. Liu
School of Physics, Liaoning University, Liaoning, China
liuxinghuixjtu@sohu.com

Y.M. Banadaki
Division of Electrical and Computer Engineering, Louisiana State University,
Baton Rouge, LA 70803, USA

College of Engineering, Southern University, Baton Rouge, LA 70813, USA
ymoham8@lsu.edu; banadaki_yaser@subr.edu

© Springer International Publishing Switzerland 2017
A. Todri-Sanial et al. (eds.), *Carbon Nanotubes for Interconnects*,
DOI 10.1007/978-3-319-29746-0_2

(EM) [6]. Because of its lower resistivity, higher melting point (1083 °C versus 660 °C of Al), and longer EM lifetime [7], copper (Cu) has replaced Al as an interconnect material in the 180 nm technology node [8] and beyond. But as interconnects scale down to the 45 nm and beyond technology generations, Cu interconnect is also facing similar problems with those of Al interconnects encountered, including increase in resistivity due to size effect [9], increase in power consumption [10], delay [11], and EM distress [12].

When the lateral dimension of a Cu wire can be comparable or smaller than its electron mean free path (MFP), according to Fuchas-Sondheimer (F-S) [13] and Mayadas-Shatzkes (M-S) model [14], Cu electrical resistivity increases [9, 15–18] compared to that of bulk Cu due to surface scattering [15] and grain boundary scattering mechanisms [19]. In addition, existence of diffusion barrier layer [20–22] in Cu interconnect technology also contributes to the resistance. Copper atoms can rapidly diffuse into interlayer dielectrics (ILDs) resulting in degradation of EM lifetime [20], Cu has poor adhesion to most ILDs, thus the barrier layer generally requires to be placed between Cu wires and the ILDs, preventing diffusion of Cu atoms into the ILDs and also providing interfacial mechanical strength [21]. Refractory metals such as titanium or tantalum and their nitrides are required to be added between Cu wires and ILDs [22].The resistivity of the barrier layer material is generally relatively higher than that of Cu, and the barrier layer also tends to reduce the effective cross section of Cu interconnect, this further increases the equivalent resistivity of Cu wire [11]. At the 22 nm technology node, Cu resistivity can increase to 5.8 $\mu\Omega$ cm [11], which is more than three times that of the bulk Cu (1.7 $\mu\Omega$ cm [23, 24]).

Increase in resistivity of Cu interconnect wire will inevitably cause performance degradation of VLSI. The time delay dominated by resistance and capacitance (R and C) will increase [11, 25]. In 35 nm Cu/low-k technology generation, the transistor delay can be ~1.0 ps and RC delay of a 1 mm length of Cu wire can be ~250 ps [10]. Moreover, the higher resistivity as well as longer interconnect length and higher current density due to reduction of dimension [2] will also make wire voltage-drop more significant, leading to an obviously increase in power consumption. It is estimated that approximately 50 % of microprocessor power is consumed by the interconnect at 0.13 μm technology node, and up to 80 % of power can be consumed by interconnect at 45 nm node [10].

Furthermore, the dimension reduction of Cu interconnect may shorten the EM failure time [26]. EM is a diffusion-controlled mass transport phenomenon [20, 27], whose driving force is related to the currents density passing through the wire, wire resistivity, and EM strength [28]. EM strength arises from a direct electrostatic force and electron wind force derived from the momentum exchange between charge carriers and diffusing atoms [29]. The movement of atoms acted by continuous electric field and currents may cause formation of void and hillocks in interconnects [30], where the void tends to increase the interconnect resistance even cause open circuit, while the hillocks will possibly cause short circuit to the neighboring wires. The EM lifetime of Cu interconnect is proportional to the dimension of Cu voids and inversely proportional to EM drift velocity (or void growth rate) [26, 28, 31]. With the dimension scales down, on one hand, the thinner Cu wire will cause a smaller critical void volume for failure [32]; on the other hand, the Cu grain size changes from bamboo type dominated structures to mixtures of bamboo and polycrystalline structures [26, 28]. The change of grain structure results in increase of Cu mass flow rate thus increase of the void growth rate, thus shortening the EM lifetime. Despite the technology innovation, the EM reliability is still a challenging problem with the newer technology generation [33–35].

2.2 Carbon Nanotubes and Graphene Nanoribbon Interconnects

Nanometer complementary metal oxides semiconductor (CMOS) technology especially in 22 nm and beyond is plagued due to performance degradation of conventional Cu/low-k dielectric as an interconnect material for gigascale and terascale integration. Thus the need for other novel materials possibly substituting Cu/low-k dielectric interconnections has brought forward other novel interconnect technologies for next-generation VLSI interconnects such as optical [36], radio frequency (capacitively coupling [37] or inductively coupling [38] and antennas interconnects [39]) as well as superconductor interconnects [40]. In search for novel technologies, no such material has aroused so much interest other than carbon nanotubes (CNTs) and graphene materials.

2.2.1 CNTs Interconnects

CNTs can be regarded as rolling up one or several graphene sheets into one-dimensional seamless cylindrical structure, called single-walled CNT (SWCNT) and multiple-walled CNT (MWCNT), respectively. The SWCNT and MWCNT were discovered in 1993 [41, 42] and 1991 [43], respectively. The diameter of SWCNTs may vary from 0.4 to 4 nm with a typical diameter of 1.4 nm and outer diameters of MWCNTs may vary from several nm to 100 nm, and the spacing between the adjacent walls is about 0.34 nm [44, 45]. CNT with unique atomic arrangement and strong covalent bonds among carbon atoms [46] exhibits not only unique band-structure, but also interesting physical properties [44, 45] strongly depending on their chiralities denoted by the structural indices (n, m). Statistically, about one-third of SWCNTs exhibit metallic properties (n-m = 3p, where p is an integer), while the remaining two-third behaves semiconducting (n-m = 3p ±1) [47, 48]. For MWCNT, in principle, each of shells can be metallic or semiconducting with different chiralities. However, a MWCNT with semiconducting shells may behave metallic due to charge transfer and orbital mixing [49].

CNTs have several advantages for application as an interconnect and device material: ballistic transport [50–52] and large electron MFP (1~several tens μm) [53–60], extraordinary mechanical strength ($\sim$1 TPa) [61–63], large current-carrying capability of up to 10^9 A/cm^2 [64–66], high thermal conductivity in the range of 2000–7500 W/mK [67–79] (for comparison, the thermal conductivity of Cu is 385 W/mK [10]), reduced capacitance [80, 81], low inductance [82–84], and saturation of skin effect [82, 84–86]. The large MFP can provide low resistivity in short-length interconnects [55], combining with the reduced capacitance, it also favors reduction of the time delay [80, 81]; the strong carbon bonds lead to an extraordinary mechanical strength and a very large current-carrying capacity, increasing tolerance to EM [66, 87]; high thermal conductivity (exceeds that of Cu by a factor of 20) can better dissipate heat of interconnects by heat transport to the surface [88], this will strongly favor the use of CNT for three-dimensional (3D) integration; and the negligible inductance and saturation of skin effect make CNTs interconnect suitable for application in high frequency electronics [86]. Thus, CNTs have attracted great attention as potential interconnect candidate for future VLSI applications [89–94].

The electric resistance of a SWCNT is limited by an intrinsic quantum resistance, Rq = 6.45 kΩ [53, 57]. When the length is longer than its MFP, CNT may also have an additional resistance due to phonon scattering [55]. Hence, it is necessary to keep parallel high density CNTs together to reduce the total resistance of CNT interconnects [80–83, 88, 95–97]. A densely packed CNT bundles can achieve 4–8 times reduction in power consumption compared to Cu interconnects at 22 and 14 nm technology nodes [88], respectively. The delay of MWCNT interconnects can reach as low as 15 % of Cu interconnect delay at 1000 μm global or 500 μm intermediate level interconnects [96]. For local interconnects, monolayer or multi-layer SWCNT interconnects can offer up to 50 % reduction in capacitance and power dissipation and up to 20 % improvement in signal delay compared to Cu interconnects at the 22 nm technology node if the length is smaller than 20 μm [80]. Global interconnect buses implemented with SWCNT bundles and SWCNT/MWCNT mixed bundles can also lead to performance gains over Cu global buses [97]. The reliability [98] and stability [99] of CNT bundles-based interconnects have been evaluated. The practical circuits-based silicon transistors and CNT interconnects with operating frequency both at low frequency [100] and high frequency regime [86] have also been reported, respectively.

2.2.2 Graphene Nanoribbon Interconnects

After the discovery of graphene in 2004 [101], it was soon considered graphene as one of the promising alternative materials for electronic applications. Graphene is a monolayer of carbon atoms in a two-dimensional hexagonal lattice with semi-metallic nature, in which electrons behave as massless Dirac fermions resulting in high carrier velocities and current density. However, the application of large-area graphene is limited for integrated circuits and a narrow stripes of graphene, known as graphene nanoribbon (GNR) is a promising alternative as a replacement of transistor channel [102, 103] and interconnect [104] for next-generation VLSI circuits. GNR can be fabricated from large-area graphene using high-resolution lithography like *e*-beam lithography. It shares many of the fascinating electrical [105], mechanical [106], and thermal [107, 108] properties of CNTs such as large carrier mobility and thermal conductivity [109]. The MFP of electron in GNRs with smooth edge is comparable with CNT and can reach to micrometer range [110]. In addition, GNR has a very large current conduction capacity (1000 times larger than Cu) with extraordinary mechanical strength and thermal conductivity (~5300 W/mK) which increases its EM tolerance compared to Cu/low-k interconnects [104]. In principle, the GNRs can be produced by patterning large-area graphene using more standard fabrication methods with much more controllability than CNTs, whose chiralities are statistically predetermined during the manufacture process. The lateral confinement of electron transport in GNR can open a bandgap of a few hundred eV [111, 112], which can be altered depending on atomic structure at the ribbon edge such that zigzag-edged ribbons are metallic and armchair-edged ribbons are semiconducting [113]. Thus, it looks difficult to produce fully metallic GNRs for the interconnect application. However, the generated bandgap is usually smaller than a Cu interconnect with the similar nanometer diameter, such that the first subbands of GNRs with narrow bandgap can be sufficiently populated at room temperature and thereby small bandgap of GNRs makes negligible degradation in conductance properties [114].

2.3 Challenges for CNT-Based and Graphene Nanoribbon Interconnects

2.3.1 Challenges for CNT-Based Interconnects

As mentioned above, many advantages of CNTs have shown a promising perspective toward possible application as next-generation interconnects. However, there still exists considerable limitations and challenges to be solved from fabrication process and design aspects, mainly focusing on growth of high density CNT bundles (larger than 10^{13} cm^{-2}) [115–118] on conductive layer at a low temperature (no more than 400 °C) [119–121] and orientation control [115, 116, 122–125] as well as formation of low resistance CNTs-metal contacts [126–129]. These factors need to be simultaneously satisfied to realize the practical application of CNT-based interconnects into VLSI [130–132]. In addition, selective growth of metallic CNTs [133, 134] and control of CNTs chirality [135] are also very important factors.

2.3.1.1 High Density Synthesis of CNTs-Based Via Interconnects

In order to minimize the intrinsic resistance of CNTs-based interconnects, super high density of CNT bundles are required so that the maximum number of tube shells contributes to the electrical conduction in parallel. Theoretically, an upper limit of the tube shells density can be evaluated if all aligned CNTs are ideally close-packed [116–118]. From these calculations, the density of ideally close-packed CNTs is reversely proportional to the square of the SWCNT diameter. The maximum density is close to 2×10^{14} cm^{-2} for SWCNTs with 0.4 nm diameter and 1×10^{13} cm^{-2} with 3 nm diameter [118], while for the SWCNTs with a mean diameter of 6 nm, the density upper limit reduces to 2.87×10^{12} cm^{-2} [136]. It should be noted that SWCNTs can be either semiconducting or metallic, maximization of the fraction of metallic SWCNTs in a bundle is critical. For the bundles with random distributed SWCNTs, the fraction of metallic SWCNTs is only one-third. A MWCNT is always metallic even the shells with semiconducting chiralities can also contribute to the electric conduction [49]. However, when a MWCNT side-bonded contacts to metal electrode, only the outermost shell can contribute to electrical transport [51] unless the inner shells are opened to contact properly with metal electrodes.

From the fabrication process point of view, the chemical vapor deposition (CVD) technique is of significant interest and importance because it allows aligned and selective growth of CNTs onto substrates through direct catalysts deposition [137] and are also scalable to large areas [138]. In principle, an individual CNT can be grown from a single catalyst nanoparticle [139–141], but not all catalytic nanoparticles are able to nucleate during synthesis of CNTs [142]. On one hand, to increase the tube shells density, it is necessary to increase the catalyst nanoparticles density; on the other hand, the diameter of grown CNTs has a positive correlation with the size of nanoparticles [141, 143], so the diameter of CNTs can be controlled by reducing the size of catalytic nanoparticles. Combining both factors, increasing density of catalyst particles or reducing the diameter of grown CNTs is an essential procedure for synthesis of high density CNT bundles [115, 122, 139, 144]. However, the reduction of the catalysts size is usually restricted by catalytic particles aggregation [144–146], which results in entangled mats of CNTs [147] or large diameter carbon nanowires [148]. Moreover, the growth of high density of CNTs on a conductive substrate is usually more difficult than on an insulating material, because metal substrate usually has higher surface energy than insulating one, leading to less easily dewet catalysts [116, 149] thus impedes transformation of catalyst into nanoparticles [150]; the metal catalysts also tend to diffuse into conducting support and then disappear [151], leading to lose catalytic roles. So the selection and treatment of catalysts is a considerable challenge for growth of ultrahigh density vertical aligned CNT bundles on a conductive substrate. The dewettability is related to the difference of surface energy between the catalysts and conductive substrate; the higher the difference of surface energy, the better the dewettability will be [149]. The transition metals Fe [116, 152–154], Co [122, 155, 156], and Ni [121, 144, 157] are the effective catalysts and usually available for

CVD growth of high density CNTs due to their high surface energy [158]. The plasma pre-treatment [116, 151, 155] and reactive ion etching [155, 159] are usually employed to convert the catalysts layer into rough nanoparticles so as to increase carbon diffusion through catalysts.

The size and density of catalyst particles strongly depend on the initial layer thickness [157, 160, 161]. Catalyst engineering [136, 156, 162, 163] and annealing treatment [139, 155, 163, 164] are usually employed to prevent from catalysts oxidation and increase activity and effectiveness of catalyst particles. Plasma pre-treatment is considered to favor the growth of CNTs by root mode due to strongly binding the catalyst nanoparticles to the substrate [151], while the density of grown CNTs by root mode is referred to denser than that of grown CNTs by tip mode [116]. However, insufficient or excess pre-treatment of catalysts will probably result in subsequently poor growth of CNTs due to either imperfect reduction of nanoparticles or low surface diffusion of carbon atoms during CVD growth [165].

There are many reports about high density growth of CNT bundles on con-ductive substrates with via holes [115, 120, 156] and without via holes structure [136, 155, 163]. However, the tube shell density obtained in these reports is merely in the range of 1×10^{11}–1×10^{12} cm^{-2}, which is still at least one or two orders too low to meet the practical application requires. To avoid the aggregation of catalytic nanoparticles, and direct exposure to high temperature during the CVD growth of CNTs, Yamazaki et al. [165] developed a multiple-step growth method with a gradually elevating temperature procedure. Firstly, 0.5 nm thick Co as a catalyst layer was deposited on a conductive TiN/Ta barrier over a Cu support, and then H_2 or Ar plasma was applied to induce formation of high density Co nanoparticles at 25–260 °C, after that CH_4/H_2 plasma was applied for nucleation at 170–350 °C. Finally, CH_4/H_2 plasma was applied to grow vertically aligned CNTs at 450–600 °C. The density of grown CNTs was estimated to be about 10^{12} cm^{-2} and the volume occupancy was about 30–40 %, respectively. There is still much room to increase the density of CNTs. Zhong et al. [139] fabricated super-high density vertically aligned SWCNT forests by plasma-enhanced CVD (PECVD) with an $Al_2O_3/Fe/Al_2O_3$ sandwich structure catalysts engineering, in which the Fe catalyst layer thickness was 0.7 nm. An average diameter of 1.0 nm and tube wall density of 1.5×10^{13} cm^{-2}, very close to the theoretical limit of densely packed bundles, has been obtained. Unfortunately, the CNT forests were grown on an insulating substrate at temperature as high as 700 °C. The growth temperature was too high and no via hole was formed. Hisashi et al. [122] grew vertically aligned CNT bundles on Ti-coated Cu supports by CVD at 450 °C. The Co-Mo-Co was used as catalysts, in which Co was an active catalyst; Mo was used to effectively stabilize small-sized Co particles. They found that 0.8 nm thick Mo prior to depositing 2.5 nm Co favored growth of highest density CNTs bundles. The mass density of grown CNTs was about 1.6 g cm^{-3}, the maximum tube shell density obtained was 7.8×10^{12} cm^{-2}, corresponding to 88 % of the maximum filling. However, the CNT-Cu supporters have a mean resistance of 22 kΩ, which is much larger than the intrinsic resistance of individual CNT. Dijon et al. [116] grew double and triple walled small diameter CNTs in bundles by using plasma pre-treatment of the

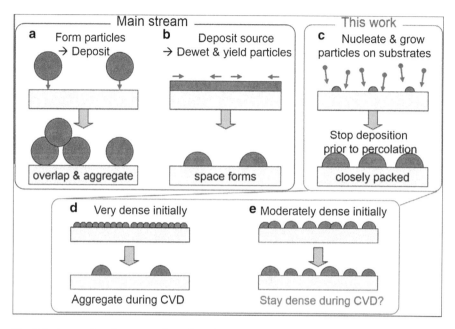

Fig. 2.1 Approaches for preparation of catalyst particle arrays on substrates. (**a**) Pre-forming catalyst particles and then depositing the particles on the substrate. (**b**) Pre-forming the catalyst layer on substrates and then transforming the layer into particles. (**c**) Directly nucleating and growing catalyst particles on substrates. (**d**, **e**) Strategy for maintaining catalyst particles at high densities during CVD. (Reprinted with permission from [144], Copyright © 2014 Elsevier Ltd)

1 nm thick Fe catalyst on doped silicon at 580 °C. CNTs with average diameter of 3.8 nm were successfully integrated in via holes with diameters ranging between 0.3 and 1 μm, corresponding to the tube shells density of more than 10^{12} cm^{-2}. They also synthesized a 70 nm diameter CNT bundle in a 140 nm diameter via hole, the tube shells density was close to 10^{13} cm^{-2} and the resistance is 44 Ω for the CNT bundle with length of 100 nm regardless of the contact resistance. Nuri et al. [144] deposited 0.6 nm thick Ni catalyst layer on a 5-nm-thick TiN conductive underlayer over the SiO$_2$/Si wafer. The flow of catalyst preparation and CNTs growth are shown in Fig. 2.1 as a comparison, two mainstream catalyst preparation methods are shown in Fig. 2.1a and b, in which Fig. 2.1a shows pre-forming the catalyst particles and then depositing the particles on the substrate, Fig. 2.1b represents pre-forming the catalyst layer on substrate and then transforming the layer into nanoparticles. Using these two methods, they encountered the catalytic particles aggregation or undesired spacing gaps between the catalyst particles, respectively. As an alternative, they directly nucleated and grew catalyst particles on the TiN substrate, and halted to deposit prior to percolation in order to avoid formation of very dense initial catalyst layer (Fig. 2.1c), which tends to cause catalyst particles aggregation during CVD (Fig. 2.1d). A combination procedure of relatively high deposition temperature (400 °C) and low deposition rate (8.1 pm s^{-1}) as well as

substrate bias voltage of -20 V was applied to maintain catalyst particles at high density. After annealing, Ni particles as dense as 2.8×10^{12} cm^{-2} can be formed on conductive TiN layer (Fig. 2.1e). Then CVD was carried out at 400 °C to grow CNTs. The role of high deposition temperature and substrate bias voltage was to enhance the surface diffusion of Ni particles and low deposition rate was used to diffuse Ni atoms for a longer period of time. They obtained the mass density of the CNTs to be as high as 1.1 g cm^{-3}, corresponding to a 1.2×10^{13} cm^{-2} of tube shell density. The measured resistances for different tube shells were between 39 and 95 kΩ.

In conclusion, some investigations have shown that synthesis of high density CNT bundles is feasible. However, the resistance value of CNT-based via is still much higher compared to that of a Cu-based via of the same diameter [10]. As low as possible CVD temperature tends to generate undesirable defects and disorder graphite in CNT structures [156] thus leads to a relatively high resistance. In addition, coexisting of metallic and semiconducting CNTs also limits the number of conduction channels in a bundle, resulting in a negatively impact on the resistance of CNTs-based vias. How to build a good electrical contact between CNT tips and metal electrode are also remaining issues. However, these issues would be gradually solved as fabrication technique further matures in the near future.

2.3.1.2 Low Temperature Synthesis of CNT-Based Via Interconnects

In order to meet the requirement to integrate CNTs into the VLSI back-end-of-line (BEOL), the maximum allowed growth temperature should be enough low (no more than 400 °C) [119–121] for future technology generations in order to protect other thin film layers [166]. The PECVD technology is well known to have an advantage in lowering the temperature of CNTs growth due to plasma-assisted dissociation of carbon source [167] and heating the substrate [168]. To alleviate the undesirable ion bombardment etching effects damage to vertically aligned CNTs, optimization process is very crucial [169]. Some alternative configurations such as the remote PECVD [120, 121, 170] and atmospheric pressure non-thermal plasma CVD [171] have been developed. For the former, plasma can be generated far away from the substrate; while for the latter, the ion bombardment can be effectively alleviated due to high collision frequency among molecules at atmospheric pressure. The synthesis of CNTs using thermal CVD usually needs a higher temperature than PECVD method [172, 173], however, some investigations have indicated that growth temperature can be controlled below 400 °C based on catalyst surface process [137] or low activation energy of catalysts [119].

A lot of investigations have managed to reduce the temperature to 350–400 °C for synthesis of vertically aligned CNT forests [137, 148, 174–177]. However, the CNTs were synthesized only on the substrate without any via hole in these works, even the CNTs were synthesized on an insulating substrate [137, 174, 175]. Some groups have reported the synthesis of CNTs via at temperatures of more than 600 °C

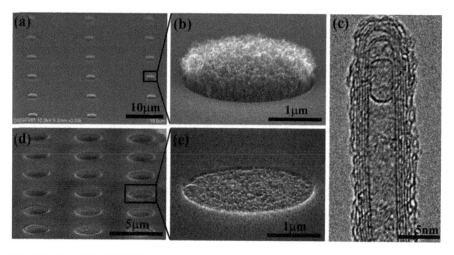

Fig. 2.2 (**a**) and 1.2 (**b**) SEM images of CNT grown at 390 °C in via holes. (**c**) High magnification TEM image of CNT grown at 390 °C before CMP. (**d**) and (**e**) SEM images of planar CNT vias after CMP. (Reprinted with permission from [120], Copyright 2007, AIP Publishing LCC)

[178–180] or around 500 °C [131, 181, 182]. It is still required to further lower the synthesis temperature in these works for CNTs application as interconnect in future technology generations.

Yokoyama et al. [120] synthesized vertically aligned MWCNTs via arrays by using remote plasma CVD at 390 °C. The initial stacking structure composed of Cu wiring (100 nm), a tantalum (Ta) diffusion barrier layer (15 nm), and a TiN contact layer (5 nm) were formed on the Si substrate by sputtering, respectively. The vias were formed by depositing 250 nm thick tetraethylorthosilicate dielectric layer and subsequent photolithography and wet-etching procedures. The average diameter of 3.8 nm Co catalyst particles were deposited over the whole substrate and CNTs were only grown selectively in via-holes. The SEM and TEM images of CNT bundles grown in a 2 μm diameter via hole for 30 min are shown in Fig. 2.2, in which Fig. 2.2a, b SEM images of CNT grown at 390 °C in via hole are shown; Fig. 2.2c high magnification TEM image of a MWCNT grown at 390 °C before CMP is shown. Figure 2.2d and e are SEM images of planarization CNT vias after CMP, respectively. The diameter of CNTs ranged from 5 to 10 nm (average of 7 nm) and the density of CNTs was about 1.6×10^{11} cm^{-2}, corresponding to one-tenth of the close-packed density of 7 nm CNTs. The measured resistance was 36 Ω for a 2 μm diameter via without CMP process, when planarization process was used to open the end caps of MWCNTs and after an annealing at 400 °C, the via resistance reduced to 0.6 Ω, which is several times higher than those of tungsten plug.

Chiodarelli et al. [121] have also successfully synthesized CNT bundles via arrays at temperature as low as 400 °C using microwave remote plasma CVD. A 1.3 nm Ni catalysts layer was deposited on whole surface of a wafer of 200 mm diameter; vertically aligned CNTs were grown only inside a series of via arrays with diameters in between 150 and 300 nm because of progress selectivity. The grown

CNTs had an average diameter of 12.1 nm and the number density of CNTs was about 1.6×10^{11} cm^{-2}. After planarization process and top Ti/Au pads deposition and subsequently anneal at 400 °C, the resistance was measured about 13.7 Ω for 576 vias in parallel, the average resistance of a single CNT (include the contributions of contact resistance at the bottom TiN/CNT and at the top Ti/CNT interfaces) was about 1.1 MΩ. The high resistance value mainly resulted from the highly defective CNTs.

Vollebregt et al. [119] have reported that synthesis of vertically aligned CNTs via interconnects at a temperature as low as 350 °C using thermal CVD method. The Co was used as a catalyst and TiN layer as a support. As a comparison, the Fe-based CNTs grown at higher temperatures (500 °C and 550 °C) were also presented. The feasibility of growing CNTs at such a low temperature by thermal CVD method was attributed to the lower activation energy of catalysts [119, 158]. The 5 nm thick Co catalysts were evaporated on TiN conductive layer in the etched chamfer on 100 mm diameter p-type Si (100) wafers. The pre-anneal and CNTs growth time was 60 min for the samples fabricated at 350 °C and 10 min for the samples fabricated at 400 °C, respectively. After CNTs growth, the wafer was covered with the thickness of 100 nm Ti and 2 μm Al for top metallization. The yield of Co-grown CNTs at 350 °C was about 86.5 %, which was comparable with Co-based CNTs at 400 °C (88.5 %) and Fe-grown CNTs fabricated at higher temperature(87.5 % for 500 °C and 83.7 % for 550 °C), respectively. However, the growth rate was found to be relatively low, the growth terminated when the length of CNTs reached about 1.5 μm at temperature of 350 °C. The average diameter of CNTs was 8 nm and the tube shell density was in the order of 5×10^{10}/cm^2. The average resistivity of a CNT-via fabricated 400 °C and 350 °C with Co catalysts was 85 mΩ cm and 139 mΩ cm, respectively. For the 500 and 550 °C Fe-grown CNT-based vias, the average resistivity was 27 mΩ cm and 22 mΩ cm, respectively.

It is challenging to grow CNTs via interconnect at a temperature of no more than 400 °C because the growth temperature highly affects the quality of the CNTs. The CNTs grown at a relatively low temperature tend to be highly defective because carbon atoms deposited at the edge of the tubes do not have enough time to diffuse [175], forming defective edges with pentagons and heptagons which will eventually induce bending in nanotubes [183]. The poor quality of CNTs as well as the lower number density of CNTs is the main contributions to large resistance. As a comparison, the MWCNTs grown using arc-discharged high temperature process exhibit fewer defects and lower resistivity (~7 μΩ cm) [66]. At a lower temperature the growth rate is also restricted due to the reduced diffusion and reaction of carbons [184]. This was demonstrated in many works [115, 119, 121, 185, 186]. Recently, Shang et al. [187] developed a method to increase the growth rates of vertically aligned CNTs at a relatively low temperature by using photo-thermal CVD technique with Ti/Fe bilayer film as catalysts. The growth temperature was 370 °C and the growth rate was up to 1.3 μm/min, which was eight times faster than the values reported by traditional thermal CVD methods. Anyway good progress has been made in growth of vertically aligned CNT-based vias in the range of temperatures compatible with VLSI process. This will provide a great hope for CNTs application as interconnects.

2.3.1.3 CNT-Based Horizontal Interconnects

Many methods have been demonstrated to grow CNT-based horizontal interconnects. One method is to route the growth direction with external forces such as electric fields [188, 189], magnetic fields [190], or gas flow [191]. Dai et al. synthesized horizontal oriented SWCNTs onto the surface of polysilicon [188] and SiO_2/Si [189] substrates based on the CVD of ethylene (900 °C) in electric fields established across patterned metal electrodes. To use this method, two important conditions should be satisfied, one is the directing ability of electric fields, and other is that the CNTs must stay away from substrate surfaces during growth and lengthening to avoid van der Waals interactions from the substrate surfaces, so that the CNTs can fully experience the aligning effect of the electric field. The horizontally controlled growth of CNTs by a magnetic field has also been proposed [190]. Magnetic field was applied in the process of adhering catalyst particles on silicon oxide substrates from dispersion, and then the iron nanoparticles fixed catalyst particles in a defined orientation under the magnetic field, which resulted in aligned growth of the CNTs. Huang et al. [191] have developed a method to control the orientation of CNTs by routing the direction of gas flow in the CVD system. The SWCNTs were synthesized on SiO_2/Si wafers by CO CVD at 900 °C with CO/H_2 mixture as feeding gas and the mono-dispersed Fe/Mo nanoparticles as catalysts. The length of grown SWCNTs can be more than 2 mm.

Ajayan et al. [123] proposed a substrate site selective growth of horizon oriented CNTs. A solution of ferrocene dissolved in xylene catalyst was firstly pre-vaporized on the patterned SiO_2/Si substrates at 150 °C, CNTs were subsequently grown by the thermal CVD at 800 °C. In this case, the CNTs were grown only on the surfaces of SiO_2 patterns, leaving the Si areas bald. The reason for site-selective partition can be related to the different adhesion energy of the iron particles on the substrate surfaces, and/or different surface diffusion rates of iron clusters on these two substrates. Later on, they [124] expanded this method to obtain tunable lengths of CNTs by selectively masking of SiO_2 patterns with a metal Au that does not catalyze CNTs growth. Seong et al. [125] obtained densely packed horizontal SWCNT arrays on a group of paralleled pattern Fe lines (thickness < 0.5 nm) on annealed quartz wafers by CVD method. After annealing the Fe at 550 °C in air to form isolated iron oxide nanoparticles with diameters about 1 nm, and then purging with hydrogen at 900 °C for 5 min, finally a flow of methane and hydrogenate was introduced at 900 °C for 1 h for growth of SWCNTs.

In these methods mentioned above, the horizontal oriented CNTs can be successfully realized. However, a common problem in these works [123–125, 188–191] lies that the CNTs are grown on insulating layers, which make the grown CNTs unsuitable for application as interconnects.

Another method to grow horizontal CNTs is to directly place the catalysts on a vertical sidewall of electrodes. Zhang et al. [192] employed physical vapor deposition to place the catalysts. Highly doped poly-Si was used as the electrode with a 50 nm SiON top layer as a hard mask. A 1 nm thick Ni film was deposited by radio frequency magnetron sputtering and rapid thermal annealing at 600 °C for 1 min in nitrogen atmosphere, followed by a wet-etching step for 2 min. Ni selectively reacted with the exposed Si to form silicide nanoparticles on the vertical sidewall, while those unreacted Ni on the hard mask and SiO$_2$ underlayer could be removed by the subsequent etching step. CNTs were grown under thermal CVD condition between 600 and 750 °C, using C$_2$H$_2$ as carbon source. The resulting yield of CNTs was found to be very low and randomly oriented due to low density of the catalyst particles deposited on the vertical sidewall of the electrode. Santini et al. [193] used electrochemical deposition method to place catalysts on the sidewalls of multiple TiN electrodes fabricated on p+ top implanted silicon substrates with a 200 nm thick SiO$_2$ layer. Ni particles were deposited only on sidewalls of the TiN electrodes due to SiON hard mask on the top surface of the TiN. CNTs were subsequently grown from the electrodeposited Ni nanoparticles by thermal CVD at temperature of 620–640 °C. The maximum density of horizontal CNTs could be up to 3×10^{11}/cm^2 by controlling the Ni catalysts sizes in the range of 15–30 nm, the resistance values are 40 Ω for the 200 nm spacing electrode. The technique described in [193] allows for a good control over CNTs horizontal direction, but too low bundle density may lead to a high resistance, and the growth temperature was also far beyond 400 °C, making the process incompatible with future application requires.

Lu et al. [194] demonstrated that horizontally suspended CNT (HSCNT) forests and bundle arrays were grown locally from selected trench sidewalls with the help of Pt thin film micro-heaters integrated on a silicon substrate. The micro-heaters supplied by a DC voltage enable to heat the substrate by the Joule effect of resistors. The iron catalysts layer was deposited and patterned onto these trench sidewalls by the tilted electron beam evaporation through a shadow mask. Local areas of the substrate were heated to about 640 °C for the CNT growth, while most other areas remained at much lower than 640 °C. The morphology of horizontal suspended CNT bundles is shown in Fig. 2.3a–d, and Fig. 2.3e indicates the I–V curves measured in the longitudinal and lateral directions of a CNT bundle. Figure 2.3f shows that the CNTs are MWCNTs and the outermost diameters are in the range of 10–30 nm. The longitudinal resistivity (along the CNTs) and lateral resistivity were, respectively, estimated to be 1.09 and 20 Ω cm, which is too large compared to that of Cu.

Chiodarelli et al. proposed an indirectly method to fabricate horizontal CNT interconnects [127]. They firstly synthesized vertically aligned CNTs inside via holes with diameters ranging from 300 to 200 nm using CVD method. The CNT bundles with tube shell density of 10^{13} cm^{-2}, lengths of 20 μm, and diameters scalable to 50 nm have been obtained. Then, the CNT bundles were flipped on the horizontal direction to serve as horizontal interconnects. Finally, to reduce the contact resistances, the symmetrical contacts were made at the tips of the CNTs in the so-called end-bonded geometry with a metallization process. The measured contact resistivity was $3.9 \times 10^{-8} \Omega$ cm^2 with Pd/Au contacts, corresponding to a resistivity of CNT of 1.1 mΩ cm, two hundred times higher than that of Cu.

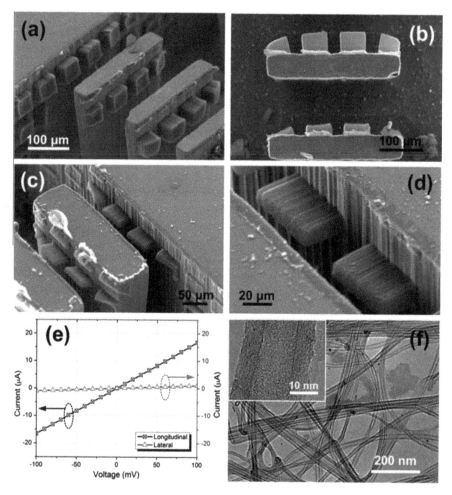

Fig. 2.3 (**a**) SEM images of HSCNT bundle arrays. (**b**) Local top view of these HSCNT bundles. (**c**) Some thin HSCNT bundles. (**d**) A local magnified view. (**e**) I–V curves taken in the longitudinal and lateral directions of a HSCNT bundle. (**f**) TEM images of HSCNTs. The insert shows a MWCNT segment. (Reprinted with permission from [194], Copyright © 2013 Elsevier Ltd)

2.3.1.4   A High Quality Contact of CNT-Metal

For CNT-metal contact, there are usually two geometries: the side-bonded contact and end-bonded contact [195]. The transport mechanism for a side-bonded contact can be explained from existence of tunneling barrier in the interface of CNT-metal [196]. The contact resistance originates from the decay of electron wave functions across the van de Waals spacing between metal and CNT [197]. The cohesion between metal and sp$_2$ carbon is inversely proportional to the metal-carbon distance [198, 199]. For the end-contact structure, there usually exists physical (physisorption) and chemical (chemisorptions) interaction at the end of the CNTs [195]. Generally speaking, the contact resistance of CNTs-metal is dependent on different electrode materials in terms of wettability [200, 201] and the work function difference [201] of electrode metal to CNTs. For those metals with low work function, surface oxides can be more dominant than the work function difference [201, 202]. The contact resistance can also results from the electrical contact between the CNTs and metal [203]. Catalysts layer with poor contact with CNTs electrically and chemically can also lead to increased resistance [204]. Impurities (defects) at near the CNTs-metal interface also contribute to the contact resistance due to physical and chemical changes [128, 199].

To effectively reduce the contact resistance between CNTs and metal electrode, various efforts have been attempted. Joule heating induced annealing [128, 205–208] was used to de-adsorb molecules in the interface region and improve the bonding between the CNTs and electrode surfaces [206] or formation of titanium carbon nitride which improves the electrical contact between the CNTs and the TiN electrodes [205]; electron beam induced deposition of carbon [209, 210] was employed to increase electronic coupling at the Fermi surface of metal-CNTs; deposition of nickel on metal electrode [211, 212] can form electrically stable bonds at the interfaces between the electrodes and CNTs; ultrasonic nanowelding [213, 214] was used to bond between the CNTs and metal electrodes and form stable Ohmic contact; and the contact resistance can also be effectively tuned by the contact configuration [215]. Though effectively reducing the CNT-metal contact resistance to some extent, the methods mentioned above are not likely to be suitable to optimize the contact between the bundles of CNT and metal electrode because they cannot be scaled up [216]. Furthermore, the CNTs would be easily destroyed due to Joule heating and contaminated by exposure to chemicals. Thermal annealing [121, 217–220] as the standard semiconductor technique is available when temperature is controlled in a certain range.

For the CNT bundle-based vertical interconnects, the large intrinsic resistances of a CNT can be reduced by paralleling densely packing aligned CNTs to increase the electrical conduction channels. In order to ensure sufficient connections between CNT's tip and top metallic contact, CMP planarization is usually required so as to maximize the number of conduction channels [120, 121, 126, 204, 221–223]. In this case, great attention is usually placed on CMP planarization treatment and selection of appropriate electrode materials. Post-CMP process such as hydrofluoric acid or plasma treatment to remove the residual oxide layer can also reduce the contact resistance [224]. Recently, a method based on densifying and then transferring CNT bundles into Via holes has also been proposed [225].

For CNTs-based horizontal interconnects, some novel methods have been proposed in order to effectively lower the contact resistance. Santini [216] has demonstrated to replace CNT-metal contacts with CNT-CNT contacts. When two suspended CNTs with diameter around 30 nm grow from opposing electrodes and touch each other at their tips with outermost shells, the contact resistivity reached about 1.4×10^{-7} Ωcm^2, which was an order of magnitude smaller than that of CNT with side-contacted TiN surface. Chiodarelli et al. [127] made symmetrical contacts

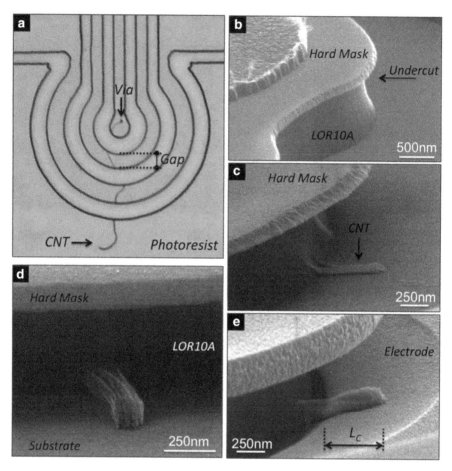

Fig. 2.4 (a) Optical image of a horizontal bundle after dry-etch. (b) Stack of materials after dry-etch. (c) Tip of a CNTs bundle emerging from under the hard mask after receding 700 nm of resin by wet-etch. (d) Front view of a CNT bundle after wet-etch. (e) Tip of a CNTs bundle after metallization (before lift-off). (Reprinted with permission from [129], Copyright 2013, AIP Publishing LCC)

at the tips of the CNTs in the so-called end-bonded geometry with metallization process. The contact resistivity was measured to be 3.9×10^{-8} Ω cm^2 with Pd/Au contacts. Later on, they expanded this method to contact CNTs at both the tip and the side of CNTs bundles in the so-called all-around geometry [129]. Firstly, they grew vertical CNTs and then flipped them horizontally by dipping the samples in isopropanol as in the "capillary folding" to obtain horizontal CNT bundles. The flow can be seen from Fig. 2.4. First, a stack of patterning materials with a 1 μm thick layer of resin, a 100 nm thick hard mask of SiO$_2$, and a 0.8 μm thick photoresist was successively deposited on the CNTs, and then the resin and the CNTs were etched in O$_2$ plasma to cut the bundle into segments, each providing one interconnect structure (Fig. 2.4a), after that, continued to wet-etch the resin only at the sidewalls with SiO$_2$

protecting the top layer (Fig. 2.4b). Consequently, the resin recedes laterally, leaving the lateral surface of the CNT bundles partially exposed (Fig. 2.4c, d). Finally, the Cr or Pd metallization was done by directional metal evaporation at a tilted angle, which defined the lateral coverage length of the CNT bundles. As samples rotated during metal evaporation, the hard mask partially shielded the deposition, so the electrode became wedge-shaped toward the borders (Fig. 2.4e). Combining end-bonded contact and "all-around" geometry contact, the measured specific contact resistivity was close to 1.6×10^{-8} Ω cm^2, which outperformed that of CNT bundles contacted at end-bonded geometry.

So far, the lowest electric resistance obtained for a CNT-based interconnect in literature is around 0.6 Ω [120, 142], which is the same order as the theoretical value of W plugs [226]. In terms of the resistivity, the typical low resistance value obtained in current technology for CNTs-based interconnects is in the range from 10^{-3} Ω cm [126–129] to 10^{-2} Ω cm [225]. As a comparison, the typical resistivity measured for a good contact between the individual MWCNT and metal electrode is in the order of 10^{-5} Ω cm [178, 227] to 7.5×10^{-4} Ω cm [228, 229]. As for Cu interconnect, the size-dependent resistivity is 5.8×10^{-6} Ω cm at 22 nm technology node [11]. The lowest resistivity achieved so far for CNTs bundles is still considerably higher for future interconnect applications.

2.3.1.5 Selective Growth of Metallic CNTs

Coexisting of metallic and semiconducting CNTs in a bundle quite restricts to lower the resistance of CNT interconnects. Increase in the ratio of metallic to semiconducting CNTs can increase the number of conduction channels of a CNT bundle, thus favors to reduce the resistance of CNTs-based interconnects.

Several research groups have pursued type-selective growth either semiconducting or metallic CNTs by optimizing process conditions. A 90 % proportion of semiconducting SWCNTs has been realized by plasma-enhanced CVD at 600 °C [230]. Two type catalysts either discrete ferritin particles with 300 Fe atoms or ~1 Å thick Fe films were coated on SiO$_2$/Si wafers, and CH$_4$/Ar gases were introduced for the growth of SWCNTs. The electrical measurements showed that 90 % of grown CNTs were semiconducting. A higher yield semiconducting SWCNTs in vertically aligned SWCNT array has been also realized by PECVD method [231]. A 0.5 nm Fe catalysts film was sputtered on a 10 nm thick Al layer pre-coated on the SiO$_2$/Si wafer. The Al was used to effectively prevent Fe nanoparticles from aggregation at the high temperature for the growth of vertically aligned SWCNTs. To start the growth of CNTs, the catalyst-coated SiO$_2$/Si wafer was quickly (<5 s) moved from a cool zone (25 °C) into the center of a radio frequency (80 W, 13.56 MHz) plasma-enhanced tube furnace heated at 750 °C under a pure gas flow of C$_2$H$_2$ at a pressure of 30 mTorr for 3 min. The measurements indicated that 96 % of semiconducting SWCNTs were obtained. The plasma and well-controlled low gas pressure were referred to play an important role in the selective growth of semiconducting vertically aligned-SWCNTs.

As for preferential growth of metallic CNTs, Wang et al. [133] used a series of monohydroxy alcohol homologues (methanol, ethanol, propanol, butanol, and pentanol) as CVD carbon source to explore the effect of the ratio of carbon atoms to oxygen atoms (RCO) in the alcohol homologues molecules on the structure of SWCNTs. They found that metallic SWCNT contents can be enriched with increase in the RCO. A maximum of 65 % metallic SWCNTs could be obtained from pentanol molecules. The selectively etching effect of hydroxyl radicals from the monohydroxy alcohol molecules and protection from formed amorphous carbon were responsible for the direct enrichment of metallic SWCNTs. Further, Harutyunyan et al. [134] have realized the fraction of tubes with metallic SWCNT up to 91 % by varying the noble gas ambient during thermal annealing of the catalysts, and in combination with oxidative and reductive species. SWCNTs were grown from Fe nanocatalyst deposited onto a SiO_2/Si support and in situ annealed in a He or Ar ambient that contained various ratios of H_2 and H_2O using methane as the carbon source at 860 °C. They achieved 91 % ratio metallic CNTs for He:H_2 (8:2) conditioning ambient in the presence of 3.5 mTorr H_2O for 1 min annealing times. They analyzed annealing effect on catalysts and found that the change in both morphology and coarsening behavior of catalyst contributes to the selective growth of CNTs.

Although progress has been made in type-selective growth of metallic CNTs, the growth mechanism is not yet very clear, and the growth temperature is relatively high compared to that of required comparable with CMOS technology. Moreover, more importantly, current fabrication techniques cannot effectively control over the selective growth of high density aligned CNT bundles. From this point of view, there is a long way to go for the process development of high selectivity of metallic type SWCNT bundles.

There have also been many studies on chirality-selective growth of CNTs [232–237]. Control of the caps type at the nucleation stage [232], catalysts type and appropriate reactive gas concentration [233], catalyst structure [234], size and energy-dependent CNT cap nucleation [235], the state of the catalysts at the point of nucleation [236], and the stability of bimetallic catalysts [237] are the key factors in determining the chirality of SWCNTs, respectively. Further, the growth of single-chirality SWCNTs has been successfully realized [135] based on a kind of W-based alloy nanoparticles as catalysts. Very high melting points and unique atomic arrangements of W are two crucial factors enabling single-chirality selective growth. High melting point ensures the crystalline structure of SWCNTs at high temperature during the growth, unique atomic arrangement can be used to control the structure of grown SWCNTs. With careful manipulation of the composition, structure, and size of the alloy catalyst nanoparticles and optimization of the CVD conditions, the yield as high as 92 % (12, 6)-dominated metallic SWCNTs was obtained. The realization of single-chirality growth of CNTs would probably provide a reference for research of type-selective growth process.

2.3.1.6 CNTs-Based Through-Silicon-Via for 3D Integration

Compared to the traditional two-dimensional integration, three-dimensional (3D) integration has advantages of high performance, low-power consumption, and high density [238]. It can present an effective candidate in meeting the challenges in the development of on-chip interconnections beyond the 32 nm technology node [239]. A critical structural component in the 3D integration is the through-silicon-via (TSV), which directly connects the stacked die structures so that the total length of interconnection is enormously shortened [240]. The TSVs in a 3D integration may be fully or partially filled by using highly conductive metals such as Cu [241, 242] and tungsten [243, 244]. However, different thermo-mechanical properties between the TSVs and the silicon may cause significant TSVs induced stress [245, 246] leading to quality and reliability issues, such as Cu extrusion [239, 247], interfacial delamination [248–250], and open-crack failure [251]. Moreover, the fill of high aspect ratio TSVs without voids through conventional deposition or electroplating process is also considered to be difficult [252]. Because of its high aspect ratio [253] and extraordinary electrical, mechanical, and thermal properties, especially high thermal conductivity, low coefficient of thermal expansion, and low thermal coefficient of resistivity, CNTs have been proposed as TSV interconnects [254–259] and some performance modeling have also been studied [260–264] recently.

So far, two main methods to fabricate CNT-based TSV have been proposed. One is direct thermal CVD method [254–256]. Xu et al. [254] fabricated CNTs bundles from the bonding TSVs with an average depth of 140 μm and diameter of 30 μm from direct thermal CVD at 700 °C, with Fe film of 2 nm thickness as the catalyst. The obtained tube density in every TSV was about 7×10^9 cm^{-2} and the diameter of the single CNT varies from 30 to 130 nm. The average electrical resistance of every CNT-based TSV was about 2 kΩ, corresponding to a resistivity was about 9.7×10^{-3} Ω cm. Wang et al. [255] synthesized CNTs in the TSV with a 50 μm diameter and different depth using thermal CVD at 700 °C, with an evaporated 1 nm thick Fe layer as catalyst. The CNT density was about 1×10^{10} cm^{-2}. The TSV top layer planarization was performed using a combined lapping and polishing process. The measured resistances from CNT-based TVSs with 50 μm diameter and 86 μm, 120 μm, and 139 μm depth are 210 Ω, 280 Ω, and 340 Ω, respectively. The high resistances in these two works probably result from the low density of CNTs and poor Ohmic contacts between the CNTs and electrodes. Xie et al. [256] fabricated CNT-based TSV using thermal CVD at 650 °C. The dip-coating FeCl$_2$ solution was adopted as catalyst. The measured resistance for a 5 μm diameter and 25 μm depth TSV was 297 Ω without considering the contact resistance. The main problem for the CNT-based TSV fabricated with thermal CVD method is that the growth temperature is too high to be compatible with future CMOS technology. On this account, a growth-densification-transfer process instead of directly growth of CNTs in the TSV has also been proposed in recent years [257–259]. The fabrication process of TSVs from Wang et al. [257] is illustrated as shown in Fig. 2.5. The aligned CNT forest was firstly synthesized on a Si substrate by thermal CVD at high temperature of 700 °C (Fig. 2.5a), and then as-grown

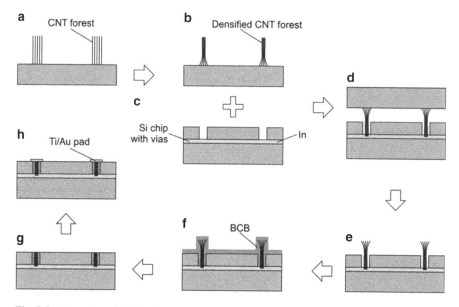

Fig. 2.5 Illustration of the fabrication process of TSVs filled with densified CNT bundles. (**a**) CNT forest grown by thermal CVD at high temperature (700 °C). (**b**) As-grown CNT forest densified through a vapor densification process. (**c**) Target chip prepared with an In layer on the bottom of the vias. (**d**) Two chips in (**b**) and (**c**) were aligned and then bonded at low temperature (200 °C). (**e**) Growth substrate separated. (**f**) BCB spin coated to fill the gaps between the CNTs and the sidewalls of the vias. (**g**) BCB-supported CNT forest planarized. (**h**) Metallic pads formed on the CNT-filled vias. (Reprinted with permission from [257], Copyright 2012 IEEE)

CNTs forest was densified to form more closely packed bundles (Fig. 2.5b) by using a densification method [225]. A Ti/Au layer was evaporated onto the top of the densified CNT bundles. The target chip was prepared with an indium layer on the bottom of the via (Fig. 2.5c), then the CNT wafer was aligned and inserted into the TSV wafer so that the Ti/Au layer contact with indium layer and bonded them at low temperature of 200 °C (Fig. 2.5d). When the growth substrate was separated from the CNT bundles (Fig. 2.5e), the finished CNT-in-via structure was then spun coated with a layer of benzocyclobutene (BCB) polymer to fill the gaps between the CNT bundles and the sidewalls of the vias (Fig. 2.5f). Finally, the planarization of CNT bundles (Fig. 2.5g) and metallic pads was formed on the CNT-filled vias (Fig. 2.5h). The resistance of the TSV with 100 μm diameter and 195 μm depth was 2.7 Ω, corresponding to a resistivity of 3.9 mΩ cm. They synthesized CNT-based TSVs with various aspect ratios [258], the measured resistance for a via of 400 μm in height and 100 μm in diameter was 1 Ω, corresponding to a resistivity of 1.6 mΩ cm. Later on, a tape-assisted transfer method and optimization of the bonding parameters were proposed to further increase the yield up to 97 % [259].

A challenging problem in current technology on CNT-based TSVs research is that the achieved resistance values are still too high for real application purposes.

As a comparison, a typical resistance is 13 mΩ for the Cu TSV with a diameter of 45 μm and depths of 95 μm [239]. Very little electrical and thermal reliability tests have been investigated so far.

2.3.2 The Challenges for Graphene Nanoribbon Interconnects

There are several limitations and challenges to implement GNRs in current technology. In the first place, wafer-scale high quality graphene is required to be synthesized on arbitrary substrates, which is suitable for patterning in the form of GNR interconnects. In general, the method of graphene preparation and the type of its substrate determine the weight of scattering sources for electrons in large-area graphene [265]. Acoustic and optical phonons in graphene, substrate polar phonons, and Coulomb scatterings due to charged impurities are the most prominent sources of scattering in graphene at a finite temperature [266].

2.3.2.1 Graphene Fabrication

A variety of methods have been introduced for graphene production such as epitaxial growth on a silicon wafer [267], direct CVD epitaxy on metal substrates [268], chemical oxidation [269], mechanical exfoliation [101], solvent exfoliation from highly oriented pyrolytic graphite (HOPG) [270], and silicon sublimation from SiC [271]. Although the mobility of suspended and annealed graphene can exceed 200,000 cm^2/V-s and demonstrate an exceptional material with highest mobility record [272, 273], it reduces to 40,000 cm^2/V-s [274] for supported graphene devices at room temperature due to trapped charges in the substrate [275]. High quality graphene on silicon can be produced by mechanical exfoliation method [276]. However, the mass production and selective placement of graphene at a specific location are almost impossible and thereby the method is not currently suitable for integrated circuits. Few atomic layers of graphene with millimeter size can be produced by silicon sublimation method, in which thin layer of SiC deposited on Si substrate followed by silicon evaporation [277]. Although this method results in the carrier mobility as high as 25,000 cm^2/V-s, it needs annealing temperature of at least 1200 °C in H$_2$ ambient condition [110] which makes it incompatible with some of the subsequent fabrication processes in manufacturing integrated circuit. The growth of large scale graphene can be achieved by ambient pressure CVD on metallic substrates such as nickel [278] and copper [279], to be transferred on arbitrary substrates by etching the metallic substrate [268], which can result in graphene flake with mobilities and sheet resistances in the range of 3700 cm^2/V-s and 280 Ω/square [268], respectively. Graphene can be produced from exfoliation of HOPG. However, the method is limited by the choice of solvent with proper surface energy thermodynamics [270] and required a chemical reduction step to recover original electronic properties similar to graphene obtained by mechanical exfoliation. Also, it usually produces monolayer graphene with low mobility in the range of 10 and 1000 cm^2/V-s [280].

2.3.2.2 Fabrication of Graphene Nanoribbon Interconnects

Graphene requires to be patterned in the form of GNR with width below 10 nm for interconnect applications, which can lead to the introduction of dangling bonds at the edges of GNR. The edge roughness is a key issue in fabrication of GNR interconnects and has crucial effects in shortening the MFP of electrons in GNR such that it can eliminate the attractive electron transport properties of graphene [281]. It increases the backscattering probability of electrons due to side wall scattering and thereby decreases the ratio of longitudinal to transverse velocity of electrons in GNR interconnects [282, 283]. Yang and Murali [284] experimentally observed the line width-dependent mobility of electrons in GNR, showing that electron mobility degrades by decreasing the GNR width below 60 nm. Edge roughness is increased by scaling down the minimum feature size due to increase in manufacturing variants of lithography and dry etching processes [285]. Thus, the efforts of most current research are to fabricate smooth-edged GNRs to preserve the superior electronic quality of graphene. Yu et al. [286] dissolved carbon atoms on nickel substrate at high temperatures and covered it with a silicon film, such that it can be patterned for GNR interconnects and transistors after removing nickel substrate. Wang et al. [287] produced smoother GNRs down to 5 nm using conventional lithography in conjunction with gas-phase etching. Dai et al. [288] showed a simple solution-based method to produce GNRs with widths down to sub-10 nm.

Another approach for the production of high quality ribbons with low disorder and smooth edge is based on unzipping the oxidized MWCNT through mechanical sonication, which can result in GNRs with approximately 20 nm length and mobility as high as 1500 cm^2/V-s [289]. Kim et al. [290] produced 45 nm width GNRs by a controlled thermally induced unwrapping of MWCNTs. Li Xie et al. [291] produced GNRs with widths between 10 and 30 nm by sono-chemical unzipping of MWCNT. An accurate control over the edge roughness of GNR can be achieved by bottom-up approach, in which the one-dimensional chains of poly-aromatic carbon precursor have been developed [292]. GNR with precisely defined width can be produced by the scalable bottom-up approach beyond the precision limit of modern lithographic approach [293]. The width and edge periphery of GNRs can be defined by the structure of precursor. However, bottom-up approaches are usually limited to some specific substrates (e.g., Au (111)) and might not be applicable for large scale production of interconnect in current technology process.

Sprinkle et al. [294] produced graphene on a template SiC substrate using a self-organized growth method and then narrowed to 40 nm width GNRs using lithography. Recently, Baringhaus et al. [295] showed that electrons in 40 nm wide GNR can have ballistic transport at room temperature for up to 16 μm length by controlling substrate geometry. GNRs are epitaxially grown on the edges of three-dimensional structures etched into silicon carbide wafers in order to produce perfectly smooth edges. As electrons flowing at the edges don't have interaction with electrons in the bulk portion of the nanoribbons, they can contribute much better than other electrons traveling in the middle and act similar to optical waveguides in optical fiber which transmits without scattering. It has been announced (www.graphenea.com/) that electron mobility reached to one million with a sheet resistance of 1 Ω/square, which are two orders of magnitude lower than two-dimensional graphene and ten times smaller than the best theoretical predictions for graphene because the production of GNR with smooth edge activates the ballistic transport of those electrons. However, the challenge comes from growing GNR on conventional substrates such as silicon and silicon oxide as SiC substrate is expensive and thereby not applicable for cost efficient integrated circuit fabrication. In addition, the growth of thin graphene films requires single crystal SiC substrate, which is not suitable for interconnects as required growth over dielectric materials. Furthermore, the back-end thermal budget in the fabrication of integrated circuits is low and thus the required high temperature in these techniques makes these not proper for producing GNR interconnects. Beside the quality and grain size of graphene produced by CVD method, the growth temperature is subject of research efforts to lower the temperature below the tolerable level ($\sim$400 °C).

2.3.2.3 Multi-Layer GNR Interconnects

Although it is experimentally demonstrated that monolayer GNR exhibits much better scalability [104], larger EM reliability [296], and current-carrying capacity ($\sim$100 A/cm^2 [104]) compared with Cu interconnect, its performance is limited by its atomically thin 2D geometry [296]. Murali et al. [285] experimentally found the average resistivity of an 18-nm wide GNR on SiO$_2$ to be 15 $\mu\Omega$ cm. The per unit length resistance of GNR interconnects can become lower than that of Cu interconnects when there exists improvement in substrate limited MFP of electrons or in the number of conduction channels as the resistance and maximum current density of an interconnect are determined by its conducting cross section [297]. Stacking graphene layers in the 3D configuration can improve the GNR performance for interconnect applications by lowering interconnect resistance [298, 299]. As stacking monolayer of graphene can resemble the graphite properties [300], multi-layer graphene nanoribbon (MLGNR) requires being electrically decoupled from each other by breaking the van der Waals force between layers graphite to preserve exceptional conducting behavior of each individual graphene monolayers as shown in Fig. 2.6 [301]. Kondo et al. [302, 303] showed that intercalation of iron chloride molecules between sheets of graphene decrease the resistivity of MLGNR interconnects by one order of magnitude and can be as low as that of Cu interconnects. By fabricating MLGNR interconnects with a width of 20 nm and stacking 11 graphene layers, they demonstrated that MLGNR interconnects have high current reliability and the variation of its resistivity by narrowing down

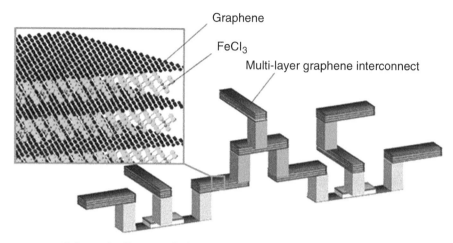

Schematic diagram of LSI using multi-layer graphene interconnects

Fig. 2.6 Schematic diagram of LSI using MLGNR interconnects (Reprinted with permission from [301]; AIST, http://www.aist.go.jp)

the width of MLGNR interconnect is negligible. The performance of stacked non-interacting multi-layer GNRs with smooth edge and Fermi energies above 0.2 eV outperforms Cu interconnects by scaling down the cross-sectional dimensions and increasing the interconnect length [289].

GNRs as local interconnects are required to be doped in order to shift the Fermi energy of electrons in graphene from Dirac point (charge neutrality point) because the Fermi level of perfect non-doped GNR is close to the charge neutrality point and thereby GNR is in highly resistive state. Miyazaki et al. [281] showed that carrier doping up to a Fermi energy close to 1 eV can reduce the resistivity of GNR interconnects to about 10 $\mu\Omega$ cm. The deposition of NH_3 and CH_4 can produce n-type graphene [304] while H_2O and NO_2 can produce p-type graphene. One challenge is the controllability and uniformity of doping on GNR surface, such that there is not significant degradation in its electron MFP, maintaining the graphene properties. Chemical doping [305] and high electric field induced carrier concentrations have minor effects in the degradation of graphene mobility, which usually guarantees the ballistic carrier transport on micrometer scale of graphene at room temperature [306]. However, the dopant charges are mostly accumulated at the interface between graphene and substrate and thereby there is a screening effect of lower GNR layers (near dopants) in MLGNR interconnects, which degrades the doping effects on upper layers. This leads to smaller contribution of upper GNR layers in total conductance of MLGNR interconnects [307]. One method is to shift the Fermi energy of upper layers by edge functionalization. However, it can convert the edges of GNR interconnect to p-type while its center remains n-type [308].

2.3.2.4 Performance and Reliability of GNR Interconnects

On-chip delay and power dissipation of interconnects are two bottle-necks for achieving high performance integrated circuits. By increasing the number of layers in the MLGNR stack, there is reduction in the equivalent resistance of the MLGNR interconnect while the equivalent capacitance increases [104]. Thus, there are optimal points in the delay and the energy-delay product (EDP) curves of MLG interconnect at some specific number of layers as shown in Fig. 2.7a, b [104]. Rakheja et al. [104] showed that MLGNR interconnects with ideal contacts or side contacts need greater number of GNR layers for reaching at minimum values of delay and EDP than the case having top contacts because the coupling between side contacts and MLGNR is stronger and thereby increase in the number of layers can leads to lower equivalent resistance of MLGNR interconnect. The optimal number of layers for minimizing the delay and EDP of the MLGNR interconnects varies by shrinking the interconnect dimension corresponding to technology node as shown in Fig. 2.7c. Figure 2.7d, e show the delay and the EDP of the MLGNR interconnects versus lengths of Cu and MLGNR interconnects at the 9.5 nm technology node. While the delay of MLGNR interconnects with ideal or side contacts (P = 0) outperforms that of Cu interconnects in absence of size

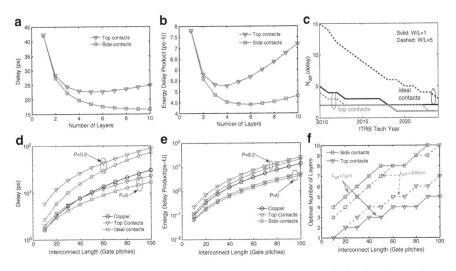

Fig. 2.7 (**a**) Delay versus number of layers in MLGNR interconnects with ideal and top contacts. (**b**) Energy-delay product versus number of layers in MLGNR interconnects with ideal and top contacts. (**c**) Optimal number of layers to minimize the delay of MLGNR interconnects as a function of the ITRS technology year. (**d**) Delay versus the lengths of Cu and MLGNR interconnects for the ideal and top contacts. (**e**) Energy-delay product of Cu and MLGNR interconnects for the ideal and top contacts. Note: (**d**) and (**e**) have been driven by 5× driver at the 9.5 nm technology node. (**f**) Optimal number of layers to minimize the delay of MLGNR interconnects versus its length for two values of mean free path 1 μm and 300 nm. (© 2013 IEEE, Reprinted with permission from [104])

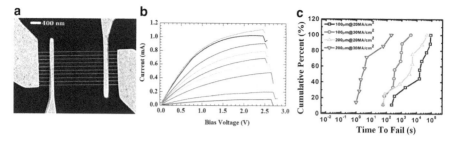

Fig. 2.8 (**a**) SEM image of ten GNRs with 21-nm-wide and 750-nm-length implemented between each electrode pair. (**b**) I–V curves of a set of parallel GNRs taken through electrical breakdown. Top I–V curve is for ten GNRs in parallel, the second curve from top for nine GNRs in parallel, and so on. (Reprinted with permission from [309], Copyright 2009, AIP Publishing LLC.) (**c**) Interconnect lifetime distribution of ten graphene interconnects for different wire lengths and stress current densities at the temperature equal to 523 K. (© 2012 IEEE, Reprinted with permission from [320])

effects, MLGNR interconnects are slower than Cu interconnects when size effect is presented. In the same scenario, the EDP of MLGNR interconnects are better than that of Cu interconnects only when the edge-scattering coefficient is very small ($P = 0$). It is shown in Fig. 2.7f that the number of layers for minimizing the delay at a given length is determined by the MFP of MLGNR interconnects [104] which needs to be considered in fabrication of MLGNR interconnects.

Another key metrics of interconnects is the maximum current-carrying capability. There exists a threshold voltage for occurring the permanent breakdown by gradually increasing voltage across the GNR interconnects [309]. Figure 2.8a shows the SEM image of ten GNRs with 21-nm wide and 750-nm length and their corresponding I–V curves is shown in Fig. 2.8b. The top I–V curve is for ten GNRs in parallel, the second curve from top for nine GNRs in parallel, and so on. The cause of failure is mostly due to Joule heating because the addition of more graphene layers in MLGNR interconnects can improve current density at breakdown. Increasing temperature results in the overpopulation of phonons, which increases the electron scattering and thereby lowers the carrier mobility in GNR interconnects. In this scenario, GNR interconnects with rough edge are more vulnerable to Joule heating due to the edges scattering, such that the threshold voltage of permanent breakdown can happen at smaller values [310, 311]. In addition, GNR interconnects have a higher degree of interaction with the substrate compared to CNT interconnects [312–315]. This can lead to smaller heat limited current degradation due to higher heat dissipation in GNR interconnects [316] while it can also work against its performance by increasing scattering due to the disorders in substrate or insulating dielectrics [317, 318]. Thus, the performance of GNR interconnects is more substrate dependent than Cu or CNT interconnects.

For MLGNR interconnects, the non-uniformity in the layers, e.g., doping profile, can be the source of degradation in its current density. Though, it would be favored for the reliability, reproducibility, and performance. For instance, the limit current density of MLGNR interconnects with five layers has been measured equal to 4×10^8 A/cm^2 [316] which is much more than Cu interconnect with 10^6 A/cm^2 [319]. Chen et al. [320] studied the degradation of graphene interconnects with time under constant current stress by exposure to air. Figure 2.8c shows the interconnect lifetime distribution of ten graphene interconnects for different wire lengths and stress current densities at the temperature equal to 523 K. They found the failure time of 6 hour for current density of 2×10^7 A/cm^2 at 250 °C. This is slightly worse than capped Cu interconnect at the same stress current density, indicating that the capping material can significantly change the performance of GNR interconnects [320]. Nasiri et al. [321] showed that the relative stability of MLGNR interconnects is increased by increasing either their length or width. Crosstalk analysis is another reliability metrics for interconnects. Das et al. [322] found that near-end crosstalk noise of GNR interconnects is larger than that of Cu and MWCNT interconnects while its far-end crosstalk noise is smaller than Cu interconnect and larger than MWCNT interconnect. Also, the gate oxide failure rate of GNR interconnects is two orders of magnitude smaller than Cu interconnects, indicating that the reliability of GNR interconnect outperform Cu interconnects [323].

2.3.2.5 GNR Interconnects in All-Graphene Circuits

In conventional technology, the material for producing devices and interconnects are usually different, which have been chosen among semiconductor and metallic materials, respectively. Fabricating GNR-based electronic circuits consisting of both transistors and interconnects in the same continuous graphene sheet would bring some release from the contact resistance of metal-to-graphene contacts [110, 324]. Kang et al. [325] proposed all-graphene circuits as shown in Fig. 2.9a–g, where both devices and interconnects are concurrently fabricated by monolithically patterning a single sheet of graphene. Tunable band gap of graphene can be adjusted for GNR interconnects by pattering it with larger width and different orientation. While all zigzag edge GNRs are metallic, GNRs with armchair edges are semi-metallic and very small bandgap if the number of dimer lines becomes (3p+1, 0), where p is an arbitrary integer [112]. In all-graphene logic gates, both the length and the bending type of GNR are critically important in using GNR as local interconnects. In order to implement GNR interconnects continuously with GNR-based transistors in an all-graphene logic gate, only one perpendicular path and two angled paths can have zigzag edges as shown in Fig. 2.9h, i. Thus, unlike conventional metallic interconnects, both the number of bending and its angle along interconnecting path can alter the delay of GNR interconnects [326]. The design and patterning of GNR interconnects in all-graphene logic circuits will require an algorithm for shortest path considering all zigzag edge bending which is challenging for design software and manufacturing process. All graphene architectures are promising designs for low-power applications due to the minimum connection to conventional metal contacts, which reduce the series contact resistance at source and drain regions together with the potential implementation of low-power

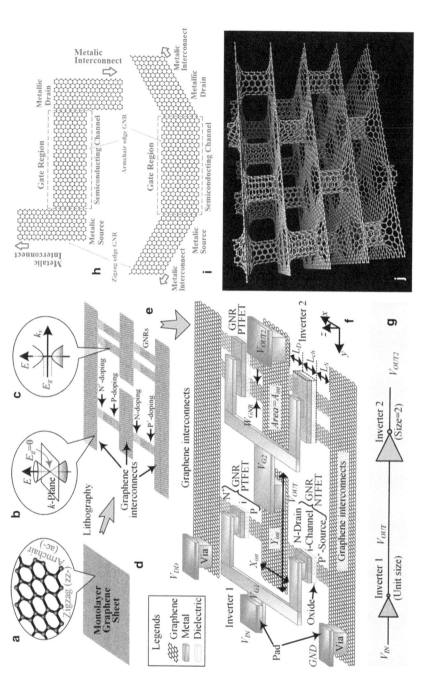

Fig. 2.9 (**a–g**) The design and fabrication steps of all-graphene integrated circuits, in which the local interconnects between GNR transistors are metallic GNR made by monolithographically patterning of large-area graphene. The figure shows the series of two all-graphene inverters (Reprinted with permission from [325], Copyright 2013, AIP Publishing LCC). (**h–i**) Metallic and semiconducting GNRs are formed by changing GNR width and chirality, such that zigzag edge GNRs can be used as metallic source and drain regions and GNR interconnects while armchair GNRs are used for the semiconducting channel. (**j**) 3D hybrid nanostructure of carbon nanotube and graphene, which shows the application of CNTs as vertical connection (via) between graphene sheets (Reprinted with permission from [332], Copyright 2015, American Chemical Society)

GNR-based transistors [327–329]. The structure can potentially reduce the complex fabrication process for local interconnect in nanoscale dimensions, leading to ultra-dense and thin integrated circuits. However, the fabrication process with atomic precision is required to implement zigzag-edged and armchair-edged GNRs for interconnects and transistors with smooth edges in order to maintain the metallic and semiconducting behaviors of GNR and prevent the reduction in MFP by edge scattering. The 1D CNT can be used as via in this structure. The 3D hybrid structure of CNTs connected perpendicularly to graphene layers has been synthesized by CVD process [330, 331] and theoretically investigated by ab inito calculation [332] as shown in Fig. 2.9j. In this approach, GNR layers are grown horizontally on cobalt film, which then disaggregates during the CVD process to form cobalt nanoparticles for the growth of vertical CNTs. However, the quality of contact between CNT via to GNR layers in this structure requires more investigation and it is a long way to use in integrated circuits.

2.4 Conclusion

For CNT-based interconnects, major efforts have been made in performance prediction, design approach, and fabrication process. However, problems such as large interconnect resistance, high deposition temperature, imperfect growth quality, and incontrollable chirality are still not solved, severely restricting the application of CNT interconnects into actual circuits. More efforts are needed in this direction. In graphene and graphene nanoribbon-based interconnects, precise control over the nature of defects and edge roughness of graphene nanoribbons are key challenges to overcome and improve electrical properties for the application in future nanoscale architectures.

References

1. Moore GE (1965) Cramming more components onto integrated circuits. Electron Mag 38:114–117
2. Dennard RH, Rideout VL, Gaensslen FH, Yu HN, Rideout VL, Bassous E, Leblanc AR (1974) Design of ion-implanted MOSFET's with very small physical dimensions. IEEE J Solid-State Circuits SC-9:256–268
3. International Technology Roadmap for Semiconductors (2013) http://www.itrs.net/link/2013ITRS/Home2013.htm
4. Vadasz LL, Grove AS, Rowe TA, Moore GE (1969) Silicon-gate technology. IEEE Spectr 6:28–35
5. Meindl JD (2003) Beyond Moore's law: the interconnect era. Comput Sci Eng 5:20–24
6. Solanki R, Pathangey B (2000) Atomic layer deposition of copper seed layers. Electrochem Solid-State Lett 3(10):479–480
7. Hu CK, Harper JME (1998) Copper interconnections and reliability. Mater Chem Phys 52:5–16

8. Yamada M, Yagi H, Sugatani S, Miyajima M, Matsunaga D, Hosoda T, Kudo H, Misawa N, Nakamura T (1999) Cu interconnect technologies in Fujitsu and problems in installing Cu equipment in an existing semiconductor manufacturing line. Proceedings of the IEEE 1999 International Interconnect Technology Conference, San Francisco, pp 115–115, 24–26 May 1999

9. Steinlesberger G, Engelhardt M, Schindlera G, Steinhogl W, Glasow AV, Mosig K, Bertagnolli E (2002) Electrical assessment of copper damascene interconnects down to sub-50 nm feature sizes. Microelectron Eng 64:409–416

10. International Technology Roadmap for Semiconductors (ITRS) (2007) http://www.itrs.net/ Links/2007ITRS/Home2007.htm

11. Koo KH, Kapur P, Saraswat KC (2009) Compact performance models and comparisons for gigascale on-chip global interconnect technologies. IEEE Trans Electron Devices 56(9):1787–1798

12. Arnaud L, Cacho F, Doyen L, Terrier F, Galpin D, Monget C (2010) Analysis of electromigration induced early failures in Cu interconnects for 45 nm node. Microelectron Eng 87:355–360

13. Sondheimer EH (1952) The mean free path of electrons in metals. Adv Phys 1:1–42

14. Mayadas AF, Shatzkes M (1970) Electrical-resistivity model for polycrystalline films: the case of arbitrary reflection at external surfaces. Phys Rev B1:1382–1389

15. Steinhoegl W, Schindler G, Steinlesberger G, Engelhardt M (2002) Size-dependent resistivity of metallic wires in the mesoscopic range. Phys Rev B 66:075414-1-4

16. Rossnagel SM, Kuan TS (2004) Alteration of copper conductivity in the size effect regime. J Vac Sci Technol B: Microelectron Nanometer Struct 22:240–247

17. Wen W, Maex K (2001) Studies on size effect of copper interconnect lines. In: 2001 Proceedings of the 6th international conference on solid-state and integrated-circuit technology (ICSICT), vol 1, pp 416–418

18. Kim CU, Park J, Michael N, Gillespie P, Augur R (2003) Study of electron-scattering mechanism in nanoscale copper interconnects. J Electron Mater 32(10):982–987

19. Ryu C, Kwon KW, Loke ALS, Lee H, Nogami T, Dubin VM, Kavari RA, Ray GW, Wong SS (1999) Microstructure and reliability of copper interconnects. IEEE Trans Electron Devices 46(6):1113–1120

20. Tokei Z, Croes K, Beyer GP (2010) Reliability of copper low-k interconnects. Microelectron Eng 87:348–354

21. Kizil H, Kim G, Chel CS, Zhao B (2001) TiN and TaN diffusion barriers in copper interconnect technology: towards a consistent testing methodology. J Electron Mater 30(4):345–348

22. Wang H, Gupta A, Tiwarii A, Zhang X, Narayan J (2003) TaN-TiN binary alloys and superlattices as diffusion barriers for copper interconnects. J Electron Mater 32(10):994–999

23. Das D, Rahaman H (2012) Modeling of single-wall carbon nanotube interconnects for different process, temperature, and voltage conditions and investigating timing delay. J Comput Electron 11:349–363

24. Moon DY, Kwon TS, Kang BW, Kim WS, Kim BM, Kim JH, Park JW (2010) Copper seed layer using atomic layer deposition for Cu interconnect. In: 2010 3rd international nanoelectronics conference (INEC), pp 450–451

25. Rakheja S, Naeemi A (2011) Modeling interconnects for post-CMOS devices and comparison with copper interconnects. IEEE Trans Electron Devices 58(5):1319–1328

26. Li B, Christiansen C, Badami D, Yang CC (2014) Electromigration challenges for advanced on-chip copper interconnects. Microelectron Reliab 54:712–724

27. Chan YC, Yang D (2010) Failure mechanisms of solder interconnects under current stressing in advanced electronic packages. Prog Mater Sci 55(5):428–475

28. Hu CK, Gignac L, Rosenberg R (2006) Electromigration of copper/low dielectric constant interconnects. Microelectron Reliab 46:213–231

29. Huntington HB, Grone AR (1961) Current-induced marker motion in gold wires. J Phys Chem Solid 20(1–2):76–87

30. Stangl M, Lipták M, Acker J, Hoffmann V, Baunack S, Wetzig K (2009) Influence of incorporated non-metallic impurities on electromigration in copper damascene interconnect lines. Thin Solid Films 517:2687–2690
31. Li B, Christiansen C, Gill J, Sullivan T (2006) Threshold electromigration failure time and its statistics for copper interconnects. J Appl Phys 100:114516-1-10
32. Hu CK, Gignac LM, Liniger E, Huang E, Greco S, McLaughlin P, Yang CC, Demarest JJ (2009) Electromigration challenges for nanoscale Cu wiring. AIP Conf Proc 1143(1):3–11
33. Galand R, Brunetti G, Arnaud L, Rouviere JL, Ciement L, Walta P, Wouters Y (2013) Microstructural void environment characterization by electron imaging in 45 nm technology node to link electromigration and copper microstructure. Microelectron Eng 106:168–171
34. Roy A, Hou Y, Tan CM (2009) Electromigration in width transition copper interconnect. Microelectron Reliab 49(9-11):1086–1089
35. Lin MH, Lin MT, Wang T (2008) Effects of length scaling on electromigration in dual-damascene copper interconnects. Microelectron Reliab 48(4):569–577
36. Mishra JK, Priye V (2014) Design of low crosstalk and bend insensitive optical interconnect using rectangular array multicore fiber. Opt Commun 331:272–277
37. Fazzi A, Magagni L, Mirandola M, Charlet B, Cioccio LD, Jung E, Canegallo R, Guerrieri R (2007) 3D capacitive interconnections for wafer-level and die-level assembly. IEEE J Solid State Circuits 42(10):2270–2282
38. Miura N, Kohama Y, Sugimori Y, Ishikuro H, Sakurai T, Kuroda T (2009) A high speed inductive-coupling link with burst transmission. IEEE J Solid State Circuits 44:947–955
39. Sasaki N, Kimoto K, Moriyama W, Kikkawa T (2009) A single-chip ultra-wideband receiver with silicon integrated antennas for inter-chip wireless interconnection. IEEE J Solid State Circuits 44(2):382–393
40. Paik KW, Mogro-Campero A (1994) Studies on the high-temperature superconductor (HTS)/metal/polymer dielectric interconnect structure for packaging applications. IEEE Trans Compon Packg Manuf Technol Part B 17(3):435–441
41. Iijima S, Ichihashi T (1993) Single-shell carbon nanotubes of 1-nm diameter. Nature 363:603–605
42. Bethune DS, Kiang CH, Beyers R (1993) Cobalt-catalyzed growth of carbon nanotubes with single-atomic layer walls. Nature 363:605–607
43. Ijima S (1991) Helical microtubes of graphitic carbon. Nature 354:56–58
44. Saito R, Dresslhaus G, Dresselhaus MS (1998) Physical properties of carbon nanotubes. Imperial College Press, London, UK
45. Dresselhaus MS, Dresselhaus G, Avouris P (2001) Carbon nanotube: synthesis, properties, structure, and applications. Springer, Berlin
46. Wilder JWG, Venema LC, Rinzler AG, Smalley RE, Dekker C (1998) Electronic structure of atomically resolved carbon nanotubes. Nature 391:59–62
47. White CT, Roberston DH, Mintmire JW (1993) Helical and rotational symmetries of nanoscale graphitic tubules. Phys Rev B 47:5485–5488
48. Saito R, Fujita M, Dresselhaus G, Dresselhaus MS (1992) Electronic structure of graphene tubules based on C60. Phys Rev B 46:1804–1811
49. Zólyomi V, Koltai J, Rusznyák Á, Kürti J, Gali Á, Simon F, Kuzmany H, Szabados Á, Surján PR (2008) Intershell interaction in double walled carbon nanotubes: charge transfer and orbital mixing. Phys Rev B 77(24):245403-1-6
50. Kong J, Yenilmez E, Tombler TW, Kim W, Dai H, Laughlin RB, Liu L, Jayanthi CS, Wu SY (2001) Quantum interference and ballistic transmission in nanotube electron waveguides. Phys Rev Lett 87:106801-1-4
51. Schönenberger C, Bachtold A, Strunk C, Salvetat JP, Forr L (1999) Interference and interaction in multiwall carbon nanotubes. Appl Phys A 69:283–295
52. Bachtold A, Fuhrer MS, Plyasunov S, Forero M, Anderson EH, Zettl A, McEuen PL (2000) Scanned probe microscopy of electronic transport in carbon nanotubes. Phys Rev Lett 84(26):6082–6085
53. White CT, Todorov TN (1998) Carbon nanotube as long ballistic conductors. Nature 393(6682):240–242

54. Li H, Srivastava N, Mao JF, Yin WY, Banerjee K (2011) Carbon nanotube vias: does ballistic electron–phonon transport imply improved performance and reliability? IEEE Trans Electron Device 58(8):2689–2701
55. McEuen PL, Fuhrer MS, Park H (2002) Single-walled carbon nanotube electronics. IEEE Trans Nanotechnol 1(1):78–85
56. Purewal MS, Hong BH, Ravi A, Chandra B, Hone J, Kim P (2007) Scaling of resistance and electron mean free path of single-walled carbon nanotubes. Phys Rev Lett 98:186808-1-4
57. Frank S, Poncharal P, Wang ZL, de Heer WA (1998) Carbon nanotube quantum resistors. Science 280:1744–1746
58. Yi W, Lu L, Zhang DL, Pan ZW, Xie SS (1999) Linear specific heat of multiwall carbon nanotubes. Phys Rev B 59:R9015–R9018
59. Li HJ, Lu WG, Li JJ, Bai XD, Gu CZ (2005) Multichannel ballistic transport in multiwall carbon nanotubes. Phys Rev Lett 95:086601-1-4
60. Appenzeller J, Martel R, Avouris P, Stahl H, Lengeler B (2001) Optimized contact configuration for the study of transport phenomena in ropes of single-wall carbon nanotubes. Appl Phys Lett 78(21):3313–3315
61. Treacy MMJ, Ebbesen TW, Gibson JM (1996) Exceptionally high Young's modulus observed for individual carbon nanotubes. Nature 381:678–680
62. Salvetat JP, Bonard JM, Thomson NH, Kulik AJ, Forr'o L, Benoit W, Zuppiroli L (1999) Mechanical properties of carbon nanotubes. Appl Phys A 69:255–260
63. Li F, Cheng HM, Bai S, Su G, Dresselhaus MS (2000) Tensile strength of single-walled carbon nanotubes directly measured from their macroscopic ropes. Appl Phys Lett 77(20):3161–3163
64. Wei B, Spolenak R, Kohler-Redlich P, Ruhle M, Arzt E (1999) Electrical transport in pure and boron-doped carbon nanotubes. Appl Phys Lett 74(21):3149–3151
65. Yao Z, Kane CL, Dekker C (2000) High field electrical transport in single-wall carbon nanotubes. Phys Rev Lett 84(13):2941–2944
66. Wei BQ, Vajtai R, Ajayan PM (2001) Reliability and current carrying capacity of carbon nanotubes. App Phys Lett 79(8):1172–1174
67. Berber S, Kwon YK, Tomanek D (2000) Unusually high thermal conductivity of carbon nanotubes. Phys Rev Lett 84(20):4613–4616
68. Cao A, Qu J (2012) Size dependent thermal conductivity of single-walled carbon nanotubes. J Appl Phys 112:013503-1-9
69. Imtani AN (2013) Thermal conductivity for single-walled carbon nanotubes from Einstein relation in molecular dynamics. J Phys Chem Solid 74:1599–1603
70. Bhattacharya S, Amalraj R, Mahapatra S (2011) Physics-based thermal conductivity model for metallic single-walled carbon nanotube interconnects. IEEE Electron Device Letts 32(2):203–205
71. Che J, Cagin T, Goddard WA III (2000) Thermal conductivity of carbon nanotubes. Nanotechnology 11:65–69
72. Mingo N, Broido DA (2005) Length dependence of carbon nanotube thermal conductivity and the "problem of long waves". Nano Lett 5(7):1221–1225
73. Qiu B, Wang Y, Zhao Q, Ruan X (2012) The effects of diameter and chirality on the thermal transport in free-standing and supported carbon-nanotubes. Appl Phys Lett 100:233105-1-4
74. Hata T, Kawai H, Ohto T, Yamashita K (2013) Chirality dependence of quantum thermal transport in carbon nanotubes at low temperatures: a first-principles study. J Chem Phys 139:044711-1-8
75. Yu C, Shi L, Yao Z, Li D, Majumdar A (2005) Thermal conductance and thermopower of an individual single-wall carbon Nanotube. Nano Lett 5(9):1842–1846
76. Pop E, Mann D, Wang Q, Goodson K, Dai H (2006) Thermal conductance of an individual single-wall carbon nanotube above room temperature. Nano Lett 6(1):96–100
77. Wang ZL, Tang DW, Li XB, Zheng XH, Zhang WG, Zheng LX, Zhu YT, Jin AZ, Yang HF, Gu CZ (2007) Length-dependent thermal conductivity of an individual single-wall carbon nanotube. Appl Phys Lett 91:123119-1-3

78. Fujii M, Zhang X, Xie H, Ago H, Takahashi K, Ikuta T, Abe H, Shimizu T (2005) Measuring the thermal conductivity of a single carbon nanotube. Phys Rev Lett 95:065502-1-4

79. Kim P, Shi L, Majumdar A, McEuen PL (2001) Thermal transport measurements of individual multiwalled nanotubes. Phys Rev Lett 87(21):215502-1-5

80. Naeemi A, Meindl JD (2007) Design and performance modeling for single-walled carbon nanotubes as local, semiglobal, and global interconnects in gigascale integrated systems. IEEE Trans Elec Dev 54(1):26–37

81. Pu SN, Yin WY, Mao JF, Liu QH (2009) Crosstalk prediction of single- and double-walled carbon-nanotube (SWCNT/DWCNT) bundle interconnects. IEEE Trans Elec Dev 56(4):560–568

82. Plombon JJ, O'Brien KP, Gstrein F, Dubin VM (2007) High-frequency electrical properties of individual and bundled carbon nanotubes. Appl Phys Lett 90:063106-1-3

83. Koo KH, Cho H, Kapur P, Saraswat KC (2007) Performance comparisons between carbon nanotubes, Optical, and Cu for future high-performance on-chip interconnect applications. IEEE Trans Electron Devices 54(12):3206–3215

84. Li H, Banerjee K (2009) High frequency analysis of carbon nanotube interconnects and implications for on-chip inductor design. IEEE Trans Electron Devices 56(10):2202–2214

85. Li H, Xu C, Srivastava N, Banerjee K (2009) Carbon nanomaterials for next-generation interconnects and passives: physics, status, and prospects. IEEE Trans Electron Devices 56(9):1799–1821

86. Close GF, Yasuda S, Paul B, Fujita S, Wong HSP (2008) A 1 GHz integrated circuit with carbon nanotube interconnects and silicon transistors. Nano Lett 8(2):706–709

87. Collins PG, Hersam M, Arnold M, Martel R, Avouris P (2001) Current saturation and electrical breakdown in multiwalled carbon nanotubes. Phys Rev Lett 86:3128–3131

88. Srivastava N, Li H, Kreupl F, Banerjee K (2009) On the applicability of single-walled carbon nanotubes as VLSI interconnects. IEEE Trans Nanotechnol 8(4):542–559

89. Graham AP, Duesberg GS, Hoenlein W, Kreupl F, Liebau M, Martin R, Rajasekharan B, Pamler W, Seidel R, Steinhoegl W, Unger E (2005) How do carbon nanotubes fit into the semiconductor roadmap. Appl Phys A 80:1141–1151

90. Srivastava A, Xu Y, Sharma AK (2010) Carbon nanotubes for next generation very large scale integration interconnects. J Nanophotonics 4:1–26

91. Xu Y, Srivastava A (2010) A model for carbon nanotube interconnects. Int J Circ Theor Appl 38:559–575

92. Naeemi A, Meindl JD (2006) Compact physical models for multiwall carbon interconnects. IEEE Electron Device Lett 27(5):338–340

93. Naeemi A, Meindl JD (2008) Performance modeling for single- and multiwall carbon nanotubes as signal and power interconnects in gigascale systems. IEEE Trans Electron Devices 55(10):2574–2582

94. Massoud Y, Nieuwoudt A (2006) Modeling and design challenges and solutions for carbon nanotube-based interconnect in future high performance integrated circuits. ACM J Emerg Technol Comput Syst 2(3):155–196

95. War JW, Nichols J, Stachowiak TB, Ngo Q, Egerton EJ (2012) Reduction of CNT interconnect resistance for the replacement of Cu for future technology nodes. IEEE Trans Nanotechnol 11(1):56–62

96. Li H, Yin WY, Banerjee K, Mao JF (2008) Circuit modeling and performance analysis of multi-walled carbon nanotube interconnects. IEEE Trans Electron Devices 55(6):1328–1337

97. Kurdahi FJ, Pasricha S, Dutt N (2010) Evaluating carbon nanotube global interconnects for chip multiprocessor applications. IEEE Trans Very Large Scale Integr Syst 18(9):1376–1380

98. Nieuwoudt A, Massoud Y (2008) On the optimal design, performance, and reliability of future carbon nanotube-based interconnect solutions. IEEE Trans Electron Devices 55(8):2097–2110

99. Haji-Nasiri S, Faez R, Moravvej-Farshi MK (2012) Stability analysis in multiwall carbon nanotube bundle interconnects. Microelectron Reliab 52:3026–3034

100. Tseng YC, Xuan P, Javey A, Malloy R, Wang Q, Bokor J, Dai H (2004) Monolithic integration of carbon nanotube devices with silicon MOS technology. Nano Lett 4(1):123–127

101. Novoselov KS, Geim AK, Morozov S, Jiang D, Zhang Y, Dubonos S (2004) Electric field effect in atomically thin carbon films. Science 306:666–669

102. Banadaki YM, Srivastava A (2015) Scaling effects on static metrics and switching attributes of graphene nanoribbon FET for emerging technology. IEEE Trans Emerg Top Comput 3:458–469

103. Obradovic B, Kotlyar R, Heinz F, Matagne P, Rakshit T, Giles MD, Stettler MA, Nikonov DE (2006) Analysis of graphene nanoribbons as a channel material for field-effect transistors. Appl Phys Lett 88:142102-1-3

104. Rakheja S, Kumar V, Naeemi A (2013) Evaluation of the potential performance of graphene nanoribbons as on-chip interconnects. Proc IEEE 101:1740–1765

105. Meric I, Han MY, Young AF, Ozyilmaz B, Kim P, Shepard KL (2008) Current saturation in zero-bandgap, top-gated graphene field-effect transistors. Nat Nanotechnol 3:654–659

106. Lee C, Wei X, Kysar JW, Hone J (2008) Measurement of the elastic properties and intrinsic strength of monolayer graphene. Science 321:385–388

107. Balandin AA, Ghosh S, Bao W, Calizo I, Teweldebrhan D, Miao F (2008) Superior thermal conductivity of single-layer graphene. Nano Lett 8:902–907

108. Banadaki YM, Mohsin K, Srivastava A (2014) A graphene field effect transistor for high temperature sensing applications. In: SPIE Smart Structures and Materials + Nondestructive Evaluation and Health Monitoring, pp 90600F-90600F-7

109. Sarma SD, Adam S, Hwang E, Rossi E (2011) Electronic transport in two-dimensional graphene. Rev Mod Phys 83:407–470

110. Berger C, Song Z, Li X, Wu X, Brown N, Naud C (2006) Electronic confinement and coherence in patterned epitaxial graphene. Science 312:1191–1196

111. Han MY, Özyilmaz B, Zhang Y, Kim P (2007) Energy band-gap engineering of graphene nanoribbons. Phys Rev Lett 98:206805

112. Banadaki YM, Srivastava A (2015) Investigation of the width-dependent static characteristics of graphene nanoribbon field effect transistors using non-parabolic quantum-based model. Solid State Electron 111:80–90

113. Nakada K, Fujita M, Dresselhaus G, Dresselhaus MS (1996) Edge state in graphene ribbons: nanometer size effect and edge shape dependence. Phys Rev B 54:17954

114. Naeemi A, Meindl JD (2008) Performance benchmarking for graphene nanoribbon, carbon nanotube, and Cu interconnects. In: 2008 International Interconnect Technology Conference, IITC 2008, pp 183–185

115. Vanpaemel J, Sugiura M, Barbarin Y, De Gendt S, Tokei Z, Vereecken PM, van der Veen MH (2014) Growth and integration challenges for carbon nanotube interconnects. Microelectron Eng 120:188–193

116. Dijon J, Fournier A, Szkutnik PD, Okuno H, Jayet C, Fayolle M (2010) Carbon nanotubes for interconnects in future integrated circuits: the challenge of the density. Diamond Relat Mater 19:382–388

117. Zhong GF, Iwasaki T, Kawarada H (2006) Semi-quantitative study on the fabrication of densely packed and vertically aligned single-walled carbon nanotubes. Carbon 44:2009–2014

118. Okuno H, Fournier A, Quesnel E, Muffato V, Poche HL, Fayolle M, Dijon J (2010) CNT integration on different materials suitable for VLSI interconnects. Comptes Rendus Physique 11:381–388

119. Vollebrgt S, Tichelaar FD, Schellevis H, Beenakker CIM, Ishihara R (2014) Carbon nanotube vertical interconnects fabricated at temperatures as low as 350°C. Carbon 71:249–256

120. Yokoyama D, Iwasaki T, Yoshida T, Kawarada H, Sato S, Hyakushima T, Nihei M, Awano Y (2007) Low temperature grown carbon nanotube interconnects using inner shells by chemical mechanical polishing. Appl Phys Letts 91:263101-1-3

121. Chiodarelli N, Li Y, Cott DJ, Mertens S, Peys N, Heyns M, De Gendt S, Groeseneken G, Vereecken PM (2011) Integration and electrical characterization of carbon nanotube via interconnects. Microelectron Eng 88(5):837–843

122. Sugime H, Esconjauregui S, Yang J, D'Arsie L, Oliver RA, Bhardwaj S, Cepek C, Robertson J (2013) Low temperature growth of ultra-high mass density carbon nanotube forests on conductive supports. Appl Phys Lett 103:073116-1-5
123. Zhang ZJ, Wei BQ, Ramanath G, Ajayan PM (2000) Substrate-site selective growth of aligned carbon nanotubes. Appl Phys Letts 77(23):3764–3766
124. Cao A, Baskaran R, Frederick MJ, Turner K, Ajayan PM, Ramanath G (2003) Direction-selective and length-tunable in plane growth of carbon nanotubes. Adv Mater 15(13):1105–1109
125. Kang SJ, Kocabas C, Ozel T, Shim M, Pimparkar N, Alam MA, Rotkin SV, Rogers JA (2007) High-performance electronics using dense, perfectly aligned arrays of single-walled carbon nanotubes. Nat Nanotechnol 2:230–236
126. Chiodarelli N, Masahito S, Kashiwagi Y, Li Y, Arstila K, Richard O, Cott DJ, Heyns M, De Gendt S, Groeseneken G, Vereecken PM (2011) Measuring the electrical resistivity and contact resistance of vertical carbon nanotube bundles for application as interconnects. Nanotechnology 22:085302-1-7
127. Chiodarelli N, Fournier A, Okuno H, Dijon J (2013) Carbon nanotubes horizontal interconnects with end-bonded contacts, diameters down to 50 nm and lengths up to 20 μm. Carbon 60:139–145
128. Yamada T, Saito T, Suzuki M, Wilhite P (2010) Tunneling between carbon nanofiber and gold electrodes. J Appl Phys 107:044304-1-5
129. Chiodarelli N, Fournier A, Dijon J (2013) Impact of the contact's geometry on the line resistivity of carbon nanotubes bundles for applications as horizontal interconnects. Appl Phys Lett 103(5):053115-1-4
130. Robertson J, Zhong G, Hofmann S, Bayer BC, Esconjauregui CS, Telg H, Thomsen C (2009) Use of carbon nanotubes for VLSI interconnects. Diamond Relat Mater 18:957–962
131. Vollebregt S, Ishihara R, Derakhshandeh J, van der Cingel J, Schellevis H, Beenakker CIM (2011) Integrating low temperature aligned carbon nanotubes as vertical interconnects in Si technology. In: 2011 11th IEEE international conference on nanotechnology, Portland Marriott, Portland, OR, USA, pp 985–990
132. Robertson J, Zhong G, Esconjauregui S, Zhang C, Hofmann S (2013) Synthesis of carbon nanotubes and graphene for VLSI interconnects. Microelectron Eng 107:210–218
133. Wang Y, Liu Y, Li X, Cao L, Wei D, Zhang H, Shi D, Yu G, Kajiura H, Li Y (2007) Direct enrichment of metallic Single-walled carbon nanotubes induced by the different molecular composition of monohydroxy alcohol homologues. Small 3(9):1486–1490
134. Harutyunyan AR, Chen G, Paronyan TM, Pigos EM, Kuznetsov OA, Hewaparakrama K, Kim SM, Zakharov D, Stach EA, Sumanasekera GU (2009) Preferential growth of single-walled carbon nanotubes with metallic conductivity. Science 326:116–120
135. Yang F, Wang X, Zhang D, Yang J, Luo D, Xu Z, Wei J, Wang JQ, Xu Z, Peng F, Li X, Li R, Li Y, Li M, Bai X, Ding F, Li Y (2014) Chirality-specific growth of single-walled carbon nanotubes on solid alloy catalysts. Nature 510:522–524
136. Zhong G, Xie R, Yang J, Robertson J (2014) Single-step CVD growth of high-density carbon nanotube forests on metallic Ti coatings through catalyst engineering. Carbon 67:680–687
137. Cantoro M, Hofmann S, Pisana S, Scardaci V, Parvez A, Ducati C, Ferrari AC, Blackburn AM, Wang KY, Robertson J (2006) Catalytic chemical vapor deposition of single-wall carbon nanotubes at low temperatures. Nano Lett 6(6):1107–1112
138. Nessim GD (2010) Properties synthesis and growth mechanisms of carbon nanotubes with special focus on thermal chemical vapor deposition. Nanoscale 2(8):1306–1323
139. Zhong G, Warner JH, Fouquet M, Robertson AW, Chen B, Robertson J (2012) Growth of ultrahigh density single-walled carbon nanotube forests by improved catalyst design. ACS Nano 6(4):2893–2903
140. Robertson J, Zhong G, Esconjauregui S, Zhang C, Fouquet M, Hofmann S (2012) Chemical vapor deposition of carbon nanotube forests. Phys Status Solidi B 249(12):2315–2322
141. Li Y, Kim W, Zhang Y, Rolandi M, Wang D, Dai H (2001) Growth of single-walled carbon nanotubes from discrete catalytic nanoparticles of various sizes. J Phys Chem B 105(46):11424–11431

142. Awano Y, Sato S, Kondo D, Ohfuti M, Kawabata A, Nihei M, Yokoyama N (2006) Carbon nanotube via interconnect technologies: size-classified catalyst nanoparticles and low-resistance ohmic contact formation. Phys Status Solidi (a) 203(14):3611–3616

143. Romo-Negreira A, Cott DJ, De Gendt S, Maex K, Heyns MM, Vereecken PM (2010) Electrochemical tailoring of catalyst nanoparticles for CNT spatial-dimension control. J Electrochem Soc 157(3):K47–K51

144. Na N, Kim DY, So YG, Ikuhara Y, Noda S (2015) Simple and engineered process yielding carbon nanotube arrays with $1.2e10^{13}$ cm^{-2} wall density on conductive underlayer at 400°C. Carbon 81:773–781

145. Liu RM, Ting JM, Huang JCA, Liu CP (2002) Growth of carbon nanotubes and nanowires using selected catalysts. Thin Solid Films 420–421:145–150

146. Seidel R, Duesberg GS, Unger E, Graham AP, Liebau M, Kreupl F (2004) Chemical vapor deposition growth of single-walled carbon nanotubes at 600 °C and a simple growth model. J Phys Chem B 108(6):1888–1893

147. Chen G, Seki Y, Kimura H, Sakurai S, Yumura M, Hata K, Futaba DN (2014) Diameter control of single-walled carbon nanotube forests from 1.3–3.0 nm by arc plasma deposition. Sci Rep 4:1–7

148. Ting JM, Liao KH (2004) Low-temperature, nonlinear rapid growth of aligned carbon nanotubes. Chem Phys Lett 396:469–472

149. Zhang C, Yan F, Allen CS, Bayer BC, Hofmann S, Hickey BJ, Cott D, Zhong G, Robertson J (2010) Growth of vertically-aligned carbon nanotube forests on conductive cobalt disilicide support. J Appl Phys 108:024311-1-6

150. Wang Y, Luo Z, Li B, Ho PS, Yao Z, Shi L, Bryan EN, Nemanich RJ (2007) Comparison study of catalyst nanoparticle formation and carbon nanotube growth: support effect. J Appl Phys 101:124310-1-8

151. Esconjauregui S, Bayer BC, Fouquet M, Wirth CT, Yan F, Xie R, Ducati C, Baehtz C, Castellarin-Cudia C, Bhardwaj S, Cepek C, Hofmann S, Robertson J (2011) Use of plasma treatment to grow carbon nanotube forests on TiN substrate. J Appl Phys 109:114312-1-10

152. Olivares J, Mirea T, Diaz-Duran B, Clement M, DeMiguel-Ramos M, Sangrador J, Frutos J, Iborra E (2015) Growth of carbon nanotube forests on metallic thin films. Carbon 90:9–15

153. Zhang C, Xie R, Chen B, Yang J, Zhong G, Robertson J (2013) High density carbon nanotube growth using a plasma pretreated catalyst. Carbon 53:339–345

154. Esconjauregui S, Xie R, Fouquet M, Cartwright R, Hardeman D, Yang J, Robertson J (2013) Measurement of area density of vertically aligned carbon nanotube forests by the weight-gain method. J Appl Phys 113:144309-1-7

155. Xie R, Zhang C, Chen B, van der Veen M, Zhong G, Robertson J (2014) Increased carbon nanotube area density after catalyst generation from cobalt disilicide using a cyclic reactive ion etching approach. J Appl Phys 115:144302-1-4

156. Liu TL, Wu HW, Wang CY, Chen SY, Hung MH, Yew TR (2013) A method to form self-aligned carbon nanotube vias using a Ta-cap layer on a Co-catalyst. Carbon 56:366–373

157. Hermann S, Ecke R, Schulz S, Gessner T (2008) Controlling the formation of nanoparticles for definite growth of carbon nanotubes for interconnect applications. Microelectron Eng 85(10):1979–1983

158. Vitos L, Ruban A, Skriver HL, Kollar J (1998) The surface energy of metals. Surf Sci 411:186–202

159. Zhang C, Yan F, Bayer BC, Blume R, van der Veen MH, Xie R, Zhong G, Chen B, Knop-Gericke A, Schlog R, Capraro BD, Hofmann S, Robertson J (2012) Complementary metal-oxide-semiconductor-compatible and self-aligned catalyst formation for carbon nanotube synthesis and interconnect fabrication. J Appl Phys 111:064310-1-6

160. Hofmann S, Cantoro M, Kaempgen M, Kang DJ, Golovko VB, Li HW, Yang Z, Geng J, Huck WTS, Jonson BFG, Robertson J (2005) Catalyst patterning methods for surface-bound chemical vapor deposition of carbon nanotubes. Appl Phys A 81:1559–1567

161. Maschmann MR, Franklin AD, Amama PB, Zakharov DN, Stach EA, Sands TD, Fisher TS (2006) Vertical single-and double-walled carbon nanotubes grown from modified porous anodic alumina templates. Nanotechnology 17:3925–3929

162. Chen Z, Cao G, Lin Z, Koehler I, Bachmann PK (2006) A self-assembled synthesis of carbon nanotubes for interconnects. Nanotechnology 17:1062–1066

163. Koji H, Furuta H, Sekiya K, Nitta N, Harigai T, Hatta A (2013) Increased CNT growth density with an additional thin Ni layer on the Fe/Al catalyst film. Diamond Relat Mater 36:1–7

164. Xie R, Zhang C, van der Ven MH, Arstila K, Hantschel T, Chen B, Zhong G, Robertson J (2013) Carbon nanotube growth for through silicon via application. Nanotechnology 24:125603-1-7

165. Yamazaki Y, Katagiri M, Sakuma N, Suzuki M, Sato S, Nihei M, Wada M, Matsunaga N, Sakai T, Awano Y (2010) Synthesis of a closely packed carbon nanotube forest by a multi-step growth method using plasma-based chemical vapor deposition. Appl Phys Express 3:055002-1-3

166. Meyyappan M, Delzeit L, Cassell A, Hash D (2003) Carbon nanotube growth by PECVD: a review. Plasma Sources Sci Technol 12:205–216

167. Bower C, Zhu W, Jin S, Zhou O (2000) Plasma-induced alignment of carbon nanotubes. Appl Phys Lett 77(6):830–832

168. Teo KBK, Hash DB, Lacerda RG, Rupesinghe NL, Bell MS, Dalal SH, Bose D, Govindan TR, Cruden BA, Chhowalla A, Amaratunga GAJ, Meyyappan M, Milne WI (2004) The significance of plasma heating in carbon nanotube and nanofiber growth. Nano Lett 4(5):921–926

169. Luo Z, Lim S, You Y, Miao J, Gong H, Zhang J, Wang S, Lin J, Shen Z (2008) Effect of ion bombardment on the synthesis of vertically aligned single-walled carbon nanotubes by plasma-enhanced chemical vapor deposition. Nanotechnology 19(25):255607-1-6

170. Zhong GF, Iwasaki T, Honda K, Furukawa Y, Ohdmari I, Kawarada H (2005) Very high yield growth of vertically aligned single-walled carbon nanotubes by point-arc microwave plasma CVD. Chem Vap Deposition 11(3):127–130

171. Nozaki T, Ohnishi K, Okazaki K, Korshagen U (2007) Fabrication of vertically aligned single-walled carbon nanotubes in atmospheric pressure non-thermal plasma CVD. Carbon 45(2):364–374

172. Juang ZY, Lai JF, Weng CH, Lee JH, Lai HJ, Lai TS, Tsai CH (2004) On the kinetics of carbon nanotube growth by thermal CVD method. Diamond Relat Mater 13(11-12):2140–2146

173. Wei S, Kang WP, Davidson JL, Huang JH (2006) Aligned carbon nanotubes fabricated by thermal CVD at atmospheric pressure using Co as catalyst with NH_3 as reactive gas. Diamond Relat Mater 15(11-12):1828–1833

174. Kyung S, Lee Y, Kim C, Lee J, Yeom G (2006) Deposition of carbon nanotubes by capillary-type atmospheric pressure PECVD. Thin Solid Films 506–507:268–273

175. Park YS, Yi J, Lee J (2013) The characteristics of carbon nanotubes grown at low temperature for electronic device application. Thin Solid Films 546:81–84

176. Yokoyama D, Iwasaki T, Ishimaru K, Sato S, Nihei M, Awano Y, Kawarada H (2010) Low-temperature synthesis of multiwalled carbon nanotubes by graphite antenna CVD in a hydrogen-free atmosphere. Carbon 48:825–831

177. Ting JM, Wua WY, Liao KH, Wua HH (2009) Low temperature, non-isothermal growth of carbon nanotubes. Carbon 47:2671–2678

178. Kreupl F, Graham AP, Duesberg GS, Steinhogl W, Liebau M, Unger E, Honlein W (2002) Carbon nanotubes in interconnect applications. Microelectron Eng 64(1):399–408

179. Li J, Ye Q, Cassell A, Ng HT, Stevens R, Han J, Meyyappan M (2003) Bottom-up approach for carbon nanotube interconnects. Appl Phys Lett 82(15):2491–2493

180. Pal SK, Talapatra S, Kar S, Ci L, Vajtai R, Borca-Tasciuc T, Schadler LS, Ajayan PM (2008) Time and temperature dependence of multi-walled carbon nanotube growth on Inconel 600. Nanotechnology 19(4):045610-1-5

181. Vollebregt S, Ishihara R, van der Cingel J, Beenakker K (2011) Low-temperature bottom-up integration of carbon nanotubes for vertical interconnects in monolithic 3D integrated circuits. In: 2011 IEEE International 3D Systems Integrated Conference (3DIC), pp 1–4
182. Vollebregt S, Banerjee S, Beenakker K, Ishihara R (2013) Thermal conductivity of low temperature grown vertical carbon nanotube bundles measured using the three-Ω method. Appl Phys Lett 102:191909-1-4
183. Choi YC, Bae DJ, Lee YH, Lee BS, Han IT, Choi WB, Lee NS, Kim JM (2000) Low temperature synthesis of carbon nanotubes by microwave plasma-enhanced chemical vapor deposition. Synth Met 108(2):159–163
184. Lee CJ, Park J, Huh Y, Lee JY (2001) Temperature effect on the growth of carbon nanotubes using thermal chemical vapor deposition. Chem Phys Lett 343(1–2):33–38
185. Wirth CT, Zhang C, Zhong G, Hofmann S, Robertson J (2009) Diffusion- and reaction-limited growth of carbon nanotube forests. ACS Nano 3(11):3560–3566
186. Zhu L, Xu J, Xiao F, Jiang H, Hess DW, Wong CP (2007) The growth of carbon nanotube stacks in the kinetics-controlled regime. Carbon 45(2):344–348
187. Shang NG, Tan YY, Stolojan V, Papakonstantinou P, Silva SR (2010) High-rate low-temperature growth of vertically aligned carbon nanotubes. Nanotechnology 21(50):505604-1-6
188. Zhang Y, Chang A, Cao J, Wang Q, Kim W, Li Y, Morris N, Yenilmez E, Kong J, Dai H (2000) Electric-field-directed growth of aligned single-walled carbon nanotubes. Appl Phys Letts 79(19):3155–3157
189. Ural A, Li Y, Dai H (2002) Electric-field-aligned growth of single-walled carbon nanotubes on surfaces. Appl Phys Lett 81(18):3464–3466
190. Lee KH, Cho JM, Sigmunda W (2003) Control of growth orientation for carbon nanotubes. Appl Phys Lett 82(3):448–450
191. Huang S, Cai X, Liu J (2003) Growth of millimeter-long and horizontally aligned single-walled carbon nanotubes on flat substrates. J Am Chem Soc 125:5636–5637
192. Zhang C, Cott D, Chiodarelli N, Vereecken P, Robertson J, Whelan CM (2008) Growth of carbon nanotubes as horizontal interconnects. Phys Status Sol (b) 245(10):2308–2310
193. Santini CA, Cott DJ, Romo-Negreira A, Capraro BD, Riva Sanseverino S, De Gendt S, Groeseneken G, Vereecken PM (2010) Growth and characterization of horizontally suspended CNTs across TiN electrode gaps. Nanotechnology 21:245604-1-9
194. Lu J, Miao J, Norford LK (2013) Localized synthesis of horizontally suspended carbon nanotubes. Carbon 57:259–266
195. Ngo Q, Petranovic D, Krishnan S, Cassell AM, Ye Q, Li J, Meyyappan M, Yang CY (2004) Electron transport through metal-multiwall carbon nanotube interfaces. IEEE Trans Nanotechnol 3(2):311–317
196. Kanda A, Ootuka Y, Tsukagoshi K, Aoyagi Y (2001) Electron transport in metal/multiwall carbon nanotube/metal structures. Appl Phys Lett 79:1354–1356
197. Tersoff J (1999) Contact resistance of carbon nanotubes. Appl Phys Lett 74:2122–2124
198. Matsuda Y, Deng WQ, Goddard WA III (2007) Contact resistance properties between carbon nanotubes and various metals from quantum mechanics. J Phys Chem C 111:11113–11116
199. Lan C, Zakharov DN, Reifenberger RG (2008) Determing the optimal contact length for a metal/multiwalled carbon nanotube interconnect. Appl Phys Lett 92(21):213112, -1-3
200. Lee S, Kahng SJ, Kuk Y. Nano-level wettings of platinum and palladium on single-walled carbon nanotubes. Chem Phys Lett 500: 82–85
201. Lim SC, Jang JH, Bae DJ, Han GH, Lee S, Yeo IS, Yeo IS, Lee YH (2009) Contact resistance between metal and carbon nanotube interconnects: effect of work function and wettability. Appl Phys Lett 95(26):264103-1-3
202. Felten A, Suarez-Martinez I, Ke X, Tendeloo GV, Ghijsen J, Pireaux JJ, Drube W, Bittencourt C, Ewels CP (2009) The role of oxygen at the interface between titanium and carbon nanotubes. Eur J Chem Phys Chem 10(11):1799–1804
203. Anantram MP (2001) Which nanowire couples better electrically to a metal contact: Armchair or zigzag nanotube. Appl Phys Lett 78:2055–2057

204. Lee S, Lim JS, Baik SJ (2011) Integration of carbon nanotube interconnects for full compatibility with semiconductor technologies. J Electrochem Soc 158(11):K193–K196
205. Santini CA, Vereecken PM, Volodin A, Groeseneken G, Gendtand, SD, Haesendonck CV (2011) A study of Joule heating-induced breakdown of carbon nanotube interconnects. Nanotechnology 22:395202-1-9
206. Dong LF, Youkey S, Bush J, Jiao J, Dubin VM, Chebiam RV (2007) Effects of local Joule heating on the reduction of contact resistance between carbon nanotubes and metal electrodes. J Appl Phys 101(2):024320-1-7
207. Ryan PM, Verhulst AS, Cott D, Romo-Negreira A, Hantschel T, Boland JJ (2010) Optimization of multi-walled carbon nanotube-metal contacts by electrical stressing. Nanotechnology 21(4):045705-1-6
208. Woo Y, Duesberg GS, Roth S (2007) Reduced contact resistance between an individual single-walled carbon nanotube and a metal electrode by a local point annealing. Nanotechnology 18(9):095203-1-7
209. Kim S, Kulkarni D, Rykaczewski K, Tsukruk HMV, Fedorov A (2012) Fabrication of an ultra low resistance ohmic contact to MWCNT-metal interconnect using graphitic carbon by electron beam induced deposition. IEEE Trans Nanotechnol 11:1223–1230
210. Rykaczewski K, Henry MR, Kim SK, Fedorov AG, Kulkarni D, Singamaneni S, Tsukruk VV (2010) The effect of the geometry and material properties of a carbon joint produced by electron beam induced deposition on the electrical resistance of a multiwalled carbon nanotube-to-metal contact interface. Nanotechnology 21(3):035202-1-12
211. Liebau M, Unger E, Duesberg GS, Graham AP, Seidel R, Kreupl F, Hoenlein W (2003) Contact improvement of carbon nanotubes via electroless nickel deposition. Appl Phys A 77:731–734
212. Seidel R, Liebau M, Duesberg GS, Kreupl F, Unger E, Graham AP, Hoenlein W, Pompe W (2003) In-situ contacted single-walled carbon nanotubes and contact improvement by electroless deposition. Nano Lett 3(7):965–968
213. Chen C, Yan L, Kong ESW, Zhang Y (2006) Ultrasonic nanowelding of carbon nanotubes to metal electrodes. Nanotechnology 17:2192–2197
214. Song X, Liu S, Gan Z, Yan H, Ai Y (2009) Contact configuration modification at carbon nanotube-metal interface during nanowelding. J Appl Phys 106:124308-1-4
215. Chen C, Zhang W, Wei L, Su Y, Hu N, Wang Y, Li Y, Zhong H, Liu Y, Liu X, Liu X, Zhang Y (2015) Investigation on nanotube-metal contacts under different contact types. Mater Lett 145:95–98
216. Santini CA, Volodin A, Haesendonck CV, Gendt SD, Groeseneken G, Vereecken PM (2011) Carbon nanotube-carbon nanotube contacts as an alternative towards low resistance horizontal interconnects. Carbon 49:4004–4012
217. Fiedler H, Toader M, Hermann S, Rodriguez RD, Sheremet E, Rennau M, Schulze S, Waechtler T, Hietschold M, Zahn DRT, Schulz SE, Gessner T (2014) Carbon nanotube based via interconnects: performance estimation based on the resistance of individual carbon nanotubes. Microelectron Eng 120:210–215
218. Yaglioglu O, Hart AJ, Martens R, Slocum AH (2006) Method of characterizing electrical contact properties of carbon nanotube coated surfaces. Rev Sci Instrum 77:095105-1-3
219. Kane AA, Sheps T, Branigan ET, Apkarian VA, Cheng MH, Hemminger JC, Hunt SR, Collins PG (2009) Graphitic electrical contacts to metallic single-walled carbon nanotubes using Pt electrodes. Nano Lett 9(10):3586–3591
220. Lee JO, Park C, Kim JJ, Kim J, Park JW, Yoo KH (2000) Formation of low-resistance ohmic contacts between carbon nanotube and metal electrodes by a rapid thermal annealing method. J Phys D Appl Phys 33(16):1953–1956
221. Katagiri M, Wada M, Ito B, Yamazaki Y, Suzuki M, Kitamura M, Saito T, Isobayashi A, Sakata A, Sakuma N, Kajita A, Sakai T (2012) Fabrication and characterization of planarized carbon nanotube via interconnects. Jpn J Appl Phys 1:05ED02–05ED04

222. Fiedler H, Toader M, Hermann S, Rennau M, Rodriguez RD, Sheremet E, Hietschold M, Zahn DRT, Schulz SE, Gessner T (2015) Back-end-of-line compatible contact materials for carbon nanotube based interconnects. Microelectron Eng 137:130–134

223. van der Veen MH, Vereecke B, Huyghebaert C, Cott DJ, Sugiura M, Kashiwagi Y, Teugels L, Caluwaerts R, Chiodarelli N, Vereecken PM, Beyer GP, Heyns MM, DeGendt S, Tökei Z (2013) Electrical characterization of CNT contacts with Cu Damascene top contact. Microelectron Eng 106:106–111

224. Lee S, Lee BJ (2012) Removal of residual oxide layer formed during chemical–mechanical-planarization process for lowering contact resistance. Surf Coat Technol 206:3142–3145

225. Jiang D, Wang T, Chen S, Ye L, Liu J (2013) Paper-mediated controlled densification and low temperature transfer of carbon nanotube forests for electronic interconnect application. Microelectron Eng 103:177–180

226. Sato S, Nihei M, Mimura A, Kawabata A, Kondo D, Shioya H, Iwai T, Mishima M, Ohfuti M, Awano Y (2006) Novel approach to fabricating carbon nanotube via interconnects using size-controlled catalyst nanoparticles. Int Interconnect Technol Conf 2006:230–232

227. Ebbesen TW, Lezec HJ, Hiura H, Bennett JW, Ghaemi HF, Thio T (1996) Electrical conductivity of individual carbon nanotubes. Nature 382:54–56

228. Chen Q, Wang S, Peng LM (2006) Establishing Ohmic contacts for in situ current–voltage characteristic measurements on a carbon nanotube inside the scanning electron microscope. Nanotechnology 17:1087–1098

229. Dai H, Wong EW, Lieber CM (1996) Probing electrical transport in nanomaterials: conductivity of individual carbon nanotubes. Science 272:523–526

230. Li Y, Mann D, Rolandi M, Kim W, Ural A, Hung S, Javey A, Cao J, Wang D, Yenilmez E, Wang Q, Gibbons JF, Nishi Y, Dai H (2004) Preferential growth of semiconducting single-walled carbon nanotubes by a plasma enhanced CVD method. Nano Lett 4(2):317–321

231. Qu L, Du F, Dai L (2008) Preferential syntheses of semiconducting vertically aligned single-walled carbon nanotubes for direct use in FETs. Nano Lett 8(9):2682–2687

232. Reich S, Li L, Robertson J (2006) Control the chirality of carbon nanotubes by epitaxial growth. Chem Phys Lett 421:469–472

233. Ghorannevis Z, Kato T, Kaneko T, Hatakeyama R (2010) Narrow-chirality distributed single-walled carbon nanotube growth from nonmagnetic catalyst. J Am Chem Soc 132:9570–9572

234. Chiang WH, Sankaran RM (2009) Linking catalyst composition to chirality distributions of as-grown single-walled carbon nanotubes by tuning Ni_xFe_{1-x} nanoparticles. Nat Mater 8:882–886

235. Fouquet M, Bayer BC, Esconjauregui S, Blume R, Warner JH, Hofmann S, Schlogl R, Thomsen C, Robertson J (2012) Highly chiral-selective growth of single-walled carbon nanotubes with a simple monometallic Co catalyst. Phys Rev B 85:235411-1-7

236. Fouquet M, Bayer BC, Esconjauregui S, Thomsen C, Hofmann S, Robertson J (2014) Effect of catalyst pretreatment on chirality-selective growth of single-walled carbon nanotubes. J Phys Chem C 118:5773–5781

237. Liu B, Ren W, Li S, Liu C, Cheng HM (2010) High temperature selective growth of single-walled carbon nanotubes with a narrow chirality distribution from a CoPt bimetallic catalyst. Chem Commun 48:2409–2411

238. Lau JH (2001) Evolution, challenge, and outlook of TSV, 3D IC integration and 3D silicon integration. In: 2011 International Symposium on Advanced Packaging Materials (APM), IEEE, pp 462–488

239. Tsai TC, Tsao WC, Lin W, Hsu CL, Hsu CM, Lin JF, Huang CC, Wu JY (2012) CMP process development for the via-middle 3D applications at 28 nm technology node. Microelectron Eng 92:29–33

240. Zhang R, Roy K, Koh CK, Janes DB (2001) Power trends and performance characterization of 3-dimensional integration for future technology generations. In: Quality electronic design, international symposium on IEEE computer society, pp 217–222

241. Du L, Shi T, Chen P, Su L, Shen J, Shao J, Liao G (2015) Optimization of through silicon via for three-dimensional integration. Microelectron Eng 139:31–38

242. Bayat P, Vogel D, Rodriguez RD, Sheremet E, Zahn DRT, Rzepka S, Michel B (2015) Thermo-mechanical characterization of copper through-silicon vias (Cu-TSVs) using micro-Raman spectroscopy and atomic force microscopy. Microelectron Eng 137:101–104

243. Koseski RP, Osborn WA, Stranick SJ, DelRio FW, Vaudin MD, Dao T, Adams VH, Cook RF (2011) Micro-scale measurement and modeling of stress in silicon surrounding a tungsten-filled through-silicon via. J Appl Phys 110:073517-1-10

244. Krauss C, Labat S, Escoubas S, Thomas O, Carniello S, Teva J, Schrank F (2013) Stress measurements in tungsten coated through silicon vias for 3D integration. Thin Solid Films 530:91–95

245. Le Texier F, Mazuir J, Su M, Castagne L, Souriau JC, Liotard JL, Saadaoui M, Inal K (2013) Effect of TSV density on local stress concentration: micro-Raman spectroscopy measurement and finite element analysis. Microelectron Eng 106:139–143

246. Ryu SK, Lu KH, Jiang T, Im JH, Huang R, Ho PS (2012) Effect of thermal stresses on carrier mobility and keep-out zone around through-silicon vias for 3-D integration. IEEE Trans Device Mater Reliab 12(2):255–262

247. Cheng EJ, Shen YL (2012) Thermal expansion behavior of through-silicon-via structures in three-dimensional microelectronic packaging. Microelectron Reliab 52(3):534–540

248. Ryu SK, Lu KH, Zhang X, Im JH, Ho PS, Huang R (2011) Impact of near-surface thermal stresses on interfacial reliability of through-silicon vias for 3-D interconnects. IEEE Trans Device Mater Reliab 11(1):35–43

249. Lu KH, Ryu SK, Qiu Z, Zhang X, Im J, Huang R, Ho PS (2010) Thermal stress induced delamination of through silicon vias in 3-D interconnects. In: Electronic components and technology conference (ECTC), 2010 proceedings of the 60th electronic components and technology conference, pp 40–45

250. Ryu SK, Jiang T, Im J, Ho PS, Huang R (2014) Thermomechanical failure analysis of through-silicon via interface using a shear-lag model with cohesive zone. IEEE Trans Device Mater Reliab 14(1):318–326

251. Liu X, Chen Q, Sundaram V, Tummala RR, Sitaraman SK (2013) Failure analysis of through-silicon vias in free-standing wafer under thermal-shock test. Microelectron Reliab 53(1):70–78

252. Kamto A, Liu Y, Schaper L, Burkett SL (2009) Reliability study of through-silicon via (TSV) copper filled interconnects. Thin Solid Films 518:1614–1619

253. Zhu L, Xu J, Xiu Y, Sun Y, Hess DW, Wong CP (2006) Growth and electrical characterization of high aspect ratio carbon nanotube arrays. Carbon 44(2):253–258

254. Xu T, Wang Z, Miao J, Chen X, Tan CM (2007) Aligned carbon nanotubes for through-wafer interconnects. Appl Phys Letts 91:042108-1-3

255. Wang T, Jeppson K, Olofsson N, Campbell EEB, Liu J (2009) Through silicon vias filled with planarized carbon nanotube bundles. Nanotechnology 20:485203-1-6

256. Xie R, Zhang C, van der Veen MH, Arstila K, Hantschel T, Chen B, Zhong G, Robertson J (2013) Carbon nanotube growth for through silicon via application. Nanotechnology 24:125603-1-7

257. Wang T, Chen S, Jiang D, Fu Y, Jeppson K, Ye L, Liu J (2012) Through-silicon vias filled with densified and transferred carbon nanotube forests. IEEE Electron Device Lett 33(3):420–422

258. Jiang D, Mu W, Chen S, Fu Y, Jeppson K, Liu J (2015) Vertically stacked carbon nanotube-based interconnects for through silicon via application. IEEE Electron Device Lett 36(5):499–501

259. Mu W, Sun S, Jiang D, Fu Y, Edwards M, Zhang Y, Jeppson K, Liu J (2015) Tape-assisted transfer of carbon nanotube bundles for through-silicon-via applications. J Electron Mater 44(8):2898–2907

260. Zhao WS, Yin WY, Guo YX (2012) Electromagnetic compatibility-oriented study on through silicon single-walled carbon nanotube bundle via (TS-SWCNTBV) arrays. IEEE Trans Electromagn Compat 54(1):149–157

261. Zhao WS, Sun L, Yin WY, Guo YX (2014) Electrothermal modelling and characterisation of submicron through-silicon carbon nanotube bundle vias for three-dimensional ICs. Micro Nano Lett 9(2):123–126

262. Qian L, Zhu Z, Xia Y (2014) Study on transmission characteristics of carbon nanotube through silicon via interconnect. IEEE Microw Wirel Compon Lett 24(12):830–832
263. Qian L, Xia Y, Liang G (2015) Study on crosstalk characteristic of carbon nanotube through silicon vias for three dimensional integration. Microelectron J 46(7):572–580
264. Majumder MK, Kumari A, Kaushik BK, Manhas SK (2014) Signal integrity analysis in carbon nanotube based through-silicon via. Active & Passive Electronic Components. Hindawi Publishing Corporation, Cario, 524107-1-7
265. Neto AC, Guinea F, Peres N, Novoselov KS, Geim AK (2009) The electronic properties of graphene. Rev Mod Phys 81:109
266. Srivastava A, Manulanda JM, Xu Y, Sharma AK (2015) Carbon-based electronics: transistors and interconnects at the nanoscale. Pan Stanford Publishing, Singapore
267. Kim KS, Zhao Y, Jang H, Lee SY, Kim JM, Kim KS, Ahn JH, Kim P, Choi JY, Hong BH (2009) Large-scale pattern growth of graphene films for stretchable transparent electrodes. Nature 457:706–710
268. Reina A, Jia X, Ho J, Nezich D, Son H, Bulovic V, Dresselhaus MS, Kong J (2008) Large area, few-layer graphene films on arbitrary substrates by chemical vapor deposition. Nano Lett 9:30–35
269. Brodie BC (1859) On the atomic weight of graphite. Philos Trans R Soc Lond 249–259
270. An X, Simmons T, Shah R, Wolfe C, Lewis KM, Washington M, Nayak SK, Talapatra S, Kar S (2010) Stable aqueous dispersions of noncovalently functionalized graphene from graphite and their multifunctional high-performance applications. Nano Lett 10:4295–4301
271. Berger C, Song Z, Li T, Li X, Ogbazghi AY, Feng R, Dai Z, Marchenkov AN, Conrad EH, First PN, de Heer WA (2004) Ultrathin epitaxial graphite: 2D electron gas properties and a route toward graphene-based nanoelectronics. J Phys Chem B 108:19912–19916
272. Bolotin K, Sikes K, Hone J, Stormer H, Kim P (2008) Temperature-dependent transport in suspended graphene. Phys Rev Lett 101:096802
273. Du X, Skachko I, Barker A, Andrei EY (2008) Approaching ballistic transport in suspended graphene. Nat Nanotechnol 3:491–495
274. Chen J-H, Jang C, Adam S, Fuhrer M, Williams E, Ishigami M (2008) Charged-impurity scattering in graphene. Nat Phys 4:377–381
275. Chen J-H, Jang C, Xiao S, Ishigami M, Fuhrer MS (2008) Intrinsic and extrinsic performance limits of graphene devices on SiO2. Nat Nanotechnol 3:206–209
276. Geim AK, Novoselov KS (2007) The rise of graphene. Nat Mater 6:183–191
277. Suemitsu M, Miyamoto Y, Handa H, Konno A (2009) Graphene formation on a 3C-SiC (111) thin film grown on Si (110) substrate. Electron J Surf Sci Nanotechnol 7:311–313
278. Gamo Y, Nagashima A, Wakabayashi M, Terai M, Oshima C (1997) Atomic structure of monolayer graphite formed on Ni (111). Surf Sci 374:61–64
279. Li X, Cai W, An J, Kim S, Nah J, Yang D, Piner R, Velamakanni A, Jung I, Tutuc E, Banerjee SK, Colombo L, Ruoff RS (2009) Large-area synthesis of high-quality and uniform graphene films on copper foils. Science 324:1312–1314
280. Gómez-Navarro C, Weitz RT, Bittner AM, Scolari M, Mews A, Burghard M, Kern K (2007) Electronic transport properties of individual chemically reduced graphene oxide sheets. Nano Lett 7:3499–3503
281. Misawa T, Okanaga T, Mohamad A, Sakai T, Awano Y (2015) Line width dependence of transport properties in graphene nanoribbon interconnects with real space edge roughness determined by Monte Carlo method. Jpn J Appl Phys 54:05EB01
282. Lin W, Moon K-S, Zhang S, Ding Y, Shang J, Chen M, Wong C (2010) Microwave makes carbon nanotubes less defective. ACS Nano 4:1716–1722
283. Ouyang Y, Wang X, Dai H, Guo J (2008) Carrier scattering in graphene nanoribbon field-effect transistors. Appl Phys Lett 92:243124
284. Yang Y, Murali R (2010) Impact of size effect on graphene nanoribbon transport. IEEE Electron Device Lett 31:237–239
285. Murali R, Brenner K, Yang Y, Beck T, Meindl JD (2009) Resistivity of graphene nanoribbon interconnects. IEEE Electron Device Lett 30:611–613

286. Yu Q, Lian J, Siriponglert S, Li H, Chen YP, Pei S-S (2008) Graphene segregated on Ni surfaces and transferred to insulators. Appl Phys Lett 93:113103
287. Wang X, Dai H (2010) Etching and narrowing of graphene from the edges. Nat Chem 2: 661–665
288. Li X, Wang X, Zhang L, Lee S, Dai H (2008) Chemically derived, ultrasmooth graphene nanoribbon semiconductors. Science 319:1229–1232
289. Jiao L, Wang X, Diankov G, Wang H, Dai H (2010) Facile synthesis of high-quality graphene nanoribbons. Nat Nanotechnol 5:321–325
290. Kim K, Sussman A, Zettl A (2010) Graphene nanoribbons obtained by electrically unwrapping carbon nanotubes. ACS Nano 4:1362–1366
291. Xie L, Wang H, Jin C, Wang X, Jiao L, Suenaga K, Dai H (2011) Graphene nanoribbons from unzipped carbon nanotubes: atomic structures, Raman spectroscopy, and electrical properties. J Am Chem Soc 133:10394–10397
292. Cai J, Ruffieux P, Jaafar R, Bieri M, Braun T, Blankenburg S, Muoth M, Seitsonen AP, Salch M, Feng X, Mullen K, Fasel R (2010) Atomically precise bottom-up fabrication of graphene nanoribbons. Nature 466:470–473
293. Hicks J, Tejeda A, Taleb-Ibrahimi A, Nevius M, Wang F, Shepperd K, Palmer J, Bertran F, Le Fevre P, Kunc J, de Heer WA, Conrad EH (2013) A wide-bandgap metal–semiconductor-metal nanostructure made entirely from graphene. Nat Phys 9:49–54
294. Sprinkle M, Ruan M, Hu Y, Hankinson J, Rubio-Roy M, Zhang B, Wu X, Berger C, de Heer WA (2010) Scalable templated growth of graphene nanoribbons on SiC. Nat Nanotechnol 5:727–731
295. Baringhaus J, Ruan M, Edler F, Tejeda A, Sicot M, Taleb-Ibrahimi A, Li AP, Jiang Z, Conrad EH, Berger C, Tegenkamp C, de Heer WA (2014) Exceptional ballistic transport in epitaxial graphene nanoribbons. Nature 506:349–354
296. Yu T, Kim E, Jain N, Xu Y, Geer R, Yu B (2011) Carbon-based interconnect: performance, scaling and reliability of 3D stacked multilayer graphene system. In: 2011 IEEE international electron devices meeting (IEDM), pp 751–754
297. Faugeras C, Nerrière A, Potemski M, Mahmood A, Dujardin E, Berger C, de Heer WA (2008) Few-layer graphene on SiC, pyrolytic graphite, and graphene: a Raman scattering study. Appl Phys Lett 92:011914
298. Yuan Q, Xu Z, Yakobson BI, Ding F (2012) Efficient defect healing in catalytic carbon nanotube growth. Phys Rev Lett 108:245505
299. Soldano C, Mahmood A, Dujardin E (2010) Production, properties and potential of graphene. Carbon 48:2127–2150
300. Cervantes-Sodi F, Csanyi G, Piscanec S, Ferrari A (2008) Edge-functionalized and substitutionally doped graphene nanoribbons: electronic and spin properties. Phys Rev B 77:165427
301. Advanced Industrial Science and Technology (AIST) (2013) Development of Technology for Producing Micro-scale Interconnect from Multi-layer Graphene. http://www.aist.go.jp/
302. Kondo D, Nakanoa H, Zhou B, Kubota I, Hayashia K, Yagi K (2013) Fabrication and evaluation of 20-nm-wide intercalated multi-layer graphene interconnects and 3D interconnects composed of graphene and vertically aligned CNTs. In: International semiconductor device research symposium (ISDRS), pp 11–13
303. Kondo D, Nakano H, Zhou B, Kubota I, Hayashi K, Yagi K, Takahashi M, Sato M, Sato S, Yokoyama N (2013) Intercalated multi-layer graphene grown by CVD for LSI interconnects. In: 2013 IEEE international interconnect technology conference (IITC), pp 1–3
304. Wei D, Liu Y, Wang Y, Zhang H, Huang L, Yu G (2009) Synthesis of N-doped graphene by chemical vapor deposition and its electrical properties. Nano Lett 9:1752–1758
305. Schedin F, Geim A, Morozov S, Hill E, Blake P, Katsnelson M, Novoselov KS (2007) Detection of individual gas molecules adsorbed on graphene. Nat Nater 6:652–655
306. Gunlycke D, Lawler H, White C (2007) Room-temperature ballistic transport in narrow graphene strips. Phys Rev B 75:085418
307. Wang H, Wu Y, Ni Z, Shen Z (2008) Electronic transport and layer engineering in multilayer graphene structures. Appl Phys Lett 92: 053504-053504-3

308. Lee EJ, Balasubramanian K, Weitz RT, Burghard M, Kern K (2008) Contact and edge effects in graphene devices. Nat Nanotechnol 3:486–490
309. Murali R, Yang Y, Brenner K, Beck T, Meindl JD (2009) Breakdown current density of graphene nanoribbons. Appl Phys Lett 94:243114
310. Cresti A, Nemec N, Biel B, Niebler G, Triozon F, Cuniberti G, Roche S (2008) Charge transport in disordered graphene-based low dimensional materials. Nano Res 1:361–394
311. Li W, Sevinçli H, Cuniberti G, Roche S (2010) Phonon transport in large scale carbon-based disordered materials: implementation of an efficient order-N and real-space Kubo methodology. Phys Rev B 82:041410
312. Pop E (2010) Energy dissipation and transport in nanoscale devices. Nano Res 3:147–169
313. Liao A, Alizadegan R, Ong Z-Y, Dutta S, Xiong F, Hsia KJ, Pop E (2010) Thermal dissipation and variability in electrical breakdown of carbon nanotube devices. Phys Rev B 82:205406
314. Mohsin KM, Srivastava A, Sharma AK, Mayberry C (2013) A thermal model for carbon nanotube interconnects. Nanomaterials 3:229–241
315. Mohsin KM, Banadaki YM, Srivastava A (2014) Metallic single-walled, carbon nanotube temperature sensor with self heating. In: Proceedings of SPIE 9060, nanosensors, biosensors, and info-tech sensors and systems, pp 906003-1-7
316. Liao AD, Wu JZ, Wang X, Tahy K, Jena D, Dai H, Pop E (2011) Thermally limited current carrying ability of graphene nanoribbons. Phys Rev Lett 106:256801
317. Hale P, Hornett S, Moger J, Horsell D, Hendry E (2011) Hot phonon decay in supported and suspended exfoliated graphene. Phys Rev B 83:121404
318. Han MY, Brant JC, Kim P (2010) Electron transport in disordered graphene nanoribbons. Phys Rev Lett 104:056801
319. Wang P-C, Filippi R (2001) Electromigration threshold in copper interconnects. Appl Phys Lett 78:3598–3600
320. Chen X, Seo DH, Seo S, Chung H, Wong H-S (2012) Graphene interconnect lifetime: a reliability analysis. IEEE Electron Device Lett 33:1604–1606
321. Haji Nasiri S, Moravvej-Farshi MK, Faez R (2010) Stability analysis in graphene nanoribbon interconnects. IEEE Electron Device Lett 31:1458–1460
322. Das D, Rahaman H (2011) Crosstalk and gate oxide reliability analysis in graphene nanoribbon interconnects. In: 2011 international symposium on electronic system design (ISED), pp 182–187
323. Yu T, Lee E-K, Briggs B, Nagabhirava B, Yu B (2010) Reliability study of bilayer graphene-material for future transistor and interconnect. In: 2010 IEEE international reliability physics symposium (IRPS), pp 80–83
324. Van Noorden R (2006) Moving towards a graphene world. Nature 442:228–229
325. Kang J, Sarkar D, Khatami Y, Banerjee K (2013) Proposal for all-graphene monolithic logic circuits. Appl Phys Lett 103:083113
326. Yan T, Ma Q, Chilstedt S, Wong MD, Chen D (2013) A routing algorithm for graphene nanoribbon circuit. ACM Trans Des Autom Electron Syst 18:61
327. Srivastava A, Banadaki YM, Fahad MS (2014) (Invited) Dielectrics for graphene transistors for emerging integrated circuits. ECS Transactions 61:351–361
328. Banadaki YM, Srivastava A (2013) A novel graphene nanoribbon field effect transistor for integrated circuit design. In: IEEE 56th international midwest symposium on circuits and systems (MWSCAS), pp 924–927
329. Johari Z, Hamid F, Tan MLP, Ahmadi MT, Harun F, Ismail R (2013) Graphene nanorib-bon field effect transistor logic gates performance projection. J Comput Theor Nanosci 10(5):1164–1170
330. Wang X, Sun G, Chen P (2014) Three-dimensional porous architectures of carbon nanotubes and graphene sheets for energy applications. Front Energy Res 2:33
331. Kondo D, Sato S, Awano Y (2008) Self-organization of novel carbon composite structure: graphene multi-layers combined perpendicularly with aligned carbon nanotubes. Appl Phys Express 1:074003
332. Dimitrakakis GK, Tylianakis E, Froudakis GE (2008) Pillared graphene: a new 3-D network nanostructure for enhanced hydrogen storage. Nano Lett 8:3166–3170

Chapter 3
Overview of Carbon Nanotube Processing Methods

Franz Kreupl

3.1 Introduction

On February 16, in year 2000, the first patent application was filed that proposed to use carbon nanotubes (CNTs) instead of metals as vertical interconnects in advanced microelectronic interconnects on semiconductor chips [1]. This patent, which has been cited in over 150 following patent applications as prior art, emphasizes that CNTs would be especially useful in vertical interconnects (vias). The quasi-ballistic current transport in CNTs would allow very efficient low resistance interconnects which can mitigate the observed reliability issues in advanced semiconductor interconnects [2].

Figure 3.1a shows an artist's view of advanced interconnect structures with various levels of horizontal copper lines and vertical connections, the vias, which connect the different levels. One major reliability failure is the formation of voids in the vias, as showcased in Fig. 3.1b, due to the required high current densities in the interconnects. The patent suggested to strengthen the vertical part of the interconnects by using the more durable CNTs as the conductor material that connects different metallization levels. Figure 3.1c shows the proposed schematic where a bundle of CNTs is vertically connecting two metal layers.

In year 2015, there are still ongoing efforts to implement CNTs as interconnects on chip, but there is not yet a single real application in this market. However, the idea to use CNTs in macroscopic interconnects has proliferated and Nanocomp Technologies in New Hampshire is using macroscopic CNT yarns to spin them into electrical cables. Their primary advantage, compared to traditional copper, is a reduced weight at equal conductance and a much increased bending fracture fatigue. Obviously a thorough knowledge about the processes and integration of CNTs is

F. Kreupl (✉)
TUM, Munich, Germany
e-mail: franz.kreupl@tum.de

© Springer International Publishing Switzerland 2017
A. Todri-Sanial et al. (eds.), *Carbon Nanotubes for Interconnects*,
DOI 10.1007/978-3-319-29746-0_3

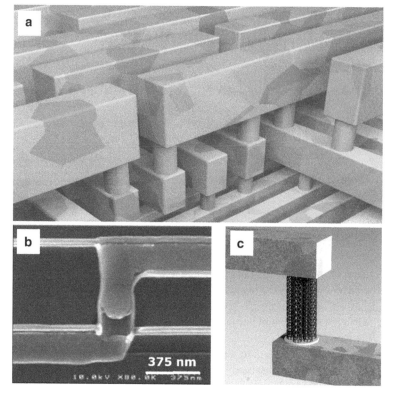

Fig. 3.1 An artist's view of the interconnects in a microelectronic circuit is shown in (**a**). The SEM picture in (**b**) shows the cross-section of a copper interconnect, where a void has been created in the vertical connection between two horizontal copper wires. The use of CNTs, as shown in (**c**), in the vertical connection could mitigate the electro-migration problem seen in (**b**)

required to achieve useful implementation and a brief discussion of these related to the topic of interconnects is the objective of this chapter—*with an emphasis on recent achievements*. CNT processing, in general, covers a much wider field and a more extensive review is given by Joselevich et al. [3].

3.2 Growth of Carbon Nanotubes

Historically, CNTs were produced by arc discharge, laser ablation, and catalyst enhanced chemical vapor deposition (CCVD). A metallic particle in high temperature environment, exposed to a carbon-containing atmosphere serves as a nucleation seed, from where CNTs start to grow. It is generally believed that the particle diameter will also determine the diameter of the CNTs. Very soon, it was discovered that almost all curved surfaces can induce the growth of CNTs by CVD. At the beginning, the transition metals Fe, Co, Ni, Mo, and bi-metallic combinations of them were used as "catalyst." Later on, other metals such as Ag, Au, Cu, Pd, Pt, and even non-metals like fullerene, diamond particles, and oxide particles (like SiO_2) were also found to enable CNT growth [4]. Even thermal SiO_2 layers are able to grow CNTs after an annealing treatment of the SiO_2 surfaces in H_2 at 950–1000 °C. The hydrogen gas is reducing the SiO_2 surface and creates defects that facilitate the growth of CNTs [5]. There is a natural fluctuation in the diameter and morphology of the catalyst particles which gives rise to variations in the diameter, chirality, and the number of walls of the CNTs. Therefore, a big step forward is now expected as chemists start to design bottom-up particles by using well defined molecules which form identical particles after reaction. This could pave the road for having chirality-selective CNT growth in the future. But by now, it looks more that the problem of accomplishing pure chirality is shifted from CNTs to the synthesis of the molecules, because the synthesis routes for bottom-up construction of catalysts are quite challenging and complicated [6].

The different possible growth processes of CNTs are illustrated in Fig. 3.2. The catalyst is often deposited as a continuous thin film by evaporation or sputtering. If the film is annealed at higher temperature, the continuous film breaks up into individual particles in order to minimize the surface energy. The small particles then serve as catalysts for the CNT growth if a hydrocarbon gas is presented. The hydrocarbon gas dissociates at the atomic steps of the catalyst and carbon starts to diffuse and recombine with other carbon atoms. The carbon may also diffuse into the bulk of the catalyst, but some reported growth speeds cannot be explained by bulk diffusion but only by surface diffusion. The base growth mode occurs if there is a strong bonding interaction of the catalyst with the substrate. The particle bonds to the substrate and a CNT starts to grow from the catalyst base. The tip growth mode can happen if the catalyst interaction with the substrate is weak. In this case,

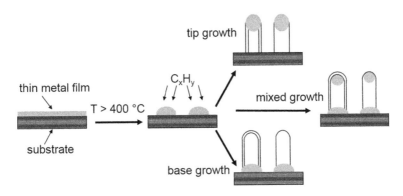

Fig. 3.2 A thin metal film breaks up into individual clusters upon annealing. After a reduction step and exposure to hydrocarbon gases, elemental carbon is formed at atomic steps of the cluster, which grows into CNTs. Two different growth modes exist, the tip growth mode and the base growth mode—a mixed growth mode can also be observed

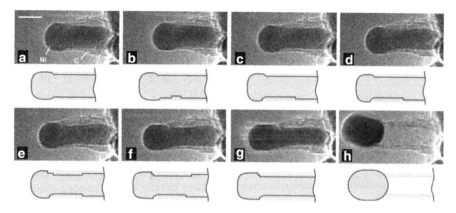

Fig. 3.3 Image sequence of the growth of a CNT with an Ni catalyst, which ended up in a tip growth mode, shown in (**h**). The pictures (**a**)–(**h**) are extracted from a video showing how a CNT starts to grow [7]. They illustrate the elongation and contraction process of the (*dark*) catalyst and the drawings below each TEM- image are a guide to the eye to show the evolving mono-atomic Ni step edges at the C-Ni interface. In image (**h**), the catalyst has lost contact to the substrate and resides on top of the carbon tubules. The scale bar is 5 nm. Reprinted with permission from Macmillan Publishers Ltd from reference [7], copyright (2004)

the formation of a nanotube lifts the catalyst from the surface and the CNTs carbon supply for growth is fed from the catalyst on the tip. Once the whole catalyst is covered completely with carbon, the growth stops. The presence of carbon etching gas species, like oxygen or hydrogen, can be very beneficial to achieve a long catalyst activity.

Figure 3.3 shows a sequence of TEM images, labelled (a)–(h), where the growth of a multi-walled CNT from an Ni cluster supported on $MgAl_2O_4$ is observed [7]. The catalyst is exposed to a 1:1 mixture of H_2 and CH_4 at a pressure of 2 mbar and a temperature of 536 °C. The initial droplet shape of the Ni catalyst is transformed in an elongated shape by the building of graphene sheets on the sidewall. In the beginning, the Ni is still attached to the support layer, but soon the cluster minimizes its energy by converting to a more ball-shaped form and the bonding to the substrate is given up in favor of a tip growth mode.

Processing of carbon nanotubes for interconnects is intimately connected to the approach how to incorporate them in a specific application. While it is possible to produce CNTs in bulk quantities in powder form [8], the immediate question arises how to bring the CNT powder and arrange it in a microelectronic via or form it into a cable? It turns out that this is not easily achievable and the more sophisticated approach of growing them directly on the microelectronic chip is needed for a vias. Even for fabricating macroscopic CNT cables and wires, the current method is not based predominantly on CNT powder, but it is to pick up a continuously growing CNT mist in a furnace and spin it into a wire. By now, the most successful integration approach for interconnects is to grow CNTs directly at the location where CNTs are needed by chemical vapor deposition (CVD) growth.

3.3   Chemical Vapor Deposition Growth

Chemical vapor deposition CNT growth is the easiest way to get to a high yield and good quality of CNTs with very few by-products. In its simplest form, CVD is carried out at atmospheric pressure and a typical reactor setup is depicted in Fig. 3.4. The substrate with the catalyst is put into a quartz tube, which resides in a tube furnace. The input and output of the exhaust need to be connected leak-tight to a gas supply at the entry and to a bubbler with perfluorinated polyether (PFPE) oil at the exit. The bubbler ensures that no atmosphere is streaming back into the furnace, which could induce an explosion with the hydrogen and/or hydrocarbon gases at high temperatures. PFPE oil has a very low vapor pressure and will give the lowest contamination risks. Some people are using water or silicone oil for the bubbler, but by back-diffusion into the furnace this contamination might impair—or even facilitate—the growth condition. After insertion of the substrate and attachment of gas entry and exhaust, it needs to be carefully checked that the whole reactor setup is leak-tight, otherwise the risk of an explosion is high. This might be facilitated by shutting of the exhaust and establishing a light overpressure with inert gas in the tube. By monitoring the pressure drop over time a leak rate can be estimated. After the leak test, the tube is flushed with inert gas for a couple of minutes to remove air components. Nitrogen should not be used as it is not inert gas at higher temperatures and might induce nitridation of catalyst and substrates. Nitrogen also interferes with the reaction kinetics during CNT growth.

During the heat-up phase of the furnace, the tube is purged with inert gas where hydrogen for the reduction of the catalyst can already be added in this step to the inert gas. Once the furnace has reached its desired temperature ($\sim$750 °C), a co-flow with hydrogen is established for a limited time, in case, hydrogen was not added during the heat-up phase. Hydrogen supply is then shut off and a small amount of hydrocarbon source (C_2H_2, C_2H_4, CH_4, or even liquids like C_2H_6O [9]) is added to

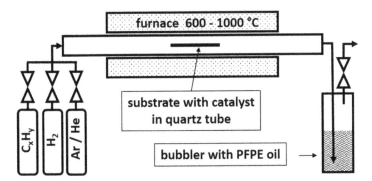

Fig. 3.4   A typical setup of an atmospheric pressure CVD reactor. A quartz tube with the substrate residing in a furnace is connected leak-tight to different gases and the exhaust is isolated from the atmosphere by a bubbler

the inert gas. Depending on the purity of the gases an addition of a small amount of water is beneficial to enhance the growth. It has been shown that the choice of inert gas, like argon or helium, and composition of background gases, like hydrogen and water, can shift the chirality prevalence of the grown CNTs from semiconducting to metallic [10]. After the desired growth time, the furnace is cooled down under inert gas flow. Every now and then, the tube can be cleaned from carbon by-products by purging it at high temperature with air or oxygen.

While the atmospheric CVD is cost-effective in terms of equipment, as no vacuum pump is needed, the consumption of inert gases like argon or helium is cost intensive in the long run. The use of a vacuum pump reduces gas consumption and gives access to a wider parameter space, where more control about background pressure, process pressure, and residence time of the gas species can be obtained. For safety reasons, the use of a vacuum pump filled with PFPE oil is highly recommended and a leak-check of the system after insertion of a new sample is also mandatory. The use of a vacuum pump allows also the application of a remote plasma at low pressures to modify the gas chemistry [11]. The inductive coil for the rf-excitation can be placed at the cool region next to the entrance of the furnace [11] (Fig. 3.4).

3.4   Vertical Alignment of Carbon Nanotubes

If the catalyst density is high enough and almost every catalyst particle induces the growth of a CNT, the nanotubes attach to one another and as the growth continues, they grow actually perpendicular to the substrate surface. In case of a horizontal flat substrate, this leads to a CNT film whose CNTs are vertically aligned to the substrate. This would be ideal for use of vertical interconnects; however, the best developed recipes rely predominately on a catalyst stack consisting of metal with an oxide as underlayer. The prominent super-growth recipe of the Hata group, for instance, uses a Fe/Al_2O_3 combination ($Fe:\sim1.0–2.0$ nm on $Al_2O_3:\sim40$ nm) [12]. For the application in interconnects, the use of oxides, which in most cases are insulators, is counterproductive as it would increase the contact resistance to a metallic interconnect severely. The material below the catalyst needs to be highly conductive and at the same time needs to allow clustering of the catalyst, prevent diffusion and alloying of the catalyst with the support layer, while giving high yield and best quality CNT growth. The natural choice is to look at refractive materials and their nitrides. The additional requirement would include compatibility with copper technology. Therefore, Ta, TaN, as well as TiSiN which all act as diffusion barrier for copper diffusion would be an optimal choice. Tantalum reacts at high temperatures with hydrocarbons and forms TaC which has a reasonable low resistivity of ~12 $\mu\Omega$ cm. The use of silicides is discussed in detail in a different chapter in this book. Recently, Robertson's group has shown that it is possible to grow high density CNTs on TiSiN as a diffusion barrier [13]. A catalyst consisting of 0.4 nm Fe/0.1 nm Al has been used to grow very dense vertically aligned CNTs in

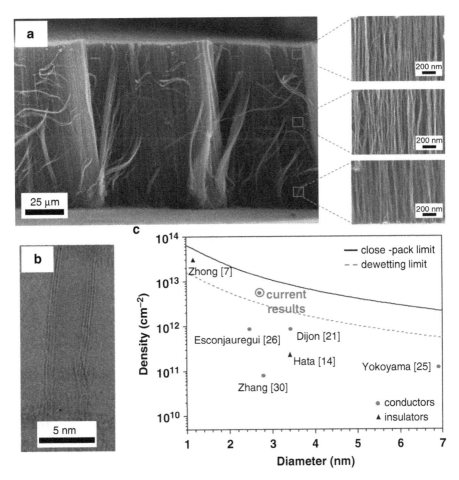

Fig. 3.5 A side-view SEM picture with vertically aligned CNT growth on a conductive TiSiN substrate is shown in (**a**), together with a TEM image of a CNT (**b**) and a comparison of the achieved area density with values from the literature (**c**). Reprinted with permission from [13], copyright 2015, AIP Publishing LLC

base growth mode at 600 °C in an atmospheric pressure process. Figure 3.5 depicts a cross-section of the CNT array and very homogeneous growth along the complete height of $\sim$100 µm, grown in 15 min. It also compares the obtained CNT density of $\sim$5*10^{12} tubes per cm^2 with other recent achievements obtained on insulating and conductive underlayers.

The requirements are even more rigorous, if the temperature budget for growth is restricted to the tolerated back-end-of-line temperature of chip manufacturing which is below 400–450 °C. Exposing a chip, which has already metallization layers, to a higher temperature induces diffusion of copper and/or other metals and silicides in the chip, which eventually can lead to a reduced reliability of the chip. This reduced

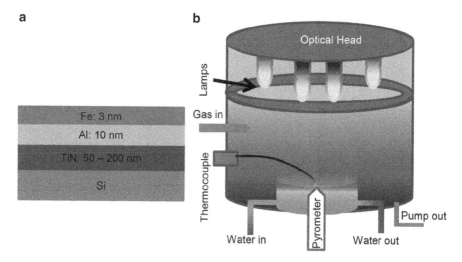

Fig. 3.6 The catalyst stack to grow high quality CNTs at 350–440 °C is shown (**a**). The low temperature growth uses the photo-thermal reactor setup depicted in (**b**). Reprinted with permission from [14]. Copyright 2015 WILEY-VCH Verlag GmbH & Co. KGaA

temperature requirement for growing CNTs is detrimental for the quality of CNTs as their structural integrity usually increases with the increase of growth temperature.

Recent growth process advances in the group of Silva have resulted in the ability to grow high quality vertically aligned CNTs at substrate temperatures between 350 and 440 °C [14]. The catalyst stack consisted of 3 nm Fe on 10 nm Al and the conductive underlayer TiN with thicknesses between 50 and 200 nm was central to focus the thermal energy onto the catalyst particle. A photo-thermal CVD reactor, as depicted in Fig. 3.6, was used to heat up only the surface of the wafer, while the silicon bottom side was connected to a water-cooled chuck. When the Si-wafer is exposed to the photonic heat source from the top, it heats only up to 350–440 °C. The catalyst on the top gets additional heating from the reflected photons and may even reach surface temperatures ranging between 550 and 715 °C. The quality of the CNTs was judged on Raman analysis, where the ratio of the intensity of the D-peak and G-peak (I_D/I_G) was used to determine a better quality. The highest structural quality of CNTs with $I_D/I_G = 0.13$ was achieved for a 100 nm thick TiN film at a substrate growth temperature of 420 °C.

3.5 Hidden Growth Parameter

Usually, it is kind of difficult to repeat a recipe for CNT growth, which is published in a paper in its own lab. The reason for this are hidden parameters of the experiments, which are not disclosed or mentioned in the paper. CNT CVD processes are characterized by pressure, flow rate of gases, heat rate, temperature, and reactor geometry. In addition, the annealing time, temperature, pressure, and atmosphere for the catalyst are important. Reviewers of publications dealing with CNT-growth often fail to demand the availability of those growth parameters in a published paper. As water or oxygen were found to play a very important role in the growth, it is strange to see that some recipes do not include the addition of water or oxygen explicitly. To understand this, it is relevant to know the background pressure of the system after the reactor has been opened and the sample has been inserted. In this context, the humidity level in the lab atmosphere is as important as the fact whether the reactor was hot during insertion of the sample or not. There will be a partial pressure of 12 Torr of water at 50 % relative humidity at 23 °C, which will absorb onto the surface of the open reactor and desorb during the process or during the purge. Viton O-rings are the most common sealing material, and it is heavily loaded with water due to the way it is manufactured. Viton O-rings will outgas water until they have been under vacuum for weeks or months. Once they are exposed to air they also will absorb water again from the air. In an atmospheric pressure CVD, the water background will depend on the purge time and flow of inert gas—a value which is not often specified in papers. Background pressure contamination control can be much better achieved with vacuum operation.

Another source of water is the quality of the used gas. Even a 5.0 grade gas can contain up to 3–5 ppm of water and doing something wrong during the hook up of the gas cylinder can contaminate the whole gas cylinder to a much lower grade. And the use of 5.0 grade gas at atmospheric pressure means that there is a partial pressure of water in the order of 0.02 mbar from the gas alone. Cheap flow meters are also often not very tight so that air can diffuse into the gas line. The contamination level then depends on the purge time and equipment use.

Using ethanol as a source for carbon carries even more risks for unknown parameters than gas. Although anhydrous ethanol is available, once the bottle is opened in normal air atmosphere, the bottle will get contaminated by the hygroscopic nature of ethanol, which will attract and dissolve water as well as oxygen and carbon dioxide. Handling ethanol with a Schlenk line or in a glove-box filled with inert gases is required to have a well understood ethanol source.

The reactor geometry is important to estimate the residence time of the gas molecules considering overall gas flow, gas expansion due to temperature in one zone, two zone or three zone furnace. Some gases like methane or ethanol need to decompose into other products before they are able to form sp^2-bonding carbon, which makes the estimation of residence times even more challenging. As the gas is heating up as it travels along the furnace, the exact sample location in the furnace also contributes to an uncertainty in the growth conditions as the composition of gases along the gas flow is changing. An attractive option to get homogeneous growth over large diameters is the use of a gas shower head, which supplies more consistently the catalyst with the same gas composition.

Another important parameter is the exact thickness of the used catalyst. Especially the growth of small diameter single-walled CNTs requires sub-monolayer deposition of catalyst, which is often achieved by sputtering or evaporation techniques. In order to have some control about this very low thicknesses, the deposition rate is reduced to very low values, so that the deposition time is in the range of 10–20 s. At very low deposition rate, the catalyst material can react with background gases of the deposition chamber and the quartz balance thickness measurement unit will effectively measure more catalyst mass than what was deposited. Another point is that the surface of sample which has topography is much bigger than a completely flat sample. Therefore, the sample with topography needs much more catalyst thickness to get to the same surface coverage as a flat sample. To conclude, a process for CNT growth needs to be established in most cases for every different reactor and infrastructure. Some hidden growth parameters might be the facilitators of a certain CNT growth recipe.

3.6 Horizontal Alignment of Carbon Nanotubes

Figure 3.1 suggests also the use of CNTs in horizontal interconnects, especially as the mean free path ($\sim$1 μm) of the charge carriers in CNTs is much longer than the typical via height. The use of CNTs in interconnects that are longer than the mean free path should be more effective in terms of reducing the resistance. This would require that CNTs are arranged horizontally on the wafer surface. For monolayer coverage of single CNTs, this might be achievable by substrate and gas flow engineering, but thick and bulky CNT layers are not attainable by the same approach [15]. Dense nanotube arrays can be grown perpendicular on any suitable substrate that supports CNT growth. This has been used to grow CNTs from the sidewall of three-dimensional structures, where catalyst particles have been deposited by tilted angle evaporation [16]. However, the quality of the produced CNT structure was poor and not very dense.

Another promising way to achieve horizontally aligned CNTs was developed by Hayamizu et al. [17] and the process is schematically depicted in Fig. 3.7. At lithographically defined locations, arrays of vertically aligned CNTs are synthesized which are several hundred micrometers in height. After CNT growth, the wafer is moved through an isopropyl alcohol solution and as the wafer is drawn out of the solution, the CNT arrays are horizontally aligned by the capillary forces of the liquid. The CNTs are densified to a compact substance after drying and can be processed with conventional semiconductor processing. By using photo-lithography and dry-etching with argon–oxygen plasmas, the CNT wafers can be structured into arbitrary designed objects. The liquid densification achieves a compression ratio of 14–19 and an initially 8 μm wide CNT array yields a 500 nm thick compact CNT solid on the wafer.

The highly anisotropic nature of the material is also reflected in the evaluated specific resistivity of the horizontally aligned and densified CNT solid. The resistivity along the direction of the CNTs is 8 mΩ cm and in the perpendicular direction only 0.2 Ω cm. The resistivity is 4000 times worse than that of copper and two times higher than highly doped polysilicon, which, with a silicide on top, can also be used as local interconnect material.

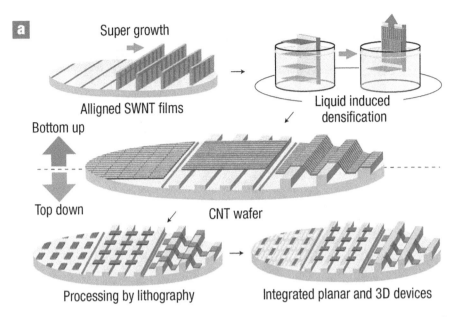

Fig. 3.7 Schematic illustration of the fabrication of horizontally aligned CNTs. First, arrays of vertically aligned CNTs are grown on a wafer. Subsequently, the wafer is submersed into a liquid and when drawn out of the liquid, the CNTs are aligned horizontally by the surface tension of the liquid. The aligned CNTs can be further processed on flat as well as on 3D substrates. Reprinted with permission from Macmillan Publishers Ltd from reference [17], copyright (2008)

In order to achieve good alignment in the liquid densification and aligning process, the width of the CNT array needs to be much wider than the thickness. This limits somehow the thickness of the densified CNT solid. In order to get thicker CNT solids, an approach has been recently developed by Li et al. [18, 19] where multiple CNT arrays have been overlaid to realize a CNT solid thickness that is multiple times higher. Figure 3.8a–d demonstrates an example where three and four CNT arrays have been overlaid by liquid flow to accomplish a much thicker horizontal CNT solid. The alignment is not perfect, but straight portion can be etched out by plasma etching. The approach shown in Figs. 3.7 and 3.8a–d is suitable for preparing horizontal CNT connections in one particular direction. However, in real chip design, the interconnects follow a 2-D Manhattan layout, where wires are also running in an orthogonal direction in order to route the wires in the shortest direction. By properly designing the catalyst and fluidic alignment process, 2-D Manhattan structures can be achieved [18, 19]. In Fig. 3.8e–g, this technique, which is central for building angled interconnects for interconnects and passive devices, like inductors, is illustrated.

While CNTs can be arranged horizontally in this way, it is less obvious how to make high quality connections to the underlying vias, or at least at the end of the CNT bundle. On one side, the CNTs are still anchored to the substrate with the Al_2O_3/Fe catalyst stack and on the other end they are opened. The CNT solid can be

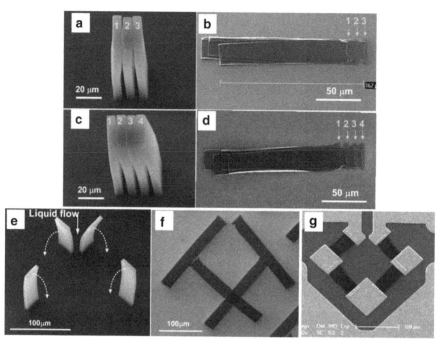

Fig. 3.8 Schematic illustration of the fabrication of horizontally aligned CNT with enhanced thickness and 2-D Manhattan alignment. (**a**) shows three CNT arrays which give a stack of triple thickness if flattened (**b**). The same technique with four layers is demonstrated in (**c**) and (**d**). The creation of a 2-D Manhattan arrangement is shown in (**e**) and (**f**), while (**g**) displays the structure in (**f**) after etching and metallization. Reprinted with permission from Li et al. [18]

trimmed on both ends or at specific location with plasma etching to allow to make a metallic contact on both the vias and the sidewalls of the CNT bundle. Extreme care needs to be taken to contact all CNTs and Li et al. have developed a two-layer resist approach employing tilted and rotated samples during metallization deposition to achieve good sidewall coverage of the CNTs with the metal [18, 19].

3.7 Carbon Nanotubes: Copper Composite Interconnects

The analysis of the experimentally observed resistivity of CNT structures demonstrates that they fail to achieve the theoretical predicted benefit of CNTs by more than a factor of thousand. A simple fact that can contribute to an explanation is the low density of 0.46 g cm^{-3} for the fabricated, horizontally aligned, and densified CNTs, discussed in the preceding chapter. The value is much less than the density of 2.2 g cm^{-3} for densely packed graphite. There is simply too much hollow space inside and between the CNTs, and also the density of states needs to be increased by doping. In order to benefit from the high current density carrying capacity of the CNTs, the resistivity needs to be decreased too, because the power density is given by $j^2 \cdot \rho$, where j is the current density and ρ is the specific resistivity of the material.

There was a considerable progress in possible ampacity (current-carrying capacity) in a new CNT–copper composite reported recently, where the supported current density reached a value of 600 mA cm^{-2} and at the same time, the resistivity was in the range of copper [20]. The composite material has an increased activation energy for Cu diffusion and blocked the typical diffusion path of Cu, leading to the one hundred fold increased ampacity. The new property is not only useful on the nanoscale in nanoelectronic application, but also in the implementations on a more macroscopic scale like in power electronic devices.

The CNT–Cu composite was created similar to the horizontal alignment procedure that was described before, but with some major modification. The dense vertical CNT forest has been removed from the substrate to give a free standing material of vertical aligned CNTs. This can be accomplished by etching in 1 % hydrofluoric acid [11] or by using a razor blade technique [21], which lifts off the whole CNT carpet. Subsequently, the free standing CNT carpet was horizontally aligned by shear force alignment between two gliding glass slides. The clip-stabilized structure was submerged into a 2.75 mM solution of copper acetate in acetonitrile and left there for 20 min for densification. Afterwards copper was deposited by a two-step electrodeposition process. Initially, copper was deposited at a very low current density in a 2.75 mM solution of copper acetate in acetonitrile solution. This was to attach copper atoms on the hydrophobic surface of the CNTs. The low current density is required to allow the penetration of the Cu atoms into the whole porous solid and a high filling ratio with copper can be achieved. The required time to fill the structure was therefore in the order of ten hours. If the current density is too high, only the outer surface would have been coated. After the first copper nucleation the bath could be switched to an aqueous commercial copper plating bath without accelerators and suppressors to fill the CNT matrix completely. After each plating step, the sample was washed with acetonitrile and vacuum dried at elevated temperatures. The filled CNT matrix was then annealed in a hydrogen atmosphere at 250 °C for 60 min. This process reduced some copper oxides that have formed through the deposition and handling. Figure 3.9 illustrates the fabrication process of the CNT–Cu composite and highlights some of the observed properties. For the first time a high ampacity and a low specific resistivity were obtained. The low resistivity was only obtained after a burn-in of the composite with a high current density. It is currently not clear whether this reduction could have been also achieved with a higher annealing temperature and/or annealing time or the treatment with high current density is a necessity. The composite also ranks second to aluminum in terms of specific conductivity (conductivity per unit weight) beating pure copper and other metals.

However it should be emphasized that the composite has been only analyzed in the direction of the aligned CNTs and no information is currently available for the direction perpendicular to the aligned CNTs. The evaluation was only for a CNT film thickness of 700 nm for the uncoated CNT matrix and 2 μm for the coated

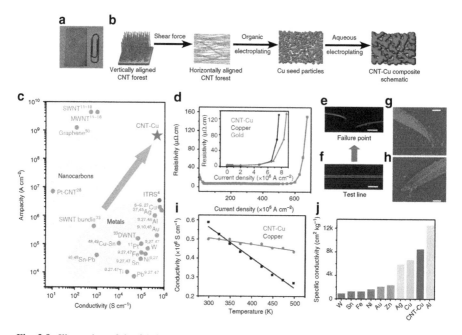

Fig. 3.9 Illustration of the fabrication of CNT–copper composites with its properties. An image of a CNT–Cu composite is shown in (**a**). The steps for forming the composite are illustrated in (**b**). The Ashby plot of ampacity versus conductivity for various materials is shown in (**c**). Measured resistivity with increasing current density for CNT–Cu composite and for copper and gold is plotted in (**d**). (**f**) SEM image of the same CNT–Cu composite and before (**f**) and after (**e**) failure. The scale bar is 4 mm. The EDX mapping for Cu-lines of the failed structure is shown in (**g**) and (**h**) with a scale bar of 500 nm. The temperature versus conductivity plot for CNT–Cu (*red*) and Cu (*black*) is shown in (**i**). A comparison chart of specific conductivity with different metals and CNT–Cu composite is plotted in (**j**). Reprinted with permission from Macmillan Publishers Ltd from reference [20], copyright (2013)

composite. Therefore, the scalability of the approach to thicker cooper lines and the achievement of an isotropic high ampacity are open questions.

3.8 Macroscopic Carbon Nanotube Interconnects: Cables and Wires

CNTs have not only been considered as interconnects in the nano-world but also in the macroscopic world. The idea is to make real-world macroscopic objects, like cables, out of nano-sized CNTs. Getting this would have far reaching applications in wiring, cables, and general power distribution. Until recently, the efforts to assemble macroscopic fibers were limited to direct spinning from the CVD growth zone [22] or spinning them from substrates covered with vertically aligned CNTs, because the obtained CNT length (up to 1 mm) enabled the fabrication of fibers with very good performance [23]. The more scalable approach of using wet chemistry to spin fibers from CNT powders was hampered by the fact that only short CNTs can be handled efficiently in the fluid and the solubility of CNTs is limited.

In a pioneering effort, Behabtu et al. have demonstrated that strong and ultrahigh conductive fibers can be created even with premade CNTs as short as 5 μm [23]. This would allow to scale up the wet processing of CNTs to industrial levels. Key to this achievement was the use of chlorosulfonic acid as the only known true solvent for CNTs. The dissolution of CNTs is based on the electrostatic stabilization through protonation of the CNT sidewalls [24]. Interestingly, CNTs whose sidewalls are highly defective did not protonate appropriately and, therefore, did not dissolve [24]. The protonation induces a delocalized positive charge on the CNT sidewalls, leading to a repulsive force between the CNTs, which counteracts the attractive van der Waals force. When the concentration exceeds a threshold of 2–6 weight (wt) %, liquid crystals are formed even with very long CNTs. Unfortunately, the handling of chlorosulfonic acid is not easy and requires outmost care, as the acid is highly corrosive and breaks down into hydrochloric and sulfuric acids if gets into contact with trace amounts of water. Chlorosulfonic acid is also incompatible with many other common reagents and special design and handling requirements apply. Nevertheless, Behabtu et al. were able to create a spinnable liquid crystal at a concentration of 2–6 wt%. They created a highly aligned CNT fiber by extruding the liquid crystal through a spinneret, removing the acid, and winding up the fiber on a drum for further processing. The resulting fibers had a tensile strength of $\sim$1 GPa and the average modulus was $\sim$120 GPa. They displayed a very low electrical resistivity of $\sim$35 $\mu\Omega$ cm. Doping the fiber by iodine decreased the resistivity further down to 17.5 $\mu\Omega$ cm. However, the stability of this resistivity upon exposure to high ampacity is questionable, as the annealing at 600 °C alone led to a deterioration of the resistivity up to 240 $\mu\Omega$ cm—increasing almost by an order of magnitude. But this is only speculation, as the current-imprint experiment has not been performed and an increase in conductivity might also be possible [12, 20]. These fibers could form the basis to braid and yarn much thicker wires.

Another successful approach, which is also used by the company Nanocomp Technologies on a bigger scale, is to wind CNT yarns and felts directly from the hot reaction zone of a CVD reactor. The floating catalyst chemical vapor deposition (FC-CVD) can be used to grow very long ($\sim$1 mm) CNTs without having a support layer directly in the gas phase [22]. The FC-CVD injects, as shown in Fig. 3.10, a catalyst precursor, like ferrocene, a carbon source, like ethanol, and thiophene as a catalyst activator into a high temperature furnace with hydrogen as carrier gas. Proper gas flow and exhaust handling is required because the main carrier gas is hydrogen and the furnace is prone to air inflow from the gas exhaust side, which could create an explosive atmosphere.

The ferrocene decomposes to catalyst particles, and CNTs grow on the floating metallic nanoparticles in the gas phase. The gas flow carries the CNT aerogel to the exit, where it can be wound up to a drum for further processing. The exact composition of the feedstock varies with furnace geometry, but for a small reactor like in reference [22], the following observations have been made. Ethanol mix

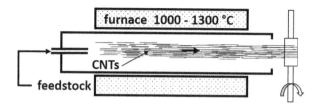

Fig. 3.10 The floating catalyst chemical vapor deposition (FC-CVD) is schematically shown. The liquid feedstock consists of ethanol mixed with ferrocene and thiophene. It is injected into the furnace with hydrogen as carrier gas. Catalyst particles are formed from ferrocene and CNTs grow from these floating particles. They are carried by the gas flow to the exit, where they are wound up

with unknown water content and with 0.23–2.3 wt% ferrocene and 1.0–4.0 wt% thiophene was used. The solution was injected at a flow of 0.08–0.25 ml/min into the flow (400–800 ml/min) of a hydrogen carrier gas. In this setup, multi-walled CNTs were formed with a thiophene concentration ranging between 1.5–4.0 wt% at a hydrogen flow rate of 400–800 ml/min, and furnace temperature between 1100° and 1180 °C. Single-walled CNTs grew at a reduced concentration of thiophene of ∼0.5 wt% and a considerably increased hydrogen flow rate of ∼1200 ml/min, at a furnace temperature of 1200 °C [22]. More recently the standard operating condition for a 65 mm inner diameter reactor has been improved and it mentions a flow of 0.8 l/min hydrogen, 80 ml/min methane, 0.35 mg/min ferrocene, and 1.3 mg thiophene as feedstock [25]. The overall atomic ratio was 0.004 for S:C and 7.65 for S:Fe. The addition of sulphur as well as an early carbon supply seems to be critical for achieving continuous production of clean aerogel with minimal unwanted by-products [25].

Nanocomp Technologies can produce large format CNT sheets of up to 1.3×2.4 m and continuous yarns by using an up-scaled furnace design [26]. It uses the approach to innovate applications such as protective clothing, energy storage, and wiring and cables predominately for aerospace, where the reduced weight at equal conductance and a much increased bending fracture fatigue of the new material are of advantage.

The draw rate of the CNTs from the furnace will enhance the alignment and determine the linear density or Tex (g/km) of the CNT roving. The roving can be spun into a yarn and plied, braided, or woven into the desired form. The physical properties of the roving can be improved through many processing techniques. One option draws the roving through a solvent, such as n-methyl pyrrolidinone (NMP). Another option is stretching the roving by a dual roller setup, where one roller moves a little fast than the other roller. During the stretching polymer or acids can be added before or after stretching. Before spinning the roving into a yarn, rinsing and baking can be applied to remove excess polymers, acids, and solvents.

As the resistance of the CNT yarn is still high compared to copper, hybrid materials, as shown in Fig. 3.11, are being developed as well. The hybrid CNT/Cu structure in Fig. 3.11a has 57 wt% Cu, is 230 μm in diameter, has an overall low density of 2 g cm^{-3}, and a resistivity of 13 $\mu\Omega$ cm. Another hybrid approach is

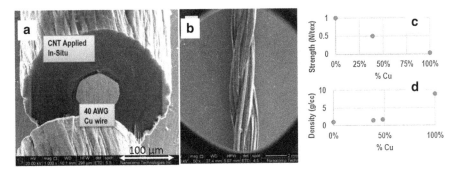

Fig. 3.11 An SEM image of a cross-section of a CNT/Cu hybrid system consisting of a copper core (40 AWG Cu wire) surrounded by a low density (~0.75 g cm^{-3}) CNT material is shown in (**a**). The SEM image in (**b**) displays a CNT/Cu hybrid yarn with 9 strands of chemical stretched yarn (CSY) and 3 strands of 40 AWG copper wire. The strength of hybrid CNT/Cu wires depending on the Cu wt is plotted in (**c**). The density of hybrid CNT/Cu systems depending on the Cu wt% is plotted in (**d**). Adapted and reprinted with permission from Schauer et al. [26]

shown in Fig. 3.11b, where 9 CNT yarns are combined with 3 strands of copper to give a wire with the properties of 415 Tex, 890 μm diameter, 1.4 g cm^{-3} density, 24 μΩ cm resistivity, and 34 % copper by weight.

Jarosz et al. used CNT sheets and ribbons to devise viable replacements for metallic conductors in wires and cables [27]. The main benefits are reduced mass of the cable, higher flexure tolerance, and stability against environmental impact. They observed a bending cycle resilience of over 200,000 bend cycles without loss of conductance or mechanical stability, whereas a normal aerospace wire started to deteriorate at 8000 bend cycles. Jarosz et al. constructed different applications, shown in Fig. 3.12, like wiring of solar cells, USB cable, an Ethernet cable, and coaxial cable—all made of CNTs and all functioning within their specifications [27].

The CNT cables showed no loss in conductivity after 80 days exposure to a highly corrosive hydrochloride environment and their resistivity does not change with temperature. The biggest drawback by now is the conductivity, which is still less than copper. Therefore, further densification and doping strategies for the CNT dope need to be investigated. Jarosz et al. improved the conductivity of a CNT paper by 400 % by immersing it into a solution of KAuCl$_4$. Nanocomp achieved a resistivity of 130 μΩ cm by doping with 5 M HNO$_3$, but care needs to be taken in the comparison, because the density of the material needs also be taken into account. Also the long-term stability of the doping, especially if exposed to environmental chemicals and humidity matters, has not been studied in case of a conductance improvement by doping. Also the consequences of high current densities flowing in the doped materials are not yet investigated.

Besides conductivity improvements, other challenges need to be solved as classical connection schemes, like in metals, will not work. The key challenge is how to make a good contact to the CNT wire. Just making a sidewall connection might not prove to be effective due to the highly anisotropic nature of the current transport in CNTs. It would be important to feed in the current in a parallel fashion

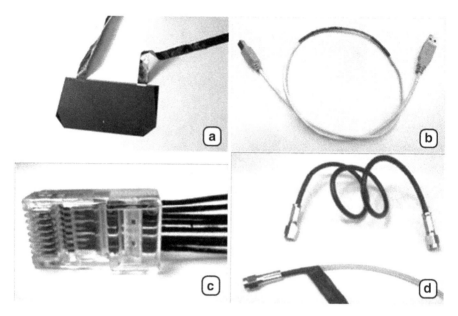

Fig. 3.12 A solar cell with CNT ribbon cable is shown in (**a**). A USB cable equipped with CNT wires is displayed in (**b**). An Ethernet cable fabricated with CNT wires is presented in (**c**) and a coaxial cable made of CNT sheets for the outer conductor in (**d**). Adapted from Jarosz et al. [27] and reprinted with permission of The Royal Society of Chemistry

to all CNTs in the yarn, which will not be easily achieved, at least in the field. Jarosz et al. also investigated ultrasonic bonding of CNT yarns to a copper foil [27]. The resulting contact resistance was evaluated to be in the range of 0.01 Ω cm^2, which is almost six orders of magnitude bigger than a metallic Cu-Cu contact resistance. Pressure-based connections like crimping are omnipresent in both data and power cables, and initial demonstrations have also been investigated by Jarosz et al. for CNTs wires [27]. One major development was to electroplate the ends of the CNT wires with a KAuCl$_4$ solution before crimping, which improved the resistance by a factor of 10. Electroplating in combination with a mild heat treatment could be a general route for achieving a better contact resistance.

3.9 Outlook

Up to now CNTs have failed to deliver their promise of a better interconnect material if one compares them with copper. Compared with tungsten and poly-Si, which both are being used as via fill material in semiconductor chips, CNTs are competitive if doped (for W), or pristine (for poly-Si). Major improvements in the understanding of growing and integrating CNTs have been developed in the past decade. The problem of low mass density translates into the fact that there are simply too few conducting

sp^2-bonds per square nanometers to compete with the conductivity of a solid metal. As a benchmark, graphite with a carbon-based mass density of 2.2 g cm^{-3} would deliver an in-plane resistivity as low as 50 $\mu\Omega$ cm and it can be brought down to 1 $\mu\Omega$ cm by doping [28]. This should be an upper limit to be obtained also with CNTs. The most promising progress could be witnessed in the development of CNT yarns, where manipulation with dopants can be more easily achieved. However, excessive reliability studies on doped yarns are still missing. Another severe problem that needs to be solved is the development of techniques that make connections with low contact resistance between yarns and metals. Due to the high specific conductivity these CNT cables could have a bright future where weight matters, for instance, in aerospace and also in power transmission cables, which are suspended above ground and under high mechanical tension.

References

1. Engelhardt M, Hönlein W, Kreupl F (2000) Electronic component comprising an electrically conductive connection consisting of carbon nanotubes and a method for producing the same. US Patent 7,321,097 B2, 16 Feb 2000
2. Kreupl F, Graham AP, Duesberg GS, Steinhögl W, Liebau M, Unger E, Hönlein W (2002) Carbon nanotubes in interconnect applications. Microelectron Eng 64(1):399–408
3. Joselevich E, Dai H, Liu J, Hata K, Windle AH (2008) Carbon nanotube synthesis and organization. In: Jorio A, Dresselhaus G, Dresselhaus MS (eds) Carbon nanotubes. Springer, Berlin, Heidelberg, pp 101–165
4. Kumar M, Ando Y (2010) Chemical vapor deposition of carbon nanotubes: a review on growth mechanism and mass production. J Nanosci Nanotechnol 10(6):3739–3758
5. Liu H, Takagi D, Chiashi S, Homma Y (2010) The growth of single-walled carbon nanotubes on a silica substrate without using a metal catalyst. Carbon 48(1):114–122
6. Wang H, Yuan Y, Wei L, Goh K, Yu D, Chen Y (2015) Catalysts for chirality selective synthesis of single-walled carbon nanotubes. Carbon 81:1–19
7. Helveg S, Lopez-Cartes C, Sehested J, Hansen PL, Clausen BS, Rostrup-Nielsen JR, Abild-Pedersen F, Nørskov JK (2004) Atomic-scale imaging of carbon nanofibre growth. Nature 427(6973):426–429
8. Zhang Q, Huang JQ, Zhao MQ, Qian WZ, Wei F (2011) Carbon nanotube mass production: principles and processes. ChemSusChem 4(7):864–889
9. Kimura H, Goto J, Yasuda S, Sakurai S, Yumura M, Futaba DN, Hata K (2013) The infinite possible growth ambients that support single-wall carbon nanotube forest growth. Sci Rep 3
10. Harutyunyan AR, Chen G, Paronyan TM, Pigos EM, Kuznetsov OA, Hewaparakrama K, Kim SM, Zakharov D, Stach EA, Sumanasekera GU (2009) Preferential growth of single-walled carbon nanotubes with metallic conductivity. Science 326(5949):116–120
11. Zhang G, Mann D, Zhang L, Javey A, Li Y, Yenilmez E, Wang Q, McVittie JP, Nishi Y, Gibbon J, Dai H (2005) Ultra-high-yield growth of vertical single-walled carbon nanotubes: hidden roles of hydrogen and oxygen. Proc Natl Acad Sci U S A 102(45):16141–16145
12. Matsumoto N, Oshima A, Sakurai S, Yumura M, Hata K, Futaba DN (2015) Scalability of the heat and current treatment on SWCNTs to improve their crystallinity and thermal and electrical conductivities. Nanoscale Res Lett 10(1):1–7
13. Yang J, Esconjauregui S, Robertson AW, Guo Y, Hallam T, Sugime H, Zhong G, Duesberg GS, Robertson J (2015) Growth of high-density carbon nanotube forests on conductive TiSiN supports. Appl Phys Lett 106(8):083108

14. Ahmad M, Anguita JV, Stolojan V, Corless T, Chen JS, Carey JD, Silva SRP (2015) High quality carbon nanotubes on conductive substrates grown at low temperatures. Adv Funct Mater. doi:10.1002/adfm.201501214

15. Ma Y, Wang B, Wu Y, Huang Y, Chen Y (2011) The production of horizontally aligned single-walled carbon nanotubes. Carbon 49(13):4098–4110

16. Yan F, Zhang C, Cott D, Zhong G, Robertson J (2010) High-density growth of horizontally aligned carbon nanotubes for interconnects. Physica Status Solidi (b) 247(11–12): 2669–2672

17. Hayamizu Y, Yamada T, Mizuno K, Davis RC, Futaba DN, Yumura M, Hata K (2008) Integrated three-dimensional microelectromechanical devices from processable carbon nanotube wafers. Nat Nanotechnol 3(5):289–294

18. Li H, Liu W, Cassell AM, Kreupl F, Banerjee K (2013) Low-resistivity long-length horizontal carbon nanotube bundles for interconnect applications—part I: process development. IEEE Trans Electron Devices 60(9):2862–2869

19. Li H, Liu W, Cassell AM, Kreupl F, Banerjee K (2013) Low-resistivity long-length horizontal carbon nanotube bundles for interconnect applications—part II: characterization. IEEE Trans Electron Devices 60(9):2870–2876

20. Subramaniam C, Yamada T, Kobashi K, Sekiguchi A, Futaba D N, Yumura M, Hata K (2013) One hundred fold increase in current carrying capacity in a carbon nanotube-copper composite. Nat Comm 4:1–7

21. Craddock JD, Weisenberger MC (2015) Harvesting of large, substrate-free sheets of vertically aligned multiwall carbon nanotube arrays. Carbon 81:839–841

22. Li YL, Kinloch IA, Windle AH (2004) Direct spinning of carbon nanotube fibers from chemical vapor deposition synthesis. Science 304(5668):276–278

23. Behabtu N, Young CC, Tsentalovich DE, Kleinerman O, Wang X, Ma AW, Amram Bengio E, ter Waarbeek RF, de Jong JJ, Hoogerwerf RE, Fairchild SB, Ferguson JB, Maruyama B, Kono J, Talmon Y, Cohen Y, Otto MJ, Pasquali M (2013) Strong, light, multifunctional fibers of carbon nanotubes with ultrahigh conductivity. Science 339(6116):182–186

24. Parra-Vasquez ANG, Behabtu N, Green MJ, Pint CL, Young CC, Schmidt J, Kesselman E, Goyal A, Ajayan PM, Cohen Y, Talmon Y, Hauge RH, Pasquali M (2010) Spontaneous dissolution of ultralong single-and multiwalled carbon nanotubes. ACS Nano 4(7):3969–3978

25. Gspann TS, Smail FR, Windle AH (2014) Spinning of carbon nanotube fibres using the floating catalyst high temperature route: purity issues and the critical role of sulphur. Faraday Discuss 173:47–65

26. Schauer MW, White MA (2015) Tailoring industrial scale CNT production to specialty markets. In: Proceedings of the MRS, vol 1752. Cambridge University Press, Cambridge, pp 103–109

27. Jarosz P, Schauerman C, Alvarenga J, Moses B, Mastrangelo T, Raffaelle R, Ridgley R, Landi B (2011) Carbon nanotube wires and cables: near-term applications and future perspectives. Nanoscale 3(11):4542–4553

28. Kreupl F (2008) Carbon nanotubes in microelectronic applications. In: Hierold C (ed) Carbon nanotube devices: properties, modeling, integration and applications, vol 8. Wiley, London

Chapter 4
Electrical Conductivity of Carbon Nanotubes: Modeling and Characterization

Antonio Maffucci, Sergey A. Maksimenko, Giovanni Miano, and Gregory Ya. Slepyan

4.1 Introduction

The main reasons why carbon nanotubes (CNTs) have been extensively explored as electronic materials reside in their intrinsic advantages for high-performance applications, due to their outstanding properties [1–4].

Indeed, in the last decade several nanoelectronics applications have been suggested for CNTs, such as nano-interconnects, nano-antennas, lumped passives, diodes and transistors, memories, plastic and transparent devices, e.g., [5–15]. Recently, their properties have also suggested their use for the THz technology, e.g., [16].

One of the main reasons of this success is to be found in their novel electromagnetic properties, due to many quantum phenomena involved in such 1D nanostructures. In general, they are related to the spatial confinement of the charge-carriers motion to sizes comparable with the de Broglie wavelength, what thereby produces a discrete spectrum of energy states in one or several directions. As a

A. Maffucci (✉)
Department of Electrical and Information Engineering, University of Cassino and Southern Lazio, Via Di Biasio 43, 03043 Cassino, Italy

INFN—LNF, Via Enrico Fermi, 00044 Frascati, Italy
e-mail: maffucci@unicas.it

S.A. Maksimenko
Institute for Nuclear Problem, Belarusian State University, Bobruiskaya 11, 220030 Minsk, Belarus

G. Miano
Department of Electrical Engineering and Information Technology, University of Naples Federico II, Via Claudio 21, 80125 Naples, Italy

G.Ya. Slepyan
School of Electrical Engineering, Tel Aviv University, 6997801 Tel-Aviv, Israel

© Springer International Publishing Switzerland 2017
A. Todri-Sanial et al. (eds.), *Carbon Nanotubes for Interconnects*,
DOI 10.1007/978-3-319-29746-0_4

result, specific dispersion laws manifest themselves in nanostructures providing their unusual electromagnetic response. This is the reason why, in view of all the above-mentioned potential applications, it is of great interest to fully understand the behavior of such carbon materials.

This chapter is devoted to modeling the electrical conductivity of CNTs, in view of their above-mentioned potential applications discussed in the other chapters. A short introduction to the electronics property of CNTs is given in Sect. 4.2, which allows further introduction of a conductivity model, given in Sect. 4.3 for isolated CNTs and in Sect. 4.4 for bundled CNTs. Such a model describes accurately the electrical transport up to the frequency range of some THz, where only intraband transitions are allowed. In this frequency range, novel phenomena may occur, which are of great potential interest for applications, such as plasmon resonance or intershell tunneling: they are discussed in Sects. 4.3 and 4.4, respectively.

Finally, in Sect. 4.5 we introduce the possibility of having interband transitions, which allows extending the conductivity model to optical range.

4.2 Band Structure of Carbon Nanotubes and Energy Subbands

Let us briefly review the band structure of the graphene, whose Bravais lattice is depicted in Fig. 4.1a. The unit cell is spanned by the two basis vectors $\mathbf{a}_1$, $\mathbf{a}_2$ of length $a_0 = \sqrt{3}b$, being $b = 1.42$ Å the interatomic distance. In the reciprocal k-space depicted in Fig. 4.1b, the graphene is characterized by the unit cell Σ_g, spanned by the two vectors $\mathbf{b}_1$, $\mathbf{b}_2$ of length $b_0 = 4\pi\sqrt{3}a_0$ [1, 3].

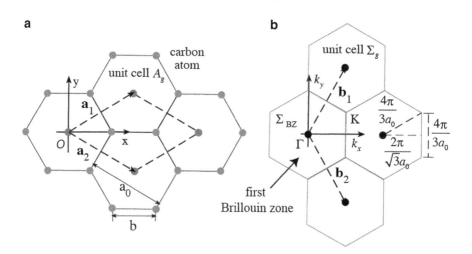

Fig. 4.1 Structure of the graphene. (**a**) Bravais lattice. (**b**) Reciprocal lattice

The graphene possesses four valence electrons for each carbon atom. Three of these (the so-called σ-*electrons*) form tight bonds with the neighboring atoms in the plane and do not play a part in the conduction phenomenon. The fourth electron (the so-called π-*electron*), instead, may move freely between the positive ions of the lattice.

In the nearest-neighbors tight-binding approximation, the energy dispersion relation of the π-electrons is given by

$$\varepsilon^{(\pm)}(\mathbf{k}) = \pm\gamma\left[1 + 4\cos\left(\frac{\sqrt{3}k_x a_0}{2}\right)\cos\left(\frac{k_y a_0}{2}\right) + 4\cos^2\left(\frac{k_y a_0}{2}\right)\right]^{1/2}, \quad (4.1)$$

where $\varepsilon^{(\pm)}$ is the energy, the $+$ sign denotes the conduction band, the $-$ sign denotes the valence band, and $\gamma = 2.7\text{eV}$ is the carbon–carbon interaction energy. The valence and conduction bands touch themselves at the sixth vertex of each unit cell, the so-called *Fermi points*. In the neighborhood of each Fermi point the energy dispersion relation may be approximated as

$$E^{(\pm)} \approx \pm\hbar v_F |\mathbf{k} - \mathbf{k}_0|, \quad (4.2)$$

where $\mathbf{k}_0$ is the wavenumber at a Fermi point, $v_F \approx 0.87 \cdot 10^6$ m/s is the Fermi velocity of the π-electrons, and $\hbar$ is the Planck constant.

In the ground state the valence band of the graphene is completely filled by the π-electrons. In general, at equilibrium the energy distribution function of π-electrons is given by the Dirac–Fermi function:

$$F\left[\varepsilon^{(\pm)}\right] = \frac{1}{e^{\varepsilon^{(\pm)}/k_B T_0} + 1}, \quad (4.3)$$

where k_B is the Boltzmann constant and T_0 is the absolute temperature.

As known, a CNT is obtained by rolling up a graphene layer. Its unit cell in the direct space is the cylindrical surface generated by the chiral vector $\mathbf{C} = n\mathbf{a}_1 + m\mathbf{a}_2$ and the translational vector $\mathbf{T} = t_1\mathbf{a}_1 + t_2\mathbf{a}_2$, where n and m are integers, and its circumference is given by $C = a_0\sqrt{n^2 + nm + m^2}$. The first Brillouin zone of an SWCNT is the set of N parallel segments s_μ, $\mu = 0, ..N - 1$, of the graphene reciprocal lattice, depicted in Fig. 4.2. Each segment has a length of $T = |\mathbf{T}|$, and is orthogonal to $\mathbf{K}_1$ and parallel to $\mathbf{K}_2$, being such vectors:

$$\mathbf{K}_1 = (-t_2\mathbf{b}_1 + t_1\mathbf{b}_2)/N, \qquad \mathbf{K}_2 = (m\mathbf{b}_1 + n\mathbf{b}_2)/N. \quad (4.4)$$

The longitudinal wave vector k is almost continuous because the length of the CNT is assumed to be very large compared with the length of the unit cell.

Fig. 4.2 The first Brillouin
zone of a CNT is the set of
line segments s_μ parallel to
$\mathbf{K}_2$. Vectors $\mathbf{K}_1$ and $\mathbf{K}_2$ are
reciprocal lattice vectors to $\mathbf{C}$
and $\mathbf{T}$, respectively

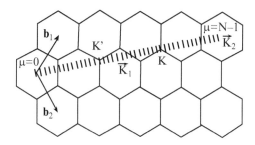

The transverse wave vector $k_\perp$ is quantized: it takes only the discrete values
$\mu \Delta k_\perp$ with $\mu = 0, 1, \ldots, N - 1$. In the zone-folding approximation the dispersion
relation for the SWCNT is given by [1, 17]:

$$\varepsilon_\mu^{(\pm)}(k) = \varepsilon^{(\pm)} \left(k \frac{\mathbf{K}_1}{|\mathbf{K}_2|} + \mu \mathbf{K}_1 \right) \quad \text{for} \quad -\frac{\pi}{T} < k < \frac{\pi}{T} \quad \text{and} \quad \mu = 0, 1, \ldots N - 1,$$

(4.5)

where $\varepsilon^{(\pm)}$ is given by (4.1).

Nanotubes are denoted as *zig-zag* if $n = 0$ or $m = 0$, *armchair* if $n = m$,
and *chiral* in all other cases. The general condition for a CNT to be metallic is
$|n - m| = 3q$, where $q = 0, 1, 2, ..$, therefore, armchair CNTs are always metallic,
whereas zig-zag CNTs are metallic only if $m = 3q$. Figure 4.3 shows the typical
band structures for a metallic and a semiconducting CNT.

4.3 Electrical Conductivity of Isolated CNTs, from DC to THz Range

4.3.1 Transport Equation

In view of modeling the CNT conductivity, we must investigate the interaction of
π-electrons with electromagnetic field. Within the quasi-classical approximation,
these electrons are modeled as a 1D electron Fermi-gas which is in the equilibrium
state, in the absence of electromagnetic field. Under the action of the field, the π-
electrons execute intraband motion, with negligible interband transitions. This limits
the mode to frequency range up to Terahertz (THz) or Far infrared (FIR) ranges.

Under the above conditions, electron gas in CNT is characterized by a distribution function of π-electrons, whose intraband motion is described by the Boltzmann
equation for the distribution function. Elimination of interband transitions does not
mean overall neglecting quantum effects, what is outlined by the adjective "quasi-
classical." Indeed, electrons in CNT are not classical particles; their quantum nature
manifests itself even in intraband motion. First of all, π-electrons are *fermions*

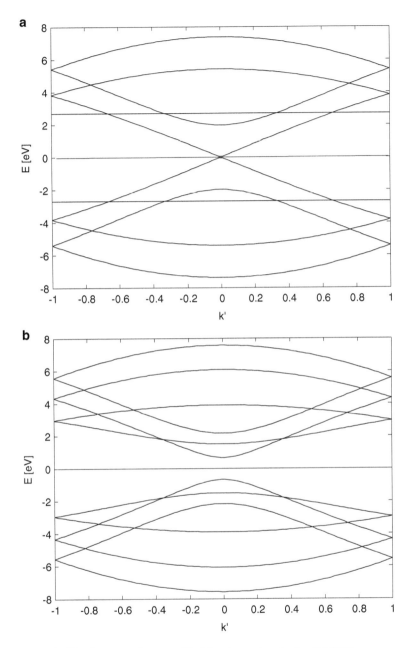

Fig. 4.3 Typical band structure for a metallic (**a**) and a semiconducting (**b**) CNT

and, therefore, their quantum nature establishes their equilibrium distribution as the Fermi distribution (4.3). Furthermore, π-electrons are not free electrons: their quantum-mechanical interaction with the CNT crystalline lattice is responsible for their particular dispersion law.

Assuming a transversally symmetric surface wave interacting with the π-electrons, the distribution function $f(\mathbf{p}, z, t)$ satisfies the following Boltzmann kinetic equation [18]:

$$\frac{\partial f}{\partial t} + v_z \frac{\partial f}{\partial z} + eE_z \frac{\partial f}{\partial p_z} = \nu \left[F(\mathbf{p}) - f \right], \tag{4.6}$$

where e is the electron charge, $\mathbf{p}$ is the two-dimensional quasi-momentum, tangential to the CNT surface, p_z is the projection of $\mathbf{p}$ on the CNT axis z, $v_z = d\varepsilon/dp_z$ is the velocity, and $F(.)$ is the Fermi distribution (4.3). The relaxation term (collision integral) in the r.h.s. of (4.6) is assumed to be momentum-independent [19], thus described by the *collision frequency* ν, i.e., the reciprocal of the time between two collisions $\nu = v_z/\lambda$, being λ the electron mean free path.

The solution of (4.6) can be linearized in the weak-field regime. Assuming $E_z = \mathrm{Re}\left[E_z^0 \exp\left(i\left(\omega t - kz \right) \right) \right]$, putting $f = F + \mathrm{Re}\left[\delta f \exp\left(i\left(\omega t - kz \right) \right) \right]$, and taking only the linear terms, we obtain

$$\delta f = -i \frac{\partial F}{\partial p_z} \frac{eE_z^0}{\omega - kv_z + i\nu}. \tag{4.7}$$

The axial surface current density $J_z = \mathrm{Re}\left[J_z^0 \exp\left(i\left(\omega t - kz \right) \right) \right]$, is then given by

$$J_z^0(\omega, k) = \tilde{\sigma}_{zz}(\omega, k) E_z^0(\omega, k), \tag{4.8}$$

where the axial conductivity in the wavenumber domain is expressed as

$$\tilde{\sigma}_{zz}(\omega, k) = -i \frac{2e^2}{(2\pi\hbar)^2} \iint\limits_{1stBZ} \frac{\partial F}{\partial p_z} \frac{v_z d^2 \mathbf{p}}{\omega - kv_z + i\nu}; \tag{4.9}$$

here, the integral spans over the first Brillouin zone (1st BZ).

To compute (4.9) we should refer to a given electron dispersion relation E(**p**). Assuming zig-zag CNTs (4.9) would become

$$\tilde{\sigma}_{zz}(\omega) = -2ie^2 n_0 \frac{1}{(\omega + i\nu) F_0} \iint\limits_{1stBZ} v_z^2(\mathbf{p}) \frac{\partial F}{\partial \varepsilon} d^2 \mathbf{p}, \tag{4.10}$$

where

$$F_0 = \iint\limits_{1stBZ} F(\mathbf{p}) d^2 \mathbf{p}, \tag{4.11}$$

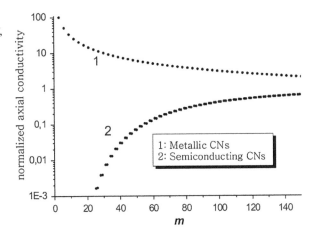

Fig. 4.4 Normalized axial conductivity for zig-zag CNT, vs chiral number m (i.e., vs CNT radius). The conductivity is normalized to that of the graphene (4.12) [18]

and n_0 is the surface density of conduction-band electrons in graphene. Note that, for infinite CNT radius, the dynamic axial conductivity of a CNT becomes locally equivalent to the dynamic conductivity of semi-metallic graphene:

$$\sigma_\infty = i\frac{2\ln 2}{\pi\hbar^2}\frac{e^2 k_B T_0}{\omega + i\nu}. \tag{4.12}$$

The numerically evaluated axial conductivity (4.10) is reported in Fig. 4.4 as a function of the chiral index m (with $n = 0$), assuming $\nu = 0.33 \cdot 10^{12}$ s^{-1}, and $T_0 = 264$ K. Since the radius of a zig-zag CNT is given by $R_c = \sqrt{3}mb/2\pi$ [1], Fig. 4.4 actually shows the dependence of $\tilde{\sigma}_{zz}$ on the cross-sectional radius. As R_c increases ($m > 300$), $\tilde{\sigma}_{zz}$ approaches σ_∞. The results show a sharp increase corresponding to $m = 3q$ (where q is an integer), i.e., to metallic condition for zig-zag CNTs. For the metallic case, it results

$$\tilde{\sigma}_{zz} = i\frac{2\sqrt{3}}{m\pi\hbar^2}\frac{e^2\gamma}{\omega + i\nu}. \tag{4.13}$$

To understand such a behavior, let us note that only energy subbands that pass through or are close to the CNT Fermi level contribute significantly to the CNT axial conductivity. In the Brillouin zones in Fig. 4.2, there are only two inequivalent graphene Fermi points contained by or are nearest to some segments s_μ, indicated with K and K'. Figure 4.5 shows the segments around K: for metallic CNTs, there is one segment passing through K, resulting in the drastic growth of the numerical value of the integral in (4.10). For semiconducting CNTs, no segment passes through K. Note that the segments that contribute significantly to $\tilde{\sigma}_{zz}$ are those falling in a circle c_k of radius $k_{eff} \approx 5k_B T_0/\hbar v_F$: this provides a criterion to evaluate the limited number of subbands which significantly contributes to the conduction. Note that this also implies the dependence of $\tilde{\sigma}_{zz}$ on temperature.

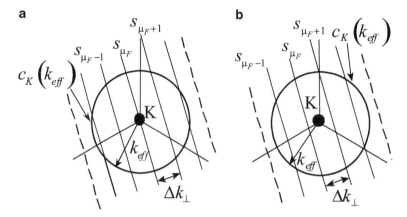

Fig. 4.5 Segments s_μ around the Fermi point K for: (**a**) a metallic and (**b**) a semiconducting CNT. Only the full line segments contribute effectively to the CNT axial conductivity

Fig. 4.6 Normalized axial conductivity for armchair CNT, vs chiral number m (i.e., vs CNT radius). The conductivity is normalized to that of the graphene (4.12) [18]

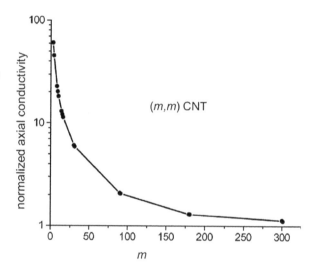

The case of armchair CNTs may be treated with the same approach, using the corresponding dispersion relation $\varepsilon(\mathbf{p})$ to solve (4.10). The numerically evaluated conductivity $\tilde{\sigma}_{zz}$ is shown in Fig. 4.6, vs the chiral number m. For small CNT radius ($m < 50$), $\tilde{\sigma}_{zz}$ may be approximated as in (4.13), divided by $\sqrt{3}$. Unlike for zig-zag CNTs, for armchair CNTs the dependence is monotonic. Physically, this follows from the fact that armchair CNTs are conductors for any m. Again, for $m \to \infty$ (i.e., $R_c \to \infty$), it is $\tilde{\sigma}_{zz} \to \sigma_\infty$.

Finally, in the general case of chiral CNTs ($n \neq m > 0$), an applied axial electric field would induce the current to flow along a helical line. Therefore, besides the axial conductivity, we should also consider an azimuthal component. However, in the linear regime such a component was shown to be negligible [20–22]. For a metallic chiral CNT of radius r_c, $\tilde{\sigma}_{zz}$ may be approximated as

$$\tilde{\sigma}_{zz} = i\frac{3e^2 b\gamma}{\pi^2 \hbar^2 r_c} \frac{1}{\omega + i\nu}. \tag{4.14}$$

4.3.2 Equivalent Parameters for an Isolated CNT

In the frequency domain, taking into account Boltzmann equation (4.6) and constitutive relation (4.8), the relation between current, charge, and field can be written as follows:

$$\left(\frac{i\omega}{\nu} + 1\right) J_z + \frac{v_F^2}{\nu}\frac{\partial \rho_s}{\partial z} = \sigma_c E_z, \tag{4.15}$$

where $\rho_s(z, \omega)$ is the surface charge density, and σ_c is the long wavelength static limit for the axial conductivity:

$$\sigma_c = \frac{v_F}{\pi r_c \nu R_0} M. \tag{4.16}$$

In (4.16) we introduced the *quantum resistance* $R_0 = \pi\hbar/e^2 \cong 12.9\,\text{k}\Omega$, and the *equivalent number of conducting channels M*. This latter parameter takes into account the contribution to the conductivity given by those segments s_μ passing through the two circles in Fig. 4.5, and may be evaluated by [17, 23]:

$$M = \frac{2\hbar}{v_F} \sum_{\mu=0}^{N-1} \int_0^{\pi/T} v_F^2 \left(\frac{dF}{dE_\mu^+}\right) dk. \tag{4.17}$$

Let us now further investigate the physical meaning of the coefficients of the linear transport equation (4.15). In the considered frequency range (up to THz), at the generic abscissa z, the current intensity and charge densities are almost uniform over the CNT circumference C, therefore we can assume

$$i(z, t) \approx 2\pi r_c j_z(z, t), \quad q(z, t) \approx 2\pi r_c \rho_s(z, t). \tag{4.18}$$

By using (4.18), transport equation (4.15) may be rewritten in time domain as

$$L_k \frac{di(z, t)}{dt} + Ri(z, t) + \frac{1}{C_q}\frac{\partial q(z, t)}{\partial z} = E_z(z, t), \tag{4.19}$$

where the electrical parameters L_k, R, and C_q have the dimensions of per-unit-length (p.u.l.) inductance, resistance, and capacitance, respectively, given by

$$L_k = \frac{R_0}{2v_F M}, \quad R = vL_k, \quad C_q = \frac{v_F^2}{L_k}. \tag{4.20}$$

These parameters are associated with three major phenomena exhibited by the CNT electrodynamics: L_k is also known as the *kinetic inductance* of the CNT, which adds to the classical magnetic inductance. Usually, L_k is 3–4 orders of magnitude greater than the magnetic one. From a physical point of view, L_k is associated with the kinetic energy of the π-electrons. The parameter C_q is, instead, the *quantum capacitance* of the CNT, associated from a physical point of view with the internal energy of the π-electrons at the ground state. Finally, the resistance R takes into account all the scattering mechanisms associated with the electron transport along the CNT lattice, since $v = v_F/\lambda$. An approximation for the mean free path λ is given by Matthiessen rule [24]:

$$\lambda^{-1} = \lambda_{AC}^{-1} + \lambda_{OP,ems}^{-1} + \lambda_{OP,abs}^{-1}, \tag{4.21}$$

which includes elastic electron scattering with AC phonons (λ_{AC}), and inelastic electron scattering by optical phonons emission ($\lambda_{OP,ems}$) and absorption ($\lambda_{OP,abs}$). Experimental evidences and theoretical models indicate that λ varies with CNT chirality, diameter, and temperature (e.g., [25]). All these parameters are strongly dependent on the number of conducting channels M (4.17), which in turn depends on CNT chirality radius and temperature. In fact, as the CNT radius increases, $\Delta k_\perp = 1/r_c$ decreases. In addition, as the absolute temperature T_0 increases, the radius k_{eff} increases too. As a consequence, M is an increasing function of both CNT radius and temperature, as shown in Fig. 4.7, for metallic and semiconducting CNTs.

For SWCNTs, with radius of some nm, it is $M \approx 2$ ($M \approx 0$) for the metallic (semiconducting) case, and thus only the metallic SWCNTs contribute to the conduction. Instead, for MWCNT shells, whose radius is typically of the order of tens of nm, the contributions of semiconducting CNTs are not negligible. The following piecewise linear fitting may be adopted for the generic CNT shell of diameter D, whose parameters are reported in Table 4.1 [23].

$$M_{shell}(D; T) \cong \begin{cases} M_0 & \text{for } D < D_0(T) = d_0/T \\ a_1 DT + a_2 & \text{for } D > D_0(T) = d_0/T \end{cases}. \tag{4.22}$$

For small diameter metallic CNTs it is $M \approx 2$, hence parameters (4.20) become

$$L_k = \frac{R_0}{4v_F} \approx 3.6 \text{ nH}/\mu\text{m}, \quad C_q = \frac{R_0 v_F}{4} \approx 350 \text{ aF}/\mu\text{m}. \tag{4.23}$$

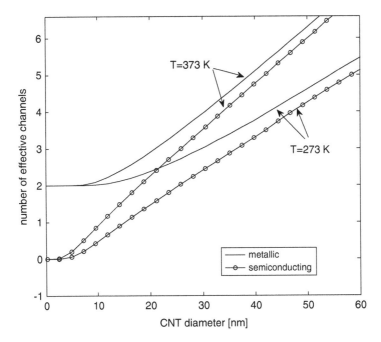

Fig. 4.7 Equivalent number of conducting channels vs CNT diameter, at $T = 273$ K and $T = 373$ K

Table 4.1 Fitting values for formula (4.22)

	SWCNT (metallic)	SWCNT (semiconducting)	MWCNT shell
M_0	2	0	2/3
a_1 (nm^{-1} K^{-1})	3.26×10^{-4}	3.26×10^{-4}	3.26×10^{-4}
a_2	0.15	-0.20	-0.08
d_0 (nm K)	5600	600	1900

As the CNT diameter increases, the kinetic inductance increases and the quantum capacitance decreases, as shown in Fig. 4.8.

4.3.3 Plasmon Resonances in CNTs

According to the results presented in previous sections, the nanotubes allow the propagation of slowly decaying surface waves (plasmons), with almost frequency-independent wavenumber and the phase velocity much smaller than the speed of light in vacuum. These properties make CNTs excellent candidates for the design of nano-antennas [5–7]. To understand this result, we can analyze the simple case of an isolated CNT, whose circuit parameters were derived in Sect. 4.3.2. In the considered frequency range, the propagation may be modeled by a simple RLC

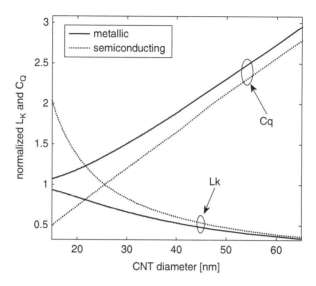

Fig. 4.8 Kinetic inductance and quantum capacitance, normalized to the values in (4.23) versus the CNT diameter, for $T = 273$K: metallic (*solid lines*) and semiconducting case (*dotted lines*)

lossy transmission line (TL) model [14], whose parameters are approximately given (assuming $L_k >> L_M$) by: R, $L \approx L_k$, and $C \approx C_E$, being C_E the electrostatic capacitance. A consequence of the huge value of the kinetic inductance is a low phase velocity, typically, $c_{CNT} \approx 10^{-2}c$, where c is the value for an ideally scaled copper interconnect of the same dimensions. The plasmon resonances are then obtained by imposing the classical resonance condition along a TL, that is, $\beta l = n\pi/2$, with $n = 1, 2, 3$.

The phenomena of plasmon resonance may be also observed in higher frequency ranges, for instance, in the FIR range. In order to investigate such a range, a rigorous evaluation of the CNT conductivity is needed, taking into account both intraband and interband transitions. This can be done, for instance, by means of the method of effective boundary conditions (EBC), as shown in [18]. The general model for conductivity in such ranges will be discussed in Sect. 4.5 and will lead to a conductivity behavior like that reported in Fig. 4.9, referring to a 200 nm-long SWCNT.

Three regions may be easily distinguished. In the low-frequency range ($\lambda > 300\,\mu$m $\rightarrow f < 1$ THz) the conductivity attains its quasi-static value, and metallic CNTs exhibit a classical Drude behavior. In the high frequency range ($\lambda < 3\,\mu$m $\rightarrow f > 100$ THz), the conductivity is dominated by the interband transitions (see Sect. 4.5 for details).

The intermediate region is the most interesting for THz applications: in such a region, the dispersion relation for the propagating surface wave reads [18]:

$$\frac{\kappa^2}{k^2}K_q(\kappa r_c)I_q(\kappa r_c) = \frac{ic}{4\pi r_c \sigma_{zz}}\left[1 - (\kappa^2 + k^2)\right],\qquad(4.24)$$

Fig. 4.9 Axial CNT conductivity vs wavelength for an SWCNT of length 200 nm

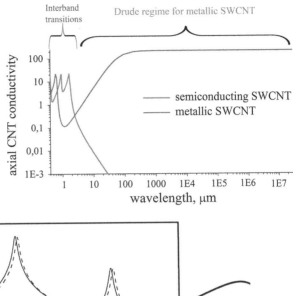

Fig. 4.10 The slow-wave coefficient β in a (9,0) metallic zig-zag SWCNT for relaxation time $\tau = 3$ ps and $T = 295$ K [18], b = 1.42 Å is the C–C bond length. 1: Re(β) and 2: Re(β)/Im(β). *Inset*: antenna resonances in polarizability of a 0.6 μm length (18,0) zig-zag SWCNT [7]

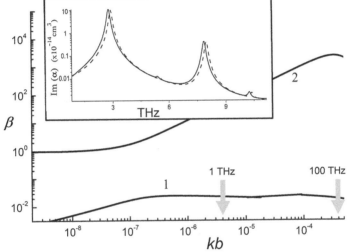

where $K_q(.)$, $I_q(.)$ are the modified Bessel functions, c is the speed of light in vacuum, $\kappa^2 = h^2 + k^2$, and h is the surface wave wavenumber. Compared to classical dispersion laws in microwave range, the presence of the CNT conductivity σ_{zz} in (4.24) is responsible for the unusual electromagnetic response of these nanostructures. In fact, Fig. 4.10 shows the wavenumber $\beta = k/h$, computed from (4.24) for an axially symmetric plasmon along a metallic zig-zag SWCNT of length 0.6 μm. In the THz and FIR ranges, the surface waves are associated with an almost frequency-independent wavenumber β, with a phase velocity given by $v_{ph} \approx 0.02c$. The positions of the plasmon resonances satisfy the above-mentioned condition: the first antenna resonance at $n = 1$ is also known as the localized plasmon resonance [26].

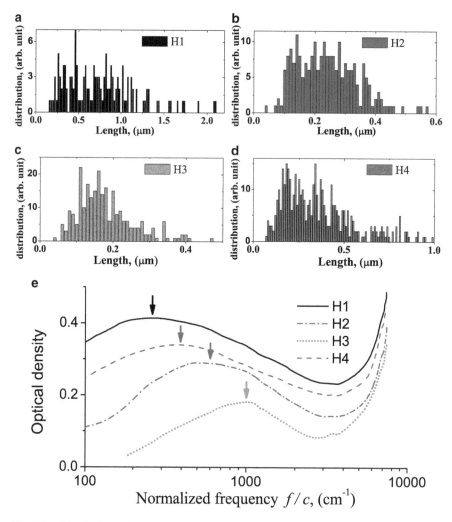

Fig. 4.11 Distributions of CNT lengths in the thin film (**a**–**d**); optical density spectra for the four sets showing the THz peaks (see *arrows*) [27]

The first experimental evidence of plasmon nature of the THz peak in SWCNTs was given in [27], where an FIR peak was found to shift to higher frequencies after decreasing the CNT length. A broad peak in the THz and FIR ranges was observed in the conductivity spectrum of composite materials containing SWCNTs [28–35]. Two hypotheses have been advanced to explain such a peak, as an effect of the finite CNT radius or of the finite CNT length. The plasmon nature of such resonances is evident from Fig. 4.11, which shows the results of an experiment carried out on a thin film containing SWCNTs with controlled lengths [36]. The length distributions of four sets are shown in Fig. 4.11a–d, whereas Fig. 4.11e shows a resonance peak in the optical density spectra which shifts to lower values for longer CNTs. The nature of plasmon excitation in SWCNTs was independently proved by alternative experiment [37], which is based on the idea that only metallic nanotubes contribute into THz peak if their nature is related to antenna (plasmon) resonance, while substitutional doping of CNTs increases "metallicity" of nanotubes. Theoretical model and experimental demonstration of the increase were reported in [38, 39], respectively. The plasmon nature of TCP was also reported in [40], on the basis of experiments analogous to that carried out in [27].

4.4 Equivalent Resistivity for a CNT Bundle from DC to THz

4.4.1 A Bundle of CNTs Without Intershell Coupling

As pointed out in this Book, CNTs are proposed as emerging research material [41] for interconnects and packages, because of their unique electrical, mechanical, and thermal properties. CNT interconnects are expected to meet many of the new requirements for bonding, molding compound, underfill, thermal interface, and die attach. The electrical performances of CNT interconnects are strongly influenced by the kinetic and quantum phenomena affecting their electrodynamics, and by the presence of a huge contact resistance at the metal/CNT contacts.

The technological solution adopted so far is the use of dense bundles of either SWCNTs or MWCNTs where each individual tube carries the same current, so that the equivalent resistance is strongly reduced by the parallel, like in the schemes reported in Fig. 4.12 for vertical (vias) or in Fig. 4.13 for horizontal interconnects.

In order to derive an equivalent resistivity of these interconnects, let us consider a bundle of length l, composed of N_b CNTs, fed in parallel.

We indicate, respectively, with I_n, V_n, and E_n, the current intensity, the voltage between the ends, and the longitudinal component of the total electric field at the surface of the n-th CNT shell with the reference directions for the current intensity

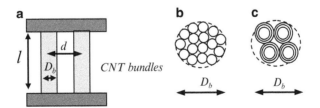

Fig. 4.12 CNT vertical interconnects (vias): (**a**) geometry; (**b**) cross section of an SWCNT via; and (**c**) cross section of an MWCNT via

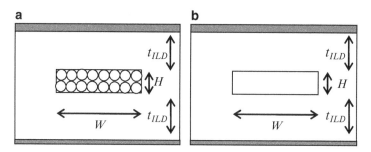

Fig. 4.13 Horizontal interconnect made by: (**a**) SWCNTs bundle and (**b**) MWCNTs bundle

and the voltage chosen according to the normal convention. Using (4.19), the equations governing the transport of the π-electrons may be expressed in matrix form as follows:

$$\left(i\omega \overset{\leftrightarrow}{\mathbf{L}}_k + \overset{\leftrightarrow}{\mathbf{R}}\right)\mathbf{I} = l\mathbf{E},\tag{4.25}$$

where $\mathbf{I}$ and $\mathbf{E}$ are column vectors ($N_b \times 1$), of components I_k and E_k, and $\overset{\leftrightarrow}{\mathbf{L}}_k$ and $\overset{\leftrightarrow}{\mathbf{R}}$ are diagonal matrices whose m-th elements are given by the kinetic inductance and the resistance of each CNT (here and hereafter the symbol $\leftrightarrow$ denotes matrices). For SWCNTs bundle, it means

$$(R)_{m,m} = \frac{R_0 l}{2M_m \lambda_m}, \qquad (L_k)_{m,m} = \frac{R_0 l}{2M_m v_F},\tag{4.26}$$

where M_m and λ_m are the number of channels and mean free path of the m-th CNT. In case of MWCNT bundles, each of them containing N_S shells, the above parameters are expressed by

$$(L_k)_{m,m} = l\left[\sum_{n=1}^{N_S}\frac{2v_F M_n}{R_0}\right]^{-1}, \qquad (R)_{m,m} = l\left[\sum_{n=1}^{N_S}\frac{2M_n \lambda_n}{R_0}\right]^{-1}.$$

In the magneto-quasi-stationary limit we have

$$i\omega \overset{\leftrightarrow}{\mathbf{L}}_m\mathbf{I} = \mathbf{V} - l\mathbf{E},\tag{4.27}$$

where $\overset{\leftrightarrow}{\mathbf{L}}_m$ is the classical magnetic inductance matrix of the bundle. By combining (4.26) and (4.27) we obtain

$$\left(i\omega \overset{\leftrightarrow}{\mathbf{L}} + \overset{\leftrightarrow}{\mathbf{R}}\right)\mathbf{I} = \mathbf{V},\tag{4.28}$$

where the total inductance is given by $\overleftrightarrow{L} = \overleftrightarrow{L}_m + \overleftrightarrow{L}_k$.

Since all the CNTs in the bundle are electrically in parallel it results $V_n = V_b$ where V_b is the voltage drop along the bundle. By solving system (4.28) with the constraint $V_n = V_b$, for $n = 1, 2 \ldots N_b$, we obtain the relation between the total current, $I_b = \sum_{n=1}^{N_b} I_n$, and the bundle voltage, expressed as

$$V_b = Z_b I_b = \left(\rho_b' + i\rho_b'' \right) \frac{l}{S_b} I_b, \tag{4.29}$$

where S_b is the bundle section and $\rho_b = \rho_b' + i\rho_b''$ may be regarded as the *bundle equivalent resistivity*. Assuming that all CNTs in the bundle are equal and that the kinetic inductance dominates over the magnetic one, as usually it is, the equivalent resistivity may be given the form [12]:

$$\rho_b' + i\rho_b'' = \rho_0 \left(1 + i\omega\tau \right), \tag{4.30}$$

where the value ρ_0 and the relaxation time τ are given by

$$\rho_0 = \frac{R_{CNT} S_b}{F_M N_b l}, \quad \tau = \frac{\lambda}{v_F}, \tag{4.31}$$

where $R_{CNT} = (R)_{m,m}$ is the resistance of the single CNT and F_M is the fraction of metallic CNTs in the bundle. Note that it is usually $F_M = 1/3$ ($F_M = 1$) for SWCNTs (MWCNTs). If we neglect the influence of the terminal contact CNT/metal, then $R_{CNT} = (R)_{m,m}$, but in realistic case we must take into account the effect of contacts, that can be modeled by adding a lumped term [14]:

$$R_{CNT} = (R)_{m,m} + \frac{R_0 + R_p}{M}, \tag{4.32}$$

where R_p is a parasitic term which goes to zero for ideal CNT/metal contact.

In DC or low-frequency conditions, the equivalent resistivity tends to the real value ρ_0. In the following we evaluate the equivalent resistivity at room temperature for several types of CNT interconnects using the geometries in Figs. 4.9 and 4.10, and considering two nanoscale technology nodes of integrated circuits: 22 and 14 nm. Their geometrical values, taken from the International Technology Roadmap of Semiconductors [41], are reported in Table 4.2. We compare the obtained results with the values of the resistivity of the corresponding copper interconnects, reported in Table 4.3, which are reported in [41] and take into account the resistivity increase with shrinking dimensions, due to electromigration, grain and boundary scattering [42].

For all the CNT realizations, we assume a metallic fraction of 1/3, a parasitic resistance $R_p = 10 \ k\Omega$, and a CNT packing density of 80%. The computed resistivity values are reported in Fig. 4.14 for vertical vias and in Fig. 4.15 for

Table 4.2 ITRS values for on-chip traces and vias for 22 and 14 nm nodes (all values are in nm)

Type		14 nm			22 nm		
		Global	Interim	Local	Global	Interim	Local
Trace	W	105	28	14	160	44	22
	H	63	28	14	96	44	44
Via	D	105	28	14	160	44	22

Table 4.3 Values for the copper resistivity at room temperature [41], as a function of the interconnect width

Width (nm)	14	28	105	Bulk
ρ ($\mu\Omega$ cm)	8.19	5.39	3.29	1.72

horizontal interconnects traces, assuming typical lengths for the local and the global levels. Due to the contact resistance, the values depend on the interconnect length: as the length increases, the effect of the contact resistance becomes more and more negligible and the resistivity of the CNT bundle outperforms that of the copper realization. This transition value is strongly dependent on the technology node and on the hierarchical level.

For high frequencies, also the imaginary part of resistivity plays a role, which can be put on evidence by defining the equivalent skin-depth, e.g., [12]:

$$\delta = \sqrt{\frac{2\rho_0}{\omega\mu}} \sqrt{\left[(\omega\tau)^2 + 1\right]\left(\sqrt{(\omega\tau)^2 + 1} - \omega\tau\right)}. \qquad (4.33)$$

Note that in classical metals the mean free path is small enough (e.g., 40 nm in copper) to have $\omega\tau << 1$, and thus (4.33) provides the classical expression for δ. For CNTs, however, larger mean free path values (hundreds of μm) induce a saturation in the skin-depth as frequency increases, according to (4.33). In other words, a CNT bundle is much more insensitive to the skin effect as compared to a copper conductor: Fig. 4.16 shows the computed high frequency distribution of the current density in a via, comparing the bulk copper to an MWCNT bundle solution (from [10]).

4.4.2   A Bundle of CNTs in Presence of Intershell Coupling

The model for CNT bundles presented in the previous paragraph does not take into account the possibility of a coupling between adjacent CNT shells. In order to introduce an intershell coupling effect, we can refer to the simple case of a double-wall CNT, as that depicted in Fig. 4.17.

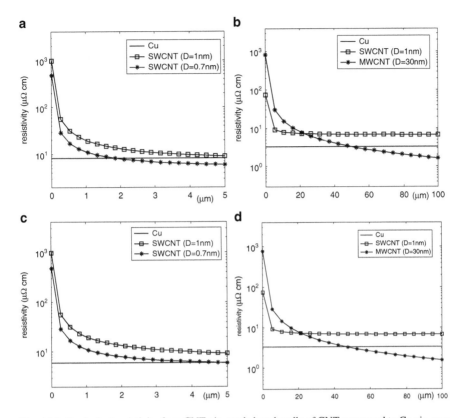

Fig. 4.14 Equivalent resistivity for a CNT via, made by a bundle of CNT compared to Cu via, as a function of the via height, for (**a**) 14-nm node, local level; (**b**) 14-nm node, global level; (**c**) 22-nm node, local level; and (**d**) 22-nm node, global level

Introducing the intershell coupling, a generalized Ohm's law may be written to express the surface density of currents on each shell in terms of the tangential components of the electrical field on each shell [11, 17]:

$$\begin{cases} \widehat{J}_{z1} = \widehat{\sigma}_{11}(\beta, \omega)\,\widehat{E}_{z1} + \widehat{\sigma}_{12}(\beta, \omega)\,\widehat{E}_{z2}, \\ \widehat{J}_{z2} = \widehat{\sigma}_{21}(\beta, \omega)\,\widehat{E}_{z1} + \widehat{\sigma}_{22}(\beta, \omega)\,\widehat{E}_{z1}, \end{cases} \tag{4.34}$$

where $\widehat{\sigma}_{11} = \widehat{\sigma}_{22} = \widehat{\sigma}_s$ are the self-conductivities given by the expressions presented in Sect. 4.3.2, whereas the mutual conductivities $\widehat{\sigma}_{12} = \widehat{\sigma}_{21} = \widehat{\sigma}_m$ may be expressed as follows [17]:

$$\widehat{\sigma}_m(\beta, \omega) \cong -i\frac{2e^2}{\pi\hbar v_F X}\frac{\omega_t\omega'}{\beta^2}\frac{1}{1-(\beta v_F/\omega')^2}\widehat{\gamma}(\beta, \omega), \tag{4.35}$$

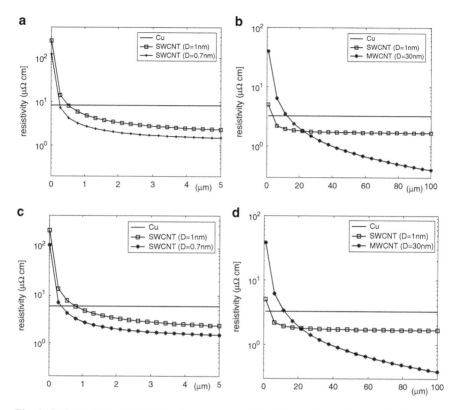

Fig. 4.15 Equivalent resistivity for a CNT trace, made by a bundle of CNT compared to Cu trace, as a function of the via height, for (**a**) 14-nm node, local level; (**b**) 14-nm node, global level; (**c**) 22-nm node, local level; and (**d**) 22-nm node, global level

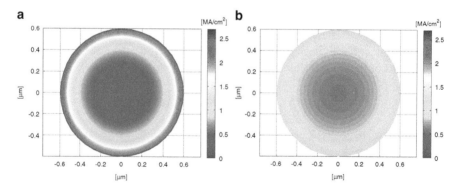

Fig. 4.16 Distribution of the current density at 200 GHz in the cross section of a chip via made by: (**a**) bulk copper and (**b**) MWCNT bundle [10]

Fig. 4.17 A double-wall
carbon nanotube

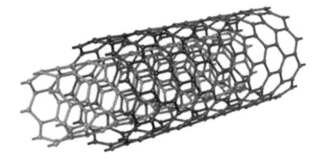

where $\omega' = \omega + i\nu$, and

$$\widehat{\gamma}(\beta, \omega) = \sum_{k=1,2} \frac{\omega'}{\beta^2} \frac{2\omega_t + (-1)^k \left[1 + (\beta v_F/\omega')^2\right]}{(2\omega_t + (-1)^k \omega') - \beta^2 v_F^2}. \tag{4.36}$$

In the above relations, a fundamental role is played by the tunneling frequency ω_t, related to the binding energy $\hbar\omega_t$ due to delocalization of the π-electrons for the tunnel effect. In a DWCNT we can assume that the shells are separated by the van der Waals distance $\delta = 0.34$ nm: for such a distance, a measured value of $\hbar\omega_t \approx 35$ meV was reported, that corresponds to a tunneling frequency of $\omega_t \approx 10^{13}$ rad/s [29].

The above binding energy is fast increasing with intershell distance, and thus to include the coupling (4.34) in the bundle model presented in Sect. 4.3.3, we can neglect the coupling between shells which are not adjacent. In this way, matrix equation (4.28) modifies as follows:

$$i\omega \overset{\leftrightarrow}{\mathbf{L}} \mathbf{I} + \overset{\leftrightarrow}{\mathbf{R}} \mathbf{I} + (i\omega + \nu) \overset{\leftrightarrow}{\mathbf{L}}_{tun} \{\mathbf{I}\} + = \mathbf{V}, \tag{4.37}$$

which means that a new term adds to the inductance and resistance terms already discussed in previous paragraph. This new term is expressed as an operator acting on the spatial distribution of the current [29]:

$$\overset{\leftrightarrow}{\mathbf{L}}_{tun} \{\mathbf{I}(z, \omega)\} = \frac{1}{4\pi} \frac{\omega_t \omega'}{\omega v_F} \left(\frac{c}{v_F}\right)^2 \overset{\leftrightarrow}{\Theta} \overset{\leftrightarrow}{\mathbf{L}}_M \int_{-\infty}^{+\infty} K_1 (z - z'; \omega') \mathbf{I}(z', \omega) \, dz', \tag{4.38}$$

where $\overset{\leftrightarrow}{\Theta}$ is the $N_b \times N_b$ matrix defined as

$$\Theta_{ij} = \begin{cases} +1 & i = j, \\ -1 & i = j \pm 2, \\ 0 & \text{elsewhere} \end{cases} \tag{4.39}$$

and the kernel is given by

$$K_1\left(z;\omega'\right) = \sin\left(\frac{2\omega_t}{v_F}|z|\right)\exp\left(-i\frac{\omega'}{v_F}|z|\right).$$ (4.40)

In order to derive an equivalent resistivity for a CNT bundle in presence of intershell coupling, we can follow the same approach as in Sect. 4.3.3, assuming all CNT shells to be fed in parallel, hence imposing on the n-th shell the same voltage $V_n = V_b$, where V_b is the total voltage drop along the bundle. However, the spatial dispersion introduced in (4.38) does not allow introducing a simple equivalent resistivity as done in (4.29), since the ratio between the total voltage V_b and the total current of the bundle $I_b = \sum_{n=1}^{N_b} I_n$, would no longer be independent from the distribution of the current itself. If we limit the CNT bundle length to values up to hundreds of nm, then up to the THz range the line is electrically short and it is possible to disregard the spatial variation of the current distribution. According to what shown in Sect. 4.3.3, such a length limit is fulfilled in typical on-chip interconnects, up to the global level. This condition, instead, is not satisfied where larger lengths are required, such as in through-silicon-vias (TSVs) in the 3D integration or as chip-to-package interconnects [43–45]. Assuming the above limitation, we can neglect the spatial dispersion introduced by (4.38) and focus on the frequency dispersion. In addition, since $\nu \approx 10^{12}$ Hz, in the THz range it is also $\omega' \approx \omega$. As a consequence, the tunneling operator (4.38) may be approximated as

$$\overset{\leftrightarrow}{\mathbf{L}}_{tun}\left\{\mathbf{I}(\omega)\right\} \approx \frac{1}{\pi}\left(\frac{c}{v_F}\right)^2\frac{\omega_t^2}{4\omega_t^2-\omega^2}\overset{\leftrightarrow}{\mathbf{L}}_M\mathbf{I}(\omega),$$ (4.41)

and the equivalent resistivity can be again defined as in (4.29).

The effect of the coupling in the equivalent resistivity is strongly related to the distribution of the feeding currents in the bundled CNTs. The properties of matrix $\overset{\leftrightarrow}{\boldsymbol{\Theta}}$ (see (4.39)) imply that the coupling is perfectly canceled out if the currents of adjacent shells are perfectly equal, $I_n = I_s$, $n = 1..N_b$. In Fig. 4.18 it is shown the frequency behavior of the computed equivalent resistivity for a DWCNT of outer diameter of 20 nm, assuming different levels of mismatch between the current flowing into the two shells: $|\Delta I|/|I| = |I_1 - I_2|/|I|$. It is evident how the coupling due to tunneling modifies the resistivity by introducing a resonance peak, whose amplitude and width increase as the mismatch increases.

4.5 Electrical Conductivity of CNTs up to the Optical Range

In order to model the CNT electrical conductivity in a wider frequency range, up to the optical frequencies, a direct solution of the quantum-mechanical equations of motion for π-electrons is needed. Consider an infinitely long rectilinear single-wall

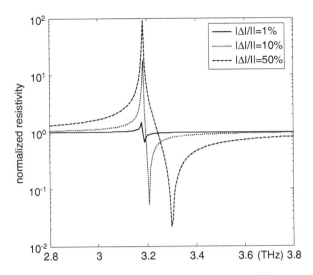

Fig. 4.18 Absolute value of the equivalent resistivity of a double-wall CNT, carrying two different currents I_1 and I_2, for different values of the current mismatch $|\Delta I|/|I|$. The results are normalized to the resistivity for $|\Delta I| = 0$.

CNT, oriented along the z axis and excited by the component of electromagnetic field polarized along this axis: $\mathbf{E}\,(\mathbf{r},t) = \mathbf{e}_z E_0 \exp{(\mathbf{k} \cdot \mathbf{r})}$, where $\mathbf{k}$ is the wave vector of the exciting field, assumed to be normal to the CNT axis. In the tight-binding approximation, the motion of electrons in the CNT crystalline lattice potential is described by the Schrödinger equation, whose solution may be written in terms of Bloch wave expansion:

$$\Psi\,(\mathbf{r},t) = \sum_q C_q\,(p_z,t)\,\Psi_q\,(\mathbf{p},\mathbf{r}), \qquad (4.42)$$

where the index q stands for the collection of quantum numbers characterizing states of π-electrons with a given quasi-momentum. In the framework of the two-band model, in the following we denote the index q with either "v" or "c," which corresponds to valence and conduction bands, respectively. The basis functions are the Bloch waves with amplitudes $u_q(\mathbf{r})$ periodic with respect to the arbitrary lattice vector $\mathbf{R}$ expressed as [46]:

$$\Psi_q\,(\mathbf{p},\mathbf{r}) = \hbar^{-1/2} \exp{(i\mathbf{p} \cdot \mathbf{r}/\hbar)}\,u_q\,(\mathbf{r}), \qquad (4.43)$$

whereas the coefficients $C_q(p_z,t)$ are solutions of the equation:

$$i\hbar\frac{\partial C_q}{\partial t} = E_q C_q - i\hbar e E_z^0 \frac{\partial C_q}{\partial p_z} - eE_z^0 \sum_{q'} C_{q'} R_{qq'}. \qquad (4.44)$$

For a zig-zag CNT of chiral number m, the matrix element $R_{qq'}$ in (4.44) may be expressed as

$$R_{vc}(p_z, s) = -\frac{b\gamma_0^2}{2E_c^2(p_z, s)}\left[1 + \cos(ap_z)\cos\left(\frac{\pi s}{m}\right) - 2\cos^2\left(\frac{\pi s}{m}\right)\right], \quad (4.45)$$

whereas for an armchair (m,m) CNT it is

$$R_{vc}(p_z, s) = -\frac{\sqrt{3}b\gamma_0^2}{2E_c^2(p_z, s)}\sin\left(\frac{ap_z}{\sqrt{3}}\right)\sin\left(\frac{\pi s}{m}\right), \quad (4.46)$$

using (4.45) or (4.46), the solution of (4.44) provides the surface density of the axial current as

$$j_z(t) = \frac{2e\gamma_0}{(\pi\hbar)^2 r_c}\sum_{s=1}^{m}\int\left\{\frac{\partial E_c}{\partial p_z}\rho_{vv} + E_c R_{vc}\mathrm{Im}(\rho_{vc})\right\}dp_z, \quad (4.47)$$

where $\rho_{qq'} = C_q C_{q'}^*$ are the elements of the matrix of density. In the weak-field regime, the Eq. (4.44) can be solved in the linear approximation: if we neglect spatial dispersion, the axial conductivity in the optical range may be put in the form [46]:

$$\sigma_{zz}(\omega) = -\frac{ie^2\omega}{\pi^2\hbar r_c}\left\{\frac{1}{(\omega+i0)^2}\sum_{s=1}^{m}\int_{1stBZ}\frac{\partial F_c}{\partial p_z}\frac{\partial E_c}{\partial p_z}dp_z + \right.$$
$$\left. -2\sum_{s=1}^{m}\int_{1stBZ}E_c|R_{vc}|^2\frac{F_c - F_v}{\hbar^2(\omega+i0)^2 - 4E_c^2}dp_z\right\}. \quad (4.48)$$

The quantities F_c, F_v are the Fermi distribution functions, obtained from (4.3) by substituting $E \rightarrow E_{c,v} - \mu_{ch}$, being μ_{ch} the electrochemical potential.

The relaxation effect can be phenomenologically incorporated in (4.48) by substituting $(\omega + i0)^2 \rightarrow \omega(\omega + i/\tau)$. The first term on the right side of (4.48) describes the intraband motion of π-electrons, thus corresponding to the quasi-classical conductivity derived in Sect. 4.3.2, based on the Boltzmann equation. The second term on the right side of (4.48) describes direct transitions between the valence and the conductivity bands.

The frequency behavior of the conductivity is reported in Fig. 4.19 for a metallic and a semiconducting zig-zag CNT. Here we assume a relaxation time of $\tau = 3$ ps and a temperature $T_0 = 295$K. The contribution of free charge-carriers to real part of the conductivity σ_{zz} is dominant at low frequencies, but is too small to be evident in both figures. The initial decay of the imaginary part of σ_{zz} with increasing frequency in Fig. 4.19a is certainly due to free charge-carriers, but the same feature is absent in Fig. 4.19b, because of the low density of free charge-carriers in semiconductors.

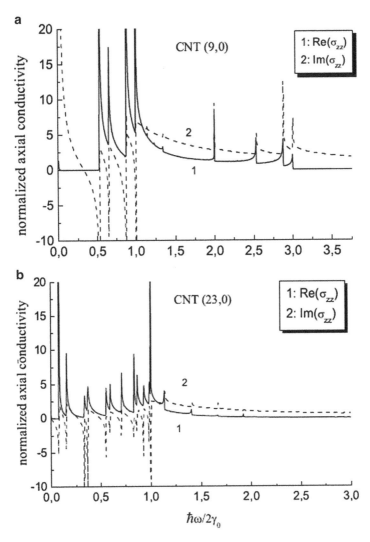

Fig. 4.19 Frequency dependence of the total axial conductivity σ_{zz} for a zig-zag CNT: (**a**) metallic (9,0) and (**b**) semiconducting (23,0). The axial conductivity is normalized to $e^2/2\pi\hbar$ [18]

As the frequency increases, $\text{Im}(\sigma_{zz})$ becomes negative. This change in sign is due to interband electronic transitions. Optical resonances appear with further increase in frequency, the frequency of the lowest resonance decreasing as the CNT radius grows. For a fixed radius, a semiconducting CNT resonates at a lower frequency than a metallic CNT.

4.6 Conclusions

This chapter presented the study of the electrical conductivity of CNTs, in view of their potential applications in nanoelectronics, based on fundamental physical principles.

The energy spectrum of π-electrons was calculated using the tight-binding approximation for hexagonal crystal lattice with quantum confinement in the azimuthal direction. The low-frequency model (from DC to THz) was obtained using the quasi-classical concept of 1D-Fermi electron gas. Its dynamic behavior in the electromagnetic field is described by the Boltzmann kinetic equation with a phenomenological collision term. The CNT conductivity was obtained by solving a linearized version of the above problem.

In the high frequency range, the conductivity was modeled by means of the quantum theory of transport, based on Schrodinger equation for $\pi-$ electrons, which accounts also for interband transitions between valence and conduction zones. In the low-frequency-limit this theory goes to the quasi-classical model mentioned above. The model of MWCNT-conductivity accounts the intershell tunneling of π-electrons between neighboring sheets.

Based on the above conductivity models, transmission line models were presented, to describe the signal propagation along single of bundled CNTs. The real values of inductance, capacitance, resistance, and characteristic impedance for metallic and semiconducting CNTs were presented and discussed, in view of using CNTs as interconnects.

The theory and experimental observation of the plasmon resonance in CNT were also described. The proposal of using CNT as a THz-nanoantenna is discussed.

Acknowledgements The authors acknowledge the support of European Commission, under Projects FP7-612285 CANTOR and FP7-318617 FAEMCAR.

References

1. Saito R, Dresselhaus G, Dresselhaus MS (2004) Physical properties of carbon nanotubes. Imperial College Press, Singapore
2. Reich S, Thomsen C, Maultzsch J (2004) Carbon nanotubes. Wiley-VCH, Weinheim
3. Anantram MP, Lonard F (2006) Physics of carbon nanotube electronic devices. Rep Prog Phys 69:507
4. Ilyinsky AS, Slepyan GYa, Slepyan AYa (1993) Propagation, scattering and dissipation of electromagnetic waves. Peter Peregrinus, London
5. Hanson GW (2005) Fundamental transmitting properties of carbon nanotube antennas. IEEE Trans Antennas Propag 53:3426–3435
6. Burke PJ, Li S, Yu Z (2006) Quantitative theory of nanowire and nanotube antenna performance. IEEE Trans Nanotechnol 5:314–334
7. Slepyan GY, Shuba MV, Maksimenko SA, Lakhtakia A (2006) Theory of optical scattering by achiral carbon nanotubes and their potential as optical nanoantennas. Phys Rev B 73:195416

8. Maffucci A, Miano G, Villone F (2008) Performance comparison between metallic carbon nanotube and copper nano-interconnects. IEEE Trans Adv Packag 31(4):692–699
9. Morris JE (2008) Nanopackaging: nanotechnologies and electronics packaging. Springer, New York
10. Close GF, Yasuda S, Paul B, Fujita S, Philip Wong H-S (2009) A 1 GHz integrated circuit with carbon nanotube interconnects and silicon transistors. Nano Lett 8(2):706–709
11. Shuba MV, Slepyan GY, Maksimenko SA, Thomsen C, Lakhtakia A (2009) Theory of multiwall carbon nanotubes as waveguides and antennas in the infrared and the visible regimes. Phys Rev B 79(15):155403–155403
12. Li H, Xu C, Srivastava N, Banerjee K (2009) Carbon nanomaterials for next-generation interconnects and passives: physics, status, and prospects. IEEE Trans Electron Devices 56(9):1799–1821
13. Ding L, Liang S, Pei T, Zhang Z, Wang S, Zhou W, Liu J, Peng L-M (2012) Carbon nanotube based ultra-low voltage integrated circuits: scaling down to 0.4 V. Appl Phys Lett 100(26):263116
14. Chiariello AG, Maffucci A, Miano G (2013) Circuit models of carbon-based interconnects for nanopackaging. IEEE Trans Compon Packag Manuf 3(11):1926–1937
15. Valitova I, Amato M, Mahvash F, Cantele G, Maffucci A, Santato C, Martel R, Cicoira F (2013) Carbon nanotube electrodes in organic transistors. Nanoscale 5:4638–4646
16. Hartmann RR, Kono J, Portnoi ME (2014) Terahertz science and technology of carbon nanomaterials. Nanotechnology 25:322001
17. Miano G, Forestiere C, Maffucci A, Maksimenko SA, Slepyan GY (2011) Signal propagation in single wall carbon nanotubes of arbitrary chirality. IEEE Trans Nanotechnol 10:135–149
18. Slepyan GYa, Maksimenko SA, Lakhtakia A, Yevtushenko O, Gusakov AV (1999) Electrodynamics of carbon nanotubes: dynamic conductivity, impedance boundary conditions, and surface wave propagation. Phys Rev B 60:17136–17149
19. Bass FG, Bulgakov AA (1997) Kinetic and electrodynamic phenomena in classical and quantum semiconductor superlattices. Nova, New York
20. Miyamoto Y, Louie SG, Cohen ML (1996) Chiral conductivities of nanotubes. Phys Rev Lett 76:2121
21. Yevtushenko OM, Slepyan GYa, Maksimenko SA, Lakhtakia A, Romanov DA (1997) Nonlinear electron transport effects in a chiral carbon nanotube. Phys Rev Lett 79:1102–1105
22. Slepyan GYa, Maksimenko SA, Lakhtakia A, Yevtushenko OM, Gusakov AV (1998) Electronic and electromagnetic properties of nanotubes. Phys Rev B 57:9485–9497
23. Forestiere C, Maffucci A, Miano G (2011) On the evaluation of the number of conducting channels in multiwall carbon nanotubes. IEEE Trans Nanotechnol 10(6):1221–1223
24. Lundstrom M (2000) Fundamentals of carrier transport. Cambridge University Press, Cambridge
25. Pop E, Mann DA, Goodson KE, Dai HJ (2007) Electrical and thermal transport in metallic single-wall carbon nanotubes on insulating substrates. J Appl Phys 101(9):093710
26. Slepyan GY, Shuba MV, Maksimenko SA, Thomsen C, Lakhtakia A (2010) Terahertz conductivity peak in composite materials containing carbon nanotubes: theory and interpretation of experiment. Phys Rev B 81:205423
27. Shuba MV, Paddubskaya AG, Plyushch AO, Kuzhir PP, Slepyan GY, Maksimenko SA, Ksenevich VK, Buka P, Seliuta D, Kasalynas I, Macutkevic J, Valusis G, Thomsen C, Lakhtakia A (2012) Experimental evidence of localized plasmon resonance in composite materials containing single-wall carbon nanotubes. Phys Rev B 85:165435
28. Bommeli F, Degiorgi L, Wachter P, Bacsa WS, De Heer WA, Forro L (1996) Evidence of anisotropic metallic behaviour in the optical properties of carbon nanotubes. Solid State Commun 99:513–517
29. Ugawa A, Rinzler AG (1999) Far-infrared gaps in single-wall carbon nanotubes. Phys Rev B 60, R11605
30. Itkis ME, Niyogi S, Meng ME, Hamon MA, Hu H, Haddon RC (2002) Spectroscopic study of the Fermi level electronic structure of single-walled carbon nanotubes. Nano Lett 2:155–159

128 A. Maffucci et al.

31. Shyu FL, Lin MF (2002) Electronic and optical properties of narrow-gap carbon nanotubes. J Phys Soc Jpn 71:1820–1823
32. Jeon T-I, Kim K-J, Kang C, Oh S-J, Son J-H, An KH, Bae DJ, Lee YH (2002) Terahertz conductivity of anisotropic single walled carbon nanotube films. Appl Phys Lett 80:3403
33. Borondics F, Kamaras K, Nikolou M, Tanner DB, Chen ZH, Rinzler AG (2006) Charge dynamics in transparent single-walled carbon nanotube films from optical transmission measurements. Phys Rev B 74:045431
34. Kampfrath T, von Volkmann K, Aguirre CM, Desjardins P, Martel R, Krenz M, Frischkorn C, Wolf M, Perfetti L (2008) Mechanism of the far-infrared absorption of carbon-nanotube films. Phys Rev Lett 101:267403
35. Akima N, Iwasa Y, Brown S, Barbour AM, Cao J, Musfeldt JL, Matsui H, Toyota N, Shiraishi M, Shimoda H, Zhou O (2006) Strong anisotropy in the far-infrared absorption spectra of stretch-aligned, single-walled carbon nanotubes. Adv Mater 18:1166–1169
36. Shuba MV, Paddubskaya AG, Kuzhir PP, Maksimenko SA, Ksenevich VK, Niaura G, Seliuta D, Kasalynas I, Valusis G (2012) Soft cutting of single-wall carbon nanotubes by low temperature ultrasonication in a mixture of sulfuric and nitric acids. Nanotechnology 23:495714
37. Zhang Q, Hároz EH, Jin Z, Ren L, Wang X, Arvidson RS, Lüttge A, Kono J (2013) Plasmonic nature of the terahertz conductivity peak in single-wall carbon nanotubes. Nano Lett 13:5991–5996
38. Nemilentsau AM, Shuba MV, Slepyan GYa, Kuzhir PP, Maksimenko SA, D'yachkov PN, Lakhtakia A (2010) Substitutional doping of carbon nanotubes to control their electromagnetic characteristics. Phys Rev B 82:235424
39. Shuba MV, Paddubskaya AG, Kuzhir PP, Slepyan GYa, Seliuta D, Kašalynas I, Valušis G, Lakhtakia A (2012) Effects of inclusion dimensions and p-type doping in the terahertz spectra of composite materials containing bundles of single-wall carbon nanotubes. J Nanophotonics 6:061707
40. Morimoto T, Joung S-K, Saito T, Futaba DN, Hata K, Okazaki T (2014) Length-dependent plasmon resonance in single-walled carbon nanotubes. ACS Nano 8:9897–9904
41. ITRS (2013) International Technology Roadmap for Semiconductors. http://public.itrs.net
42. Steinhögl W et al (2005) Comprehensive study of the resistivity of copper wires with lateral dimensions of 100 nm and smaller. J Appl Phys 97:023706
43. Naeemi A, Meindl JD (2008) Performance modeling for single- and multiwall carbon nanotubes as signal and power interconnects in gigascale systems. IEEE Trans Electron Devices 55(10):2574–2582
44. Chiariello AG, Maffucci A, Miano G (2012) Electrical modeling of carbon nanotube vias. IEEE Trans Electromagn Compat 54(1):158–166
45. Wang T, Chen S, Jiang D, Fu Y, Jeppson K, Ye L, Liu J (2012) Through-silicon vias filled with densified and transferred carbon nanotube forests. IEEE Electron Device Lett 33(3):420–422
46. Maksimenko SA, Slepyan G Ya (2002) Slepyan: electrodynamics of carbon nanotubes. J Commun Technol Electron 47:235–252

Chapter 5
Computational Studies of Thermal Transport Properties of Carbon Nanotube Materials

Leonid V. Zhigilei, Richard N. Salaway, Bernard K. Wittmaack, and Alexey N. Volkov

5.1 Introduction

Carbon nanotube (CNT) materials constitute a broad class of hierarchical materials deriving their properties from the intimate connections between the atomic structure of individual CNTs, the arrangements of CNTs into mesoscopic structural elements, such as CNT bundles and branching structures, and the structural organization of the mesoscopic elements into macroscopic network materials. Depending on the material density and production method, the CNT materials exist in many forms, from low-density aerogels and sponges with densities of $\sim$0.01 g/cm^3 [1, 2], to medium-density CNT films [3–6], "forests" [7, 8], mats, and "buckypaper" [9–11] with densities of $\sim$0.1 g/cm^3, and to high-density super-aligned CNT fibers [12–14], forests [8] and films [15] with CNT arrangements approaching the ideal packing limit. While, in general, the strong van der Waals attraction between nanotubes [16] results in their self-assembly into networks of interconnected bundles [17, 18], the degree of CNT alignment, bundle and pore size distributions, and other structural characteristics of CNT networks are not uniquely defined by the length and flexural rigidity of CNTs and material density, but can be modulated by changing the

L.V. Zhigilei (✉) • B.K. Wittmaack
Department of Materials Science and Engineering, University of Virginia, 395 McCormick Road, Charlottesville, VA 22904-4745, USA
e-mail: lz2n@virginia.edu

R.N. Salaway
Department of Mechanical and Aerospace Engineering, University of Virginia, 122 Engineers Way, Charlottesville, VA 22904-4746, USA

A.N. Volkov
Department of Mechanical Engineering, University of Alabama, 7th Avenue, Tuscaloosa, AL 35407, USA

© Springer International Publishing Switzerland 2017
A. Todri-Sanial et al. (eds.), *Carbon Nanotubes for Interconnects*,
DOI 10.1007/978-3-319-29746-0_5

parameters of the production process [18], mechanical [11] and chemical [19, 20] post-processing or radiative treatment.

The complex hierarchical organization of the CNT materials and the wide diversity of material structures give rise to a large variability of physical properties. The experimentally measured values of thermal conductivity of CNT materials, in particular, fall into an amazingly broad range of values from 0.02 to 1000 W/m/K [14, 15, 21–34] and suggest that the variation in structure and density of the CNT networks as well as the modulation of the intrinsic thermal conductivity of individual CNTs can turn CNT materials from perfect thermal conductors to insulators. Moreover, the high degree of anisotropy of thermal conductivity reported for aligned arrays of nanotubes [24, 26, 30] opens up attractive opportunities for guiding the heat transport along a desired path in thermal management applications. In order to realize these opportunities, however, a clear understanding of the key microstructural features and elementary processes that control thermal transport properties of CNT materials has to be achieved.

As reviewed in Chap. 10 of this book, the steady advancement in experimental techniques for probing thermal conductivity of individual CNTs and CNT materials has yielded important information on the connections between the structural characteristics of CNT materials and their thermal transport properties. The small size of the individual nanotubes and their propensity to form bundles and aggregate into intertwined structures, however, present a number of technical challenges that hamper a systematic experimental investigation of the dependence of thermal conductivity on the geometrical and structural parameters of CNTs (length, diameter, chirality, strain, and presence of defects) and their arrangement into network structures. The experimental challenges have motivated extensive computational efforts aimed at providing insights into the mechanisms and pathways of the heat transfer in CNT materials.

Due to the complexity and inherently multiscale nature of the structural organization of the CNT network materials, a complete analysis of the heat transfer cannot be performed within a single computational approach and a number of complementary computational techniques have to be combined for establishing the connections between the intrinsic thermal conductivity of individual CNTs, intertube thermal conductance, and the collective heat transfer through the CNT material, as schematically shown in Fig. 5.1. The atomistic molecular dynamics (MD) computational technique is well suited for simulation of the vibrational (phononic) heat transfer in individual CNTs [35–69] or heat exchange between CNTs [21, 70–81]. The information gained from the atomistic simulations can be used for parameterization of mesoscopic models that adopt simplified representations of nanotubes and are capable of simulating the collective behavior and properties of large ensembles of CNTs arranged into interconnected networks of bundles [82–99]. A detailed analysis of the results of the mesoscopic simulations of heat flow in CNT materials with realistic entangled structures [60, 79, 100, 101] can provide ideas for designing theoretical models of the heat (and charge) transfer in CNT materials and, in particular, for making a transition from the theoretical analysis of model systems composed of straight randomly arranged or aligned nanofibers

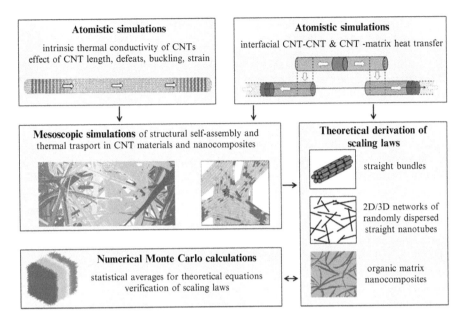

Fig. 5.1 Schematic diagram of the connections between different computational approaches aimed at providing insights into the heat transfer in CNT materials and advancing theoretical understanding of thermal transport properties of this practically important class of nanofibrous network materials

[72, 100–114] to the derivation of the scaling laws for complex network structures and nanocomposites. The statistically averaged thermal transport properties of CNT networks can be obtained with a number of theoretical approaches based on simplified geometrical models of fibrous networks or numerically in Monte Carlo simulations [100, 102–105, 109, 110, 112–114]. These theoretical and numerical approaches provide additional guidance for revealing the scaling laws that govern the structure–properties relationships of the transport properties of CNT materials. Overall, the atomistic and mesoscopic modeling plays an important role in the advancement of the theoretical understanding of the heat transfer in CNT materials and, in general, in a broad class of network nanofibrous materials.

In this chapter, we first provide a brief overview of the results of atomistic simulation studies of the thermal conductivity of individual CNTs and inter-tube contact conductance (Sects. 5.2 and 5.3), and then discuss the emerging mesoscopic computational approaches to the calculation of the effective thermal conductivity of CNT materials (Sects. 5.4 and 5.5). The interconnections between the processes occurring at different length scales are highlighted and the open questions and future research directions are discussed in Sect. 5.6.

5.2 Atomistic Modeling of Thermal Conductivity of Individual CNTs

The classical MD technique has been (and remains) the most commonly used computational method in investigations of heat transfer in CNT systems. The MD technique allows one to follow the time evolution in a system of interacting atoms by solving a set of classical equations of motion for all atoms in the system [115]. A key advantage of this technique is in its ability to fully control the conditions of the simulations and to perform systematic analysis of the dependence of the thermal conductivity on CNT length [38–41, 44–50, 54, 55, 62–68], diameter [37–39, 44, 47–49, 54, 57, 59, 62–65], elastic deformation [51–53, 58, 67], buckling [51, 56, 60, 61, 67], presence of isotope dopants [44, 45, 49, 64], vacancies and Stone–Wales defects [36, 45, 64], or chemisorbed molecules [40] in a systematic manner and in isolation from other structural parameters. Despite the full control over the simulated system, however, the MD simulation technique can hardly be considered to be suitable for quantitative predictions of the exact values of thermal conductivity of CNTs. The two main limiting factors, the classical treatment of atomic motions and the sensitivity of the results to the computational parameters, are briefly described below.

Since the Debye temperature of CNTs is on the order of 1000 K, the phonon population at room temperature is governed by quantum mechanics (Bose–Einstein statistics), and cannot be adequately represented by the classical MD technique. Quantum corrections, based on redefining the temperature so that the classical kinetic energy of the system is equal to the corresponding quantum kinetic energy of a harmonic lattice, have been developed for the description of thermodynamic and structural properties [116, 117] and extended to thermal transport properties in a number of works, e.g., [46, 50]. This extension, however, has been shown to be flawed [118] as the classical and quantum relaxation times cannot be reconciled using a system-level quantum correction. A number of more refined "quantum thermal bath" approaches accounting for quantum-mechanical effects through application of a Langevin-type thermostat with a colored noise enforcing the Bose–Einstein distribution of the vibrational energy have recently been suggested [118–123]. The complexity of these approaches as well as a number of technical issues, such as the zero-point energy leakage [124, 125], however, restricts their applications to relatively simple model systems, and leaves the challenge of fully accounting for quantum effects in MD simulations unresolved.

Beyond the missing quantum effects, the predictive power of MD simulations is affected by the sensitivity of the results to the computational procedures and interatomic potentials employed in the calculations. Even when the simulated system is nominally the same, the values of thermal conductivity predicted in different MD simulations exhibit a surprisingly large variability. An example of such inconsistency is demonstrated by the sample of MD results obtained for single-walled (10, 10) CNTs at ~300 K provided in Table 5.1. The values of thermal conductivity range from ~100 $Wm^{-1} K^{-1}$ to more than 6000 $Wm^{-1} K^{-1}$, and are

Table 5.1 A sample of room temperature values of CNT thermal conductivity, k_T, predicted in MD simulations of (10,10) CNTs performed with equilibrium MD (EMD) method, where k_T is calculated from small fluctuations of heat current using Green–Kubo relation [115], non-equilibrium MD (NEMD) method, where a steady-state temperature gradient is created between "heat bath" regions and k_T is determined from the Fourier law, and homogenous NEMD (HNEMD) method [126], in which k_T is evaluated from the response of the system to a small artificial "thermal force" applied to individual atoms and representing the effects of the heat flow

Source	k_T (Wm^{-1} K^{-1})	L_T (nm)	Potential	Method	Boundary conditions
Lukes and Zhong [46]	~20–160	5–40	Brenner-2 + LJ	EMD	Periodic and non-periodic
Ma et al. [67]	~105–175	~25–90	AIREBO	NEMD	Periodic
Pan [57]	243	~29.4	Brenner-2	NEMD	Periodic
Salaway and Zhigilei [66]	154–258	47–630	AIREBO	NEMD	Non-periodic
Xu and Buehler [51]	301	49.26	AIREBO	NEMD	Periodic
Wei et al. [58]	302	49.26	AIREBO	NEMD	Periodic
Padgett and Brenner [40]	~35–350	~10–310	Brenner-2	NEMD	Periodic
Thomas et al. [54]	~300–365	200–1000	Brenner-2	NEMD	Non-periodic
Lukes and Zhong [46]	233–375	5–10	Brenner-2 + LJ	HNEMD	Periodic
Maruyama [38]	~275–390	6–404	Brenner	NEMD	Non-periodic
Feng et al. [64]	~50–590	6.52–35	Brenner-2	NEMD	Non-periodic
Bi et al. [45]	~400–600	2.5–25	Tersoff	EMD	Periodic
Moreland et al. [41]	215–831	50–1000	Brenner	NEMD	Periodic
Cao and Qu [62]	~560–1620	100–2400	Tersoff-2	NEMD	Periodic
Ren et al. [53]	~1430	6	AIREBO	NEMD	Non-periodic
Sääskilahti et al. [68]	~160–1600	~120–4000	Tersoff-2	NEMD	Non-periodic
Grujicic et al. [42]	1730–1790	~2.5–40	AIREBO	EMD	Periodic
Che et al. [36]	~2400–3050	~2.5–40	Brenner	EMD	Periodic
Berber et al. [35]	~6600	~2.5	Tersoff	HNEMD	Periodic

The definition of the length of the CNTs, L_T, includes the length of the heat bath regions [66]. The interatomic potentials used in the simulations are the Tersoff potential in its original formulation [127, 128] (Tersoff) and with parameters suggested in [129] (Tersoff-2), the Brenner potentials with parameterizations described in [130, 131] (Brenner and Brenner-2, respectively), and the adaptive intermolecular reactive bond order (AIREBO) potential described in [132]. A description of non-bonding van der Waals interactions through addition of the Lennard-Jones (LJ) potential to the Brenner-2 potential is denoted as Brenner-2 + LJ

clearly affected by the computational setups adopted in different studies. A recent analysis of the effects of common computational parameters of non-equilibrium MD simulations [66] demonstrates that the type of boundary conditions, size and location of heat bath regions, definition of the CNT length, and the choice of interatomic potential all have a substantial influence on the predicted values of thermal conductivity. The choice of interatomic potential, in particular, is shown to be responsible for an up to fourfold variability in thermal conductivity for otherwise identical simulation conditions.

While the above discussion casts doubt on the utility of MD simulations for evaluation of accurate values of thermal conductivity of CNTs, the real power of the MD simulations is in the ability to reveal the general trends and physical mechanisms that control the heat transfer within and between the nanotubes, as well as to provide a valuable semi-quantitative information on the effect of different factors (e.g., CNT length, structural defects, and different modes of deformation) on thermal conductivity of CNTs.

The length dependence of thermal conductivity of CNTs, in particular, is commonly observed in MD simulations [33, 36, 38–44, 46–49, 54, 61, 63–68] and attributed to two main physical origins. First, when the length of a nanotube is smaller than or comparable to the phonon mean free path, phonons are capable of traveling ballistically through the CNT without being impeded by phonon–phonon scattering. As the CNT length increases, the effective length of the ballistic transport increases, which also increases the overall thermal transport and results in higher conductivity values. Second, the longest available phonon wavelength that can exist in a CNT is defined by the length of the nanotube. Thus, as the length increases, the maximum allowable phonon wavelength also increases. The additional long-wavelength phonons offer effective channels for thermal transport and can make a substantial contribution to the thermal conductivity [46, 62, 67, 68].

The CNT length that corresponds to the transition from the ballistic heat conduction regime, where the thermal conductivity increases with CNT length, to the diffusive regime, where the thermal conductivity approaches a constant value, is temperature dependent as the phonon mean free path decreases with increasing temperature. A recent review of room temperature experimental measurements performed for CNTs with length exceeding 0.5 μm [33] suggests the diffusive regime of the heat transfer. At the same time, the results of MD simulations performed for CNTs with lengths of 10–100s of nm typically exhibit a pronounced increase of k_T with increasing CNT length, that is the characteristic of the ballistic phonon transport [38–41, 44, 46–49, 54, 62, 64, 66, 68]. The length dependences predicted in different MD studies, however, vary widely for the same (10,10) CNT, with the transition to the diffusive regime (saturation of k_T) predicted for as short CNTs as 100 nm or even in some of the investigations [36, 42, 43, 63], while no saturation is observed for CNTs with length exceeding 1 μm in other studies [62, 68].

A sample set of computational results shown in Fig. 5.2 suggests the dominant contribution from the ballistic phonon transport for CNTs shorter than $\sim$200 nm,

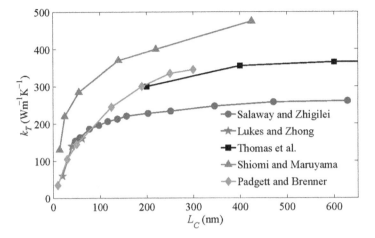

Fig. 5.2 Dependences of thermal conductivity, k_T, of (10,10) CNTs on sample length, L_C, predicted in MD simulations by Lukes and Zhong [46], Thomas et al. [54], Shiomi and Maruyama [48], Padgett and Brenner [40], and Salaway and Zhigilei [66]. The sample length is defined as the distance between the hot and cold heat bath regions

which exhibit strong length dependence. In studies where CNTs longer than 200 nm are investigated, a transition to weaker length dependence is observed, suggesting a transition to the diffusive-ballistic regime. The transition between the two regimes is observed for the nanotube length that roughly corresponds to the room temperature phonon mean free path, which has been estimated to be of the order of 200–500 nm based on experimental data [133, 134], results of MD simulations [62], and theoretical analysis [135]. Note that the extent of the transitional diffusive-ballistic regime is defined by the longest mean free paths of long-wavelength phonons (of the order of several μm [136] or even longer [68] at room temperature). Thus, while the length dependences shown in Fig. 5.2 become notably weaker as the length increases above 200 nm, the gradual increase of the thermal conductivity with increasing length may be expected up to CNT lengths of the order of tens of μm [68, 137, 138].

While the computational studies discussed above are addressing the intrinsic thermal conductivity of perfect fully relaxed CNTs, the nanotubes in a real material are likely to contain various structural defects and experience elastic deformation caused by their interaction with surrounding CNTs in an interconnected network of bundles or application of an external mechanical loading to the material. The precise control over atomic structure and strain in MD simulations makes it possible to perform a detailed analysis of the effect of mechanical deformation and defects on thermal conductivity of CNTs.

In particular, the effect of the axial deformation was investigated in non-equilibrium MD (NEMD) simulations reported in [51] and a substantial reduction of the thermal conductivity for both tension (~30 % reduction for strain of 15 %, close

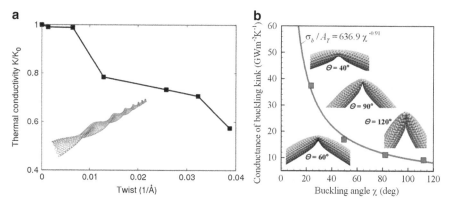

Fig. 5.3 The dependences of the nanotube thermal conductivity on torsional deformation [51] (**a**) and thermal conductance of bending buckling kink normalized by the cross-sectional area of a CNT, σ_b/A_T, on the buckling angle χ [60] (**b**) predicted for (10,10) CNTs in MD simulations. The insets show atomic configurations for a twist strain of 0.026 in (**a**) and for buckling angles $\chi = 23.6°$, $49.5°$, $82.1°$, and $112.0°$ (correspond to bending angles $\theta = 40°$, $60°$, $90°$, and $120°$, respectively) in (**b**)

to the failure limit) and compression (up to ~36 % reduction for strain of -5 %) was observed for a 49 nm long (10,10) CNT. The reduction of thermal conductivity was attributed to softening of phonon modes under tension and enhanced phonon scattering due to the transverse bending/buckling upon compression. In another study, performed for a shorter, 6 nm long (10,10) CNT [53], the maximum of thermal conductivity was observed at about 2 % elongation of the nanotube while additional elongation and compression are found to decrease the conductivity. The torsional strain is reported to have a negligible effect on thermal conductivity up to the onset of buckling of the nanotube cross-section, when the thermal conductivity exhibits a substantial drop, Fig. 5.3a [51]. A similar behavior is observed for bending deformation, where the effect of bending on thermal conductivity is hardly noticeable [52, 67] up to the onset of bending buckling, while the appearance of buckling kinks creates effective "thermal resistors" for the heat conduction along the nanotube [51, 60, 61, 67]. The results of MD simulations performed for different bending angles reveal a strong dependence of the thermal conductance of a buckling kink on the buckling angle, Fig. 5.3b [60], and enable reliable parameterization of a mesoscopic model capable of simulating heat transfer in network structures composed of thousands of CNTs (see Sect. 5.4). The mesoscopic simulations show that the effect of the finite buckling conductance is amplified by the preferential buckling of thin bundles and individual CNTs serving as interconnections between thicker bundles in the network structures, resulting in about 20 % reduction in the effective conductivity of a network material composed of 1 µm long (10,10) CNTs [60].

Overall, the results of the MD simulations of strained nanotubes suggest that homogeneous elastic deformation of nanotubes has relatively small effect on their thermal conductivity, while the onset of geometric instability and buckling greatly enhances scattering of phonons and decreases thermal conductivity. This conclusion is also consistent with the results of recent simulations of the free vibrations of 26 nm long (10,10) CNTs, where a dramatic increase in the rate of the energy dissipation of longitudinal and bending oscillations (i.e., the energy transfer from low-frequency mechanical oscillations to high-frequency vibrational modes) is observed at the onset of axial or bending buckling [90].

5.3 Atomistic Modeling of Inter-Tube Contact Conductance

Despite the high intrinsic thermal conductivity of individual CNTs, the values of the effective conductivity reported for CNT-based materials are often relatively low and exhibit large variability [14, 15, 21–34]. The weak thermal coupling between CNTs, defined by non-bonding van der Waals inter-tube interactions, is commonly assumed to be the limiting factor that controls thermal transport in CNT materials [21, 70–77, 100, 139–141]. This assumption has been put into question by the results of recent mesoscopic simulations [101], which suggest that the effective thermal conductivity of CNT network materials is to a large extent controlled by the finite values of the intrinsic thermal conductivity of the nanotubes (see Sect. 5.4). Nevertheless, the dependence of the inter-tube conductance on the density and geometrical characteristics of the CNT–CNT contacts is clearly critical for establishing the physical mechanisms of heat transfer in CNT materials.

The only direct experimental measurements of thermal conductance between individual nanotubes reported to date are the ones reported in [140, 141] for multi-walled CNTs. The results of these measurements indicate that the conductance per unit area is about an order of magnitude lower for CNTs that are aligned with each other at the contact as compared to the CNTs crossing each other at an angle [140]. The strong dependence of the interfacial conductance on the geometry of the contact suggests a high sensitivity of the effective thermal conductivity of CNT materials to their structural organization and puts into question the reliability of the estimations of the inter-tube contact conductance based on the experimental values of the effective conductivity of CNT materials [21, 34, 113, 144]. These estimations typically assume a fixed value of the inter-tube conductance and rely on analytical equations derived for idealized systems composed of randomly dispersed straight nanotubes [72, 100, 101, 111]. The arrangement of CNTs into bundles in real CNT materials, however, can result in a broad spectrum of the inter-tube contacts, which has been shown to have a dramatic effect on the inter-tube heat exchange and the effective thermal conductivity [100, 101].

Under conditions when nanoscale manipulation of CNTs and reliable measurement of the conductance between individual nanotubes still present significant challenges, computational analysis based on MD simulations has been playing the leading role in advancing the physical understanding of the thermal transport across CNT–CNT contacts [21, 70–81]. The simulations have provided important

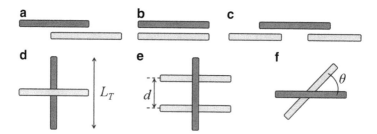

Fig. 5.4 Schematic representations of computational setups used in MD simulation studies of the inter-tube contact conductance: (**a**–**c**) parallel, partially or fully overlapping CNTs and (**d**–**f**) CNTs crossing each other at an angle. The setup in (**e**), with two cross-junctions separated by a distance d, is used for investigation of the effect of contact density on the conductance per junction

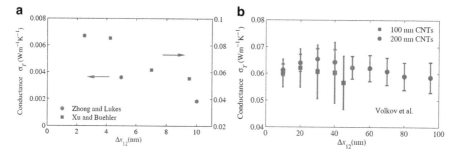

Fig. 5.5 Inter-tube conductance per unit length, σ_T, versus overlap length, Δx_{12}, predicted in MD simulations of parallel partially overlapping (10,10) CNTs by Zhong and Lukes [70], Xu and Buehler [74] (**a**), and Volkov et al. [79] (**b**). The computational setup of Fig. 5.4a with rigid boundary conditions applied to the outer ends of the two nanotubes is used by Zhong and Lukes, while the setup shown in Fig. 5.4c with periodic boundary conditions applied in the direction parallel to the axes of nanotubes is used in the other two studies. The *error bars* in (**b**) show one sample standard deviation calculated for "instantaneous" values of inter-tube conductance collected during the steady-state part of the simulation

insights into the mechanisms responsible for the inter-tube heat transfer for parallel CNTs [70, 71, 74, 75, 77, 79, 81] and CNTs crossing each other at an angle [21, 72, 73, 76–78, 80, 81]. Various configurations used in MD simulations of inter-tube conductance are schematically represented in Fig. 5.4. Similarly to the MD simulations of the intrinsic conductivity of CNTs discussed above, in Sect. 5.2, quantitative comparison of the values of inter-tube conductance predicted in different simulations is hampered by the differences in computational setups, interatomic potentials, definitions of the contact area, and length and type of the CNTs used in the simulations. Nevertheless, it may be instructive to compare the results obtained in different investigations of the well-studied (10,10) CNTs that are also used as an example in the discussion of CNT conductivity in the previous section.

Starting with parallel, partially overlapping CNTs (setups shown in Fig. 5.4a, c) and plotting the data reported in [70, 74, 79] in a uniform way, in terms of conductance per overlap length, σ_T in units of $Wm^{-1} K^{-1}$, in Fig. 5.5, two trends can be observed. In the simulations performed for short overlap length, $\Delta x_{12} \le 10$ nm, the conductance per overlap length is decreasing with increasing overlap length [70, 74]. The drop in σ_T is particularly sharp in simulations reported in [70], from $\sigma_T \approx 0.0065 \, Wm^{-1} \, K^{-1}$ at $\Delta x_{12} = 2.5$ nm, to $\sigma_T \approx 0.0034 \, Wm^{-1} \, K^{-1}$ at $\Delta x_{12} = 5$ nm, and to $\sigma_T \approx 0.0018 \, Wm^{-1} \, K^{-1}$ at $\Delta x_{12} = 10$ nm, and more moderate in simulations of [74] where σ_T decreases from ~ 0.08 to $\sim 0.05 \, Wm^{-1} \, K^{-1}$ as Δx_{12} increases from 4 to 9.5 nm. For longer overlap lengths of 10–95 nm considered in [79], however, no statistically significant variation of σ_T with the overlap length is observed. Since the area of a CNT–CNT contact is directly proportional to the overlap length, the disparate trends observed for short and long overlaps suggest that the inter-tube contact conductance is not a simple function of the contact area and may depend on particular geometry/types of the contact.

At quantitative level, the values of σ_T obtained for long overlaps in [79], $\sim 0.06 \, Wm^{-1} \, K^{-1}$, are consistent with the range of 0.05–$0.08 \, Wm^{-1} \, K^{-1}$ reported for shorter overlaps in [74]. Comparable values, 0.03 and 0.048 $Wm^{-1} \, K^{-1}$, can be calculated from the results reported for two parallel 20 nm long (10,10) CNTs [77] and two parallel 4.3 nm long (10,0) CNTs embedded into a "frozen" matrix [75],[1] respectively. On the other hand, an order of magnitude smaller values of 0.0065–$0.0018 \, Wm^{-1} \, K^{-1}$ are calculated from the data of [70]. The use of relatively short nanotubes ($L_T = 5$–40 nm as compared to $L_T = 100$–200 nm in [79] and $L_T = 25$–75 nm in [74]) in combination with fixed boundary conditions at the ends of the interacting CNTs may be responsible for both the strong overlap length dependence of σ_T and the small values of the conductance observed in [70]. Indeed, the dependence on the CNT length in [70] is especially pronounced for $L_T < 10$ nm and becomes weaker as the CNT length increases from 10 to 40 nm. The observation of the pronounced CNT length dependence for short CNTs is consistent with the results of an MD simulation study of the interfacial conductance between a (5,5) CNT and a surrounding octane liquid [139], where an increase in the interfacial conductance per CNT surface area with increasing nanotube length is observed up to a length of ~ 3.5 nm and attributed to the extinction of low-frequency phonons in short CNTs.

Further insights into the dependence of the inter-tube conductance on the contact area and the geometry of the contact can be obtained from simulations performed for CNTs crossing each other at various angles (computational setups shown in Fig. 5.4d, f). The definition of the contact area for cross-junctions, however, is not straightforward and different approaches have been adopted in different studies. In particular, the inverse thermal conductance calculated in [77] for different angles

---

[1] The values of interface conductance given in [75] in units of $Wm^{-1} \, K^{-1}$ (Figs. 6 and 7 of [75]) are about an order of magnitude larger than the actual values found in the simulations, as established through private communication with Dr. Vikas Varshney.

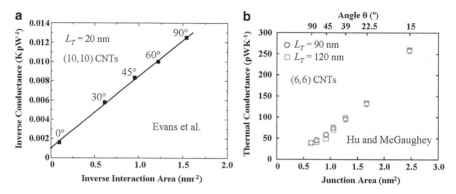

Fig. 5.6 Thermal conductance of CNT–CNT cross-junctions predicted in MD simulations by Evans et al. [77] (**a**) and Hu and McGaughey [80] (**b**). The computational setups of Fig. 5.4d, f with periodic boundary conditions applied in the direction parallel to the axes of nanotubes is used by Evans et al., while the ends of the nanotubes are kept rigid during the simulations by Hu and McGaughey. The contact area is defined through the inter-tube interaction energy in (**a**) and calculated as $D^2/\sin\theta$, where D is the nanotube diameter in (**b**)

θ is plotted in Fig. 5.6a as a function of the inverse contact area using the total inter-tube bonding energy as a measure of the contact area. The linear scaling of the inverse conductance with inverse "contact area" is discussed in this study in terms of a combined contribution of two thermal resistances placed in series, an internal resistance associated with the energy exchange between high- and low-frequency modes within the nanotubes and an external junction resistance associated with heat flow between the nanotubes mostly via low-frequency vibrational modes. The notion of the major contribution of long-wavelength phonons in the energy transfer between the nanotubes is supported by the results of wavelet analysis of thermal pulse propagation along a CNT forming perpendicular cross-junction with another nanotube [73]. Evaluation of the vibrational frequencies excited in the second CNT reveals the dominant frequencies that are relatively low (<10 THz), implying that the low-frequency modes are largely responsible for the heat transfer across CNT–CNT junctions.

An approximate linear scaling of the cross-junction conductance on the contact area is also observed in a study performed for (6,6) CNTs [80], where a simple geometrical definition of the contact area as $D^2/\sin(\theta)$, where D is the nanotube diameter, is used, Fig. 5.6b. A substantial deviation from the linear dependence, however, can be seen for the data point calculated for the smallest angle of $\theta = 15°$, where deformation of the contact geometry by the inter-tube interaction forces acting to align the nanotubes may be expected.

The combined results of several series of MD simulations performed for various inter-tube contact configurations shown in Fig. 5.4 have recently been used to formulate a general model of CNT–CNT conductance applicable to junctions of arbitrary configuration [81]. The analysis of the combined set of data reveals a non-linear dependence of the conductance on the number of interatomic inter-tube

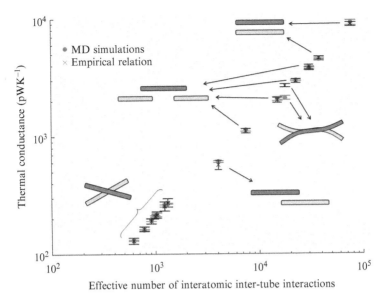

Fig. 5.7 The values of inter-tube thermal conductance obtained in MD simulations (*red circles*) and predicted by the general empirical equation (*black crosses*) for various CNT–CNT contact configurations [81]

interactions (used as a proxy for contact area) and suggests a larger contribution to the conductance from areas of the contact where the density of interatomic inter-tube interactions is smaller. An empirical relation expressing the conductance of an arbitrary contact configuration through the area of the contact region, quantified by the number of interatomic inter-tube interactions, and the density of interatomic inter-tube interactions, characterized by the average number of interatomic inter-tube interactions per atom in the contact region, is suggested based on the results of MD simulations. The empirical relation is found to provide a good quantitative description of the contact conductance for various CNT configurations, as shown in Fig. 5.7. Moreover, the empirical relation and the underlying concept of the sensitivity of the conductance to the density of interatomic inter-tube interactions reconcile the results of earlier studies of the conductance between parallel partially overlapping CNT, Fig. 5.5, where the conductance per overlap length was shown to be independent of the overlap length for long overlaps [79] but was found to exhibit a pronounced decrease with increasing length of the overlap for short overlaps [70, 74]. The general description of the conductance of an arbitrary CNT–CNT contact configuration is also suitable for incorporation into mesoscopic models capable of predicting the effective thermal transport properties of CNT network materials, as discussed below in Sect. 5.4.

Before starting the discussion of the mesoscopic modeling, we would like to mention two issues related to the inter-tube interactions in CNT materials that received a substantial, and somewhat controversial, attention in literature. First,

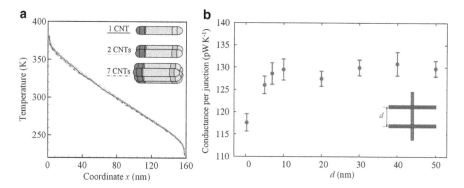

Fig. 5.8 The temperature profiles along nanotubes obtained in MD simulations performed for 160 nm long (10,10) CNTs arranged into three configurations shown in the inset [79] (**a**) and the values of the contact conductance predicted in simulations of double junctions separated by distance d [81] (**b**). In (**a**), the heat flux is applied to each CNT individually and the values of thermal conductivity are obtained from the steady-state temperature profiles

the van der Waals interactions between nanotubes in CNT bundles leading to the enhancement of phonon scattering and decrease in the intrinsic thermal conductivity of nanotubes is commonly discussed as a major factor responsible for a substantially lower thermal conductivity of CNT bundles as compared to individual CNTs [25, 31, 142, 143, 145]. The reduction of the intrinsic thermal conductivity of CNTs and other carbon structures due to the non-bonding interactions in materials that consist of multiple structural elements was suggested in [35] based on the results of atomistic Green-Kubo calculations of thermal conductivity. This suggestion, however, while echoed in a number of works, e.g., [31, 25, 143, 145, 146], is only a conjecture extrapolated from the computational prediction on the difference between the thermal conductivity of a graphene monolayer and graphite. An opposite conclusion of a weak effect of the interactions of perfect CNTs in a bundle on thermal conductivity, which can be substantially enhanced by structural defects, has been made in [149] based on the kinetic model calculations of thermal conductivity. To clarify the question on the effect of the inter-tube coupling in CNT bundles on the intrinsic thermal conductivity of the CNTs, a series of MD simulations was performed in [79] for systems of one, two, and seven 160 and 300 nm long (10,10) CNTs arranged in configurations shown as insets in Fig. 5.8a. The temperature profiles predicted in constant heat flux NEMD simulations for the three configurations are almost identical, Fig. 5.8a, suggesting that the thermal conductivities of individual CNTs, defined by the Fourier law, are not significantly affected by the interactions among the CNTs. Indeed, the values of k_T calculated for CNTs in each of the three configurations coincide with each other within the statistical error of the calculation.

The absence of any significant effect of the van der Waals inter-tube coupling in CNT bundles on the intrinsic thermal conductivity of individual CNTs is consistent with relatively small changes of the vibrational spectra of CNTs due to the inter-tube interactions and negligible contribution of inter-tube phonon modes to the thermal conductivity of bundles [149, 150]. It also suggests that three-phonon umklapp scattering involving phonons from neighboring CNTs does not play any significant role in perfect bundles consisting of defect-free CNTs. The experimental observations of the pronounced decrease of the thermal conductivity of bundles with increasing bundle thickness [31, 145, 152], therefore, is likely to have alternative explanations, such as the higher degree of CNT misalignment and increased concentration of inter-tube defects, cross-links, and foreign inclusions in larger bundles, which could result in the increase of both the phonon scattering and inter-tube contact thermal resistance.

The second effect related to the inter-tube interactions, which has been a subject of contradictory computational predictions, is the influence of the contact density on the conductance of an individual contact. The results of atomistic Green's function calculations performed for both single and double contact junctions between (10,10) CNTs (configurations shown in Fig. 5.4d, e) have predicted an order of magnitude reduction of the contact conductance in a double junction with respect to an isolated junction [21]. This dramatic reduction of inter-tube conductance in the presence of neighboring junctions, if true, has major implications for interpretation of the experimental data for high-density CNT materials. The results of recent NEMD simulations performed for single and double junctions between 100 nm long (10,10) CNTs [81], however, predict very weak sensitivity to the presence of neighboring junctions even at the smallest distances that can be realized in real materials. The dependence of the inter-tube conductance per junction on the junction separation distance plotted in Fig. 5.8b indicates that the conductance across CNT–CNT junctions is unaffected by the presence of neighboring junctions when the CNTs creating the junctions are outside the range of direct van der Waals interaction with each other. When junctions are created by CNTs separated by the minimum distance (equilibrium distance between CNTs), the conductance per junction is reduced by only $\sim$10 % with respect to an isolated CNT–CNT contact. While these results are in a sharp contrast with predictions of [21] discussed above, they are closer to the results of NEMD simulations reported in [80], where a somewhat stronger $\sim$20 % reduction in conductance per junction is observed for double junctions made up of 60 nm long (6,6) CNTs separated by relatively large distances of 10 or 20 nm, but no significant reduction in conductance is observed for neighboring junctions between 30 and 90 nm long nanotubes separated by 10 nm.

5.4 Mesoscopic Modeling of Thermal Transport in CNT Network Materials

The results of the atomistic simulations briefly reviewed above have provided a wealth of information on the intrinsic thermal conductivity of individual CNTs, its dependence on defects and elastic deformation, as well as the inter-tube contact

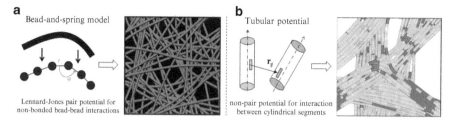

Fig. 5.9 Schematic representation of two descriptions of van der Waals inter-tube interactions adopted in the bead-and-spring model [83] (**a**) and in the mesoscopic model with tubular potential developed in [82, 84] (**b**). The *right panels* in (**a, b**) show fragments of CNT networks obtained with the two models, as reported in [85] and [87], respectively. In (**b**), the CNTs are colored by local radii of curvature, with red marking the segments adjacent to buckling kinks. Large artificial barriers inherent for the bead-and-spring model impede relative sliding of CNTs and prohibit formation of a network of bundles. In contrast, a network of bundles forms naturally in the dynamic simulation performed with the tubular potential

conductance for various contact configurations. The translation of all this information to the effective (macroscopic) thermal conductance of CNT materials, however, is far from being straightforward and requires a clear understanding of the collective heat transfer through thousands of nanotubes arranged into complex interconnected network structures. The gap between the predictions of atomistic simulations and the effective thermal transport properties of CNT materials can be bridged with the help of emerging mesoscopic computational models [82–101] capable of describing the collective behavior and properties of large groups of interacting nanotubes while still retaining the critical information on the individual nanotubes and their interactions revealed in the atomistic simulations. In this section, a brief overview of the mesoscopic models and their ability to reproduce structural self-organization of CNTs into continuous networks of bundles characteristic of real CNT materials is provided first, and is followed by a discussion of the mesoscopic calculations of thermal transport properties of the CNT network materials.

The mesoscopic models of CNT materials adopt coarse-grained representations of nanotubes, with the dynamic degrees of freedom of the models describing the motion of the nanotube segments composed of many atoms [82–84, 95]. Several alternative mesoscopic models proposed for CNT materials have similar formulations of the internal parts of the mesoscopic force fields that account for stretching, bending, and torsional deformation of individual nanotubes and are parameterized based on the results of atomistic simulations [82, 83, 86, 87, 90, 93, 95]. The different models, however, can be clearly distinguished by the computational approaches used for the description of the non-bonded van der Waals inter-tube interactions.

The most straightforward approach to the description of CNT–CNT interactions is suggested in [83] and is based on the bead-and-spring model commonly used in simulations of polymers [147]. In this model, the van der Waals inter-tube interactions are represented through the spherically symmetric pair-wise interactions between mesoscopic segments of the nanotubes (Fig. 5.9a). Due to the simplicity of this approach, it has been adopted by several groups for analysis of the structure and mechanical properties of CNT materials [89, 91–93, 96, 98, 99]. An important drawback of this approach, however, is the existence of large artificial barriers for relative displacements of neighboring CNTs introduced by the pair-wise interactions between the "beads" in the bead-and-spring model [87, 95]. The presence of these barriers casts doubt on the ability of the model to provide an adequate description of the mechanical properties of the CNT materials and may prevent long-range rearrangements of CNTs required for their self-assembly into continuous networks of bundles. The latter effect can be illustrated by a snapshot from a simulation of a layer-by-layer deposition of straight randomly oriented CNTs shown in Fig. 5.9a. The bead-and-spring model predicts interlocking of the deposited nanotubes into a stable layered structure of randomly oriented individual CNTs [85, 98] which, experimentally, can only be stabilized by chemical functionalization or charging of the CNTs [148].

More advanced descriptions of non-bonding inter-tube interactions that do not result in the artificial corrugated inter-tube interactions have been developed and include the distinct element method [94, 95, 97] and the mesoscopic model adopting the tubular potential method for evaluation of the inter-tube interactions [82, 84]. As shown in Figs. 5.9b and 5.10a, b, the same procedure of layer-by-layer deposition of straight randomly oriented CNTs simulated with the tubular potential method results in a spontaneous self-assembly of CNTs into a continuous network of bundles with partial hexagonal ordering of CNTs in the bundles (Fig. 5.10b) and structural characteristics similar to the ones observed experimentally, e.g., [1–11]. The structure of the simulated CNT networks can be to a large extent controlled by parameters of the computational procedures used for the generation of the computational samples, and can be fine-tuned to match the results of experimental characterization. In particular, two distinct CNT structures, a CNT film and a "forest" of vertically aligned carbon nanotubes (VACNT), produced in mesoscopic simulations are shown in Fig. 5.10a, c along with experimental images of corresponding structures.

The availability of the computational samples reproducing the mesoscopic structure of real CNT materials opens up a broad range of opportunities for investigation of the dependence of the thermal transport properties on various material characteristics (material density, CNT type and length, density of cross-links, pore and bungle size distributions, degree of anisotropy in CNT orientation, etc.). First mesoscopic calculations of thermal conductivity performed for CNT films with preferential in-plane nanotube orientation [60, 100, 101], VACNT forests, and close-packed bundles of CNTs [79] have already provided important insights into the mechanisms and channels of the heat transfer in CNT network materials and are briefly discussed below.

Contrary to the conventional atomic-level MD simulations, the values of thermal conductivity of mesoscopic samples cannot be directly obtained from the analysis of the dynamic behavior of the coarse-grained elements of the model, and the calculation has to rely on special rules governing the heat transfer within and

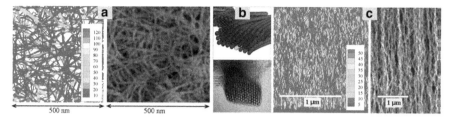

Fig. 5.10 Structure of CNT networks generated in mesoscopic simulations and observed experimentally. The experimental images of (**a**) buckypaper, (**b**) cross-section of a bundle, and (**c**) VACNT forest (all composed of single-walled CNTs) are from [8], [17], and [153], respectively. The computational images are for a CNT film consisting of 8000 200-nm-long (10,10) CNTs in (**a**) and a VACNT forest composed of 20,438 2-μm-long CNTs in (**c**). An enlarged view of a cross-section of a typical bundle is shown in (**b**). The computational samples have densities of 0.2 g/cm^3 in (**a**) and 0.02 g/cm^3 in (**c**), and dimensions of $0.5 \times 0.5 \times 0.1$ μm^3 in (**a**) and $2 \times 2 \times 2$ μm^3 in (**c**). Only a 20-nm-thick slice of the sample is shown in (**a**)

between the individual nanotubes. The complex structure of the entangled CNT networks introduces ambiguity in the concept of CNT–CNT contact area and calls for the design of a "heat transfer" function that accounts for the broad range of possible geometrical arrangements of the interacting CNT segments. As discussed above, in Sect. 5.3, a general description of the inter-tube conductance can be designed based on the results of atomistic simulations performed for various types of CNT–CNT contact configurations (see Fig. 5.7). The first mesoscopic simulations, however, were performed before the complete picture of the angular dependence of the contact conductance had emerged from the atomistic simulations and a simplified version of the heat transfer function [60, 79, 100, 101] that still ensured a continuous transition between the limiting cases of the inter-tube conductance between parallel and perpendicular CNTs was adopted in these simulations. Once the assumptions on the intrinsic thermal conductivity of individual CNTs and the CNT–CNT thermal conductance are made, the values of thermal conductivity of CNT networks generated in the dynamic mesoscopic simulations can be calculated by a method that is schematically illustrated in Fig. 5.11a. The opposing sides of a sample generated in a mesoscopic simulation are connected to the hot and cold heat reservoirs, the temperatures of all CNT segments located within the heat reservoirs are fixed at T_1 and T_2, the temperatures of all "internal" CNTs are iteratively calculated from the balance of incoming and outgoing heat fluxes in each CNT or, in general, based on the solution of a system of one-dimensional heat conduction equations describing the heat propagation along nanotubes and energy exchange between them. Once the steady-state distribution of temperature is obtained, the heat flux Q through the sample can be determined, and the value of the effective thermal conductivity k of the CNT sample can be calculated from the Fourier law.

The results of the mesoscopic calculations of in-plane and out-of-plane conductivities of CNT films with preferential in-plane orientation of nanotubes arranged into continuous networks of bundles are shown in Fig. 5.11b. A striking result

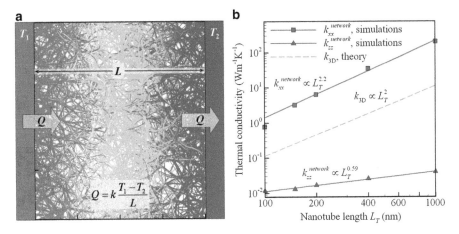

Fig. 5.11 Schematic of the method used in the calculation of thermal conductivity of CNT materials (**a**) and the values of the thermal conductivity predicted in simulations performed for systems composed of (10,10) CNTs of various length L_T and fixed material density of 0.2 g cm^{-3} [100]. The nanotubes in (**a**) are colored by their temperature and only a slice of the material with thicknesses of 50 nm is shown. The symbols in (**b**) show the results of the mesoscopic calculations, with red squares and blue triangles corresponding to the in-plane and out-of-plane conductivities of the anisotropic networks. The dashed line in (**b**) shows the prediction of an analytical scaling law obtained for a random distribution of straight CNTs (Eq. (5.1)) and discussed below, in Sect. 5.5

of these calculations is the large difference (more than two orders of magnitude) between the values of thermal conductivity predicted for the two directions in the material. The thermal conductivity within the plane of preferred orientation of CNTs, $k_{xx}^{network}$, increases from 0.8 to 205 Wm^{-1} K^{-1} as the length of CNTs increases from 100 nm to 1 μm, whereas the conductivity perpendicular to the plane, $k_{zz}^{network}$, increases from 0.01 to 0.04 Wm^{-1} K^{-1} within the same range of nanotube lengths. The strong structural dependence of the thermal conductivity predicted in the simulations provides a clue for explaining the large variability of experimental data on thermal conductivity reported in literature, with values measured for various CNT materials ranging from ~0.02 to ~1000 Wm^{-1} K^{-1} [14, 15, 21–34]. At the same time, the anisotropy of heat conduction and different scaling of the in-plane and out-of-plane conductivities with the CNT length, $k_{xx}^{network} \propto L_T^{2.2}$ and $k_{zz}^{network} \propto L_T^{0.59}$, support the feasibility of designing CNT-based materials capable of controlling and directing the heat flow along a desired path in thermal management applications.

The mesoscopic simulations discussed above and illustrated in Fig. 5.11 are done under assumption of a negligible contribution of the intrinsic thermal resistance of CNTs to the effective thermal resistance of a CNT material, which is thought to be largely defined by the inter-tube contact resistance. This assumption, commonly made in theoretical studies of heat transfer in CNT materials, e.g., [21, 72, 100], was critically evaluated in recent mesoscopic simulations that accounted for the finite intrinsic thermal conductivity of CNTs [101]. The results of the simulations,

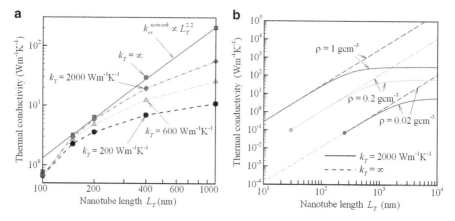

Fig. 5.12 Thermal conductivity of continuous networks of CNT bundles generated in mesoscopic simulations (**a**) and samples composed of straight randomly distributed CNTs (**b**) predicted for different values of intrinsic CNT conductivity k_T, nanotube length L_T, and material density ρ [101]. The material density of all networks in (**a**) is 0.2 g cm^{-3}. The solid and dashed lines in (**b**) show the predictions of Eq. (5.2) for $k_T = 2000$ Wm^{-1} K^{-1} and $k_T = \infty$, respectively, and a constant value of inter-tube conductance of $\sigma_c = 5 \times 10^{-11}$ WK^{-1}. The red and green circles mark the validity limit of the quadratic scaling of the conductivity with L_T as the system is approaching the percolation threshold

illustrated in Fig. 5.12a, have revealed an unexpectedly strong contribution of the intrinsic thermal resistance of CNTs that is found to make the dominant contribution to the effective thermal resistance of materials composed of CNTs that are longer than several hundreds of nanometers. Even for a relatively high intrinsic conductivity $k_T = 2000$ Wm^{-1} K^{-1} (see Table 5.1) and $L_T = 1$ μm, the value of the effective conductivity of the CNT network is ~3.5 times smaller than that predicted under assumption of negligible contribution of the intrinsic thermal resistance of CNTs, i.e., $k_T = \infty$. Further increase of the length of the nanotubes with finite values of k_T results in the saturation of the thermal conductivity of the materials at a level that is defined by the volume fraction of nanotubes (or density of the material). The physical origin of the strong effect of the intrinsic thermal resistance of the CNTs is explained by a theoretical analysis that provides scaling laws for thermal conductivity of samples composed of randomly dispersed straight CNTs [101]. The results of this analysis are illustrated in Fig. 5.12b and discussed in Sect. 5.5.

The effect of the intrinsic thermal resistance of the CNTs on the effective conductivity of CNT network materials can be amplified by the contributions from various structural defects and elastic deformation of CNTs. As discussed in Sect. 5.2 and illustrated in Fig. 5.3, the results of atomistic simulations reveal a substantial increase of the thermal resistance of nanotubes undergoing mechanical deformation and buckling. Mesoscopic modeling provides opportunities to translate the predictions of the atomistic simulations to the effective conductivity of CNT-based materials. In particular, the implications of the finite thermal conductance of the buckling kinks on the conductivity of CNT-based materials have been investigated in mesoscopic calculations performed for films composed of thousands of CNTs arranged into continuous networks of bundles [60]. The mesoscopic calculations predict a substantial contribution of the angular-dependent thermal resistance of the buckling kinks to the thermal conductivity of CNT-based materials, which is amplified by the preferential buckling of thin bundles and individual CNTs serving as interconnections between thicker bundles in the network structures. The total heat flux passing through the CNTs that are parts of the interconnections is, on average, higher than in other parts of the network structures. Consequently, the high concentration of the buckling kinks in interconnects results in a stronger impact of the buckling on the overall thermal conductivity of the films.

A recent series of calculations performed for VACNT "forests" consisting of 2 μm long (10,10) CNTs also predict a noticeable, up to 15 %, reduction of the thermal conductivity in the direction of the preferential orientation of nanotubes when the thermal resistance of the buckling kinks is accounted for in the calculations. The corresponding change in the heat flux along the nanotubes is illustrated in Fig. 5.13 for one of the simulated VACNT samples. Note that the sample used in these calculations is fully relaxed and all buckling kinks appear as a result of spontaneous self-organization of individual CNTs into bundles. The number of the buckling kinks can be expected to increase dramatically in the course of mechanical deformation, when the collective buckling is known to occur in response to compressive loading, e.g., [151, 154]. While the onset of collective buckling is likely to lead to a substantial drop in the effective thermal conductivity of the CNT material, the extent of this drop is defined by complex redistribution of the heat flux in response to the appearance of new "thermal resistors" and cannot be evaluated simply based on the average density of the buckling kinks. The mesoscopic modeling of heat transfer provides a unique opportunity for investigation of the changing pathways of the heat flow in CNT materials undergoing mechanical deformation.

5.5 Derivation of Scaling Laws and Monte Carlo Simulations

The results of the mesoscopic simulations discussed above can be related to theoretical models for thermal and electrical conductivity of nanofibrous materials [72, 100–114]. While different in details, these models are based on assumptions of straight dispersed nanotubes with point inter-tube contacts responsible for the heat exchange between CNTs. As a result, these models are not capable of quantitative description of the heat transport in continuous networks of CNTs, where the nanotubes are arranged into bundles and the interconnections between bundles are playing a prominent role. Nevertheless, the theoretical analysis performed with simplified models can be used for revealing the basic scaling laws with respect to key material parameters and can provide general guidance for designing CNT materials with improved thermal transport properties.

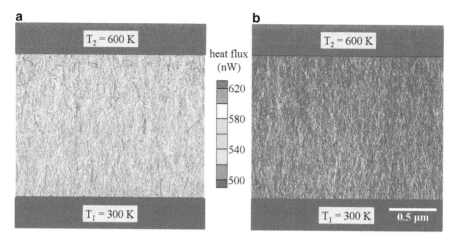

Fig. 5.13 The steady-state heat flux along the nanotubes obtained in the calculations of thermal conductivity of a VACNT sample consisting of 20,438 2-μm-long (10,10) CNTs. The same intrinsic thermal conductivity of 2000 $Wm^{-1} K^{-1}$ is assumed for buckling-free CNTs in both calculations. The thermal resistance of buckling kinks is neglected in the calculation illustrated in (**a**) and is accounted for in (**b**). The values of the effective thermal conductivity of the VACNT sample predicted in the two calculations are 16.4 and 13.9 $Wm^{-1} K^{-1}$, respectively

The capabilities and limitations of the theoretical approach to the analysis of the heat transfer in CNT network materials can be illustrated by the results of a recent study where a three-dimensional (3D) system composed of straight randomly dispersed nanotubes of length L_T and radius R_T is considered [100]. In this work, the derivation of the effective thermal conductivity of the CNT network is based on the ensemble averaging over configurational space performed under assumption of homogeneous and isotropic distribution of individual nanotubes in a system where a fixed temperature gradient is established. With an additional assumption of the dominant contribution of the inter-tube thermal contact resistance to the effective thermal resistance of the material (i.e., the intrinsic thermal resistance of CNTs is neglected and individual CNTs are considered to be isothermal), the following theoretical equation can be obtained:

$$k_{3D} = \frac{\sigma_c}{R_T} \frac{\pi \bar{n}_V^2}{36} \left(1 + 16\bar{R}_T + 80\bar{R}_T^2 + 192\bar{R}_T^3 + 153.6\bar{R}_T^4\right), \qquad (5.1)$$

where $\bar{n}_V = n_V L_T^2 R_T$ is the dimensionless density parameter, n_V is the volume number density of CNTs in the sample, $\bar{R}_T$ is the ratio of CNT radius R_T to its length L_T, $\bar{R}_T = R_T/L_T$, and σ_c is the inter-tube contact conductance (different from σ_T used in the discussion of inter-tube conductance per unit length for parallel CNTs in Sect. 5.3).

Since the material mass density, ρ, is proportional to $n_V L_T$, Eq. (5.1) predicts a quadratic dependence of the thermal conductivity on both ρ and L_T, $k_{3D} \propto \rho^2 L_T^2$.

This scaling law is in agreement with the results of Monte Carlo simulations of the electrical conductivity of 3D percolating networks dominated by contact resistance [102], where the quadratic scaling with density was observed. The scaling law $k_{3D} \propto \rho^2 L_T^2$ is also consistent with the results of a theoretical analysis of 3D systems of high aspect ratio rod-like particles [107, 108], which reveals the quadratic dependence on both rod length and density in the limit of dense systems. Recently, the quadratic scaling of thermal conductivity of the CNT films with material density was confirmed experimentally [113].

It is instructive to compare the predictions of the scaling law derived for randomly arranged straight CNTs and given by Eq. (5.1) with the results of the mesoscopic simulations performed for continuous networks of intertwined CNT bundles and discussed in Sect. 5.4. The results obtained for systems of (10,10) CNTs with the same material density of 0.2 g cm^{-3} and the value of contact conductance between the nanotubes crossing each other at a 90° angle, $\sigma_c = 5 \times 10^{-11}$ WK^{-1}, used in Eq. (5.1) are shown in Fig. 5.11b. Although the quadratic dependence on L_T predicted by Eq. (5.1) is similar to the super-quadratic dependence $k_{xx}^{network} \propto L_T^{2.2}$ observed for the random networks of CNT bundles produced in the mesoscopic simulations, the values of thermal conductivity in the networks of bundles significantly, by more than an order of magnitude, exceed the values predicted for the random arrangements of individual nanotubes. This difference can be attributed to the bundle structure of the CNT networks [100]. The tight arrangements of CNTs in the bundles result in a larger CNT–CNT contact area as compared to the random arrangements of CNTs assumed in the derivations of Eq. (5.1). Thus, we can conclude that while the model of straight individual nanotubes is commonly used in interpretation of the results of experimental measurements, the thermal conductivity of real nanotube materials is likely to be strongly enhanced by self-organization of CNTs into continuous networks of bundles. The design of the analytical description (scaling laws) capable of accounting for the realistic structure of real CNT materials is an outstanding challenge on the way of providing an adequate quantitative description of the thermal transport properties of CNT materials.

The validity of the assumption of negligible effect of the intrinsic thermal resistance of CNTs on the effective conductivity of CNT network materials is evaluated in a recent study [101] where the simplification of isothermal nanotubes is relaxed and the temperature distributions along individual nanotubes are calculated by solving one-dimensional heat conduction equations along every CNTs and accounting for both the intrinsic thermal conductivity and the heat exchange with surrounding nanotubes. In this case, the evaluation of thermal conductivity within the theoretical framework similar to that used in [100] for isothermal nanotubes results in the following equation:

$$k'_{3D} = \frac{k_{3D}}{1 + Bi_c \langle N_J \rangle /12} = \frac{k_{3D}}{1 + Bi_T/12}, \qquad (5.2)$$

where k_{3D} is the thermal conductivity of the CNT network at $k_T = \infty$, given by Eq. (5.1), $Bi_T = Bi_c \langle N_J \rangle = \sigma_c \langle N_J \rangle L_T / (k_T A_T)$ is a dimensionless parameter, $\langle N_J \rangle$ is the ensemble-averaged number of thermal contacts per CNT, and A_T is the cross-sectional area of a CNT. The parameter Bi_T is defined by the ratio of the total contact conductance $\sigma_c \langle N_J \rangle$ of a nanotube at all contacts it has with other CNTs to the intrinsic conductance of the nanotube, $k_T A_T/L_T$, and can be referred to as a Biot number for a nanotube. Similarly to the conventional definition of the Biot number in heat transfer analysis, the value of Bi_T can be used as a measure of non-isothermal distribution of temperature in an individual nanotube. The value of Bi_T, which increases with increasing material density and L_T, can be fairly large for real materials even if the ratio of the conductance in a single contact to the intrinsic conductance of the CNT, $\sigma_c L_T/(k_T A_T)$, is small.

The predictions of Eq. (5.2) are illustrated in Fig. 5.12b, where the solid curves show the dependence of k'_{3D} on L_T, for 3D samples composed of randomly oriented and distributed (10,10) CNTs for three values of material density ρ, typical of the CNT films and buckypaper. At small L_T, k'_{3D} practically coincides with k_{3D} (dashed curves in Fig. 5.12b) and scales quadratically with both ρ and L_T. An increase of L_T at a fixed ρ also increases Bi_T and results in the deviation of k'_{3D} from k_{3D} as Bi_T approaches and exceeds unity. In the limit of infinitely large Bi_T, the conductivity approaches an asymptotic value that is independent of L_T and linearly proportional to ρ. Somewhat unexpectedly, the transition between the two limiting scaling laws, $k \propto \rho^2 L_T^2$ at $Bi_T = 0$ and $k \propto \rho$ at $Bi_T \to \infty$ is observed for relatively short CNTs, on the order of hundreds of nanometers. The theoretical predictions given in Eq. (5.2) are consistent with the results of mesoscopic simulations discussed in Sect. 5.4 and illustrated in Fig. 5.12a. Thus, contrary to the common assumption of the dominant effect of the contact conductance, we can conclude that the intrinsic CNT conductivity, rather than contact conductance, is defining the overall thermal conductivity of CNT materials composed of nanotubes with the characteristic length on the order of micrometers or longer.

5.6   Concluding Remarks

The complexity of structural organization of CNT network materials is presenting a challenge for predictive modeling of their thermal transport properties and calls for a tightly integrated multiscale computational approach to this challenging problem. The applications of well-established atomistic simulation techniques and emerging mesoscopic models to the investigation of various aspects the heat transport in CNT materials have already resulted in a substantial progress in the fundamental understanding of the mechanisms and pathways of the heat transfer in different types of CNT materials. The results of atomistic simulations, in particular, have clarified a number of questions related to the heat transfer in individual CNTs and CNT–CNT heat conduction. The length dependence of the intrinsic thermal conductivity of CNTs, the conditions defining the transition from ballistic to diffusive phonon transport regimes, the effect of the structural defects and various modes of elastic and inelastic deformation on the conductivity of CNTs are among the questions that have been addressed in atomistic simulations of individual nanotubes and briefly reviewed in Sect. 5.2. The atomistic simulations of two or more interacting nanotubes have also been performed and provided important information on the inter-tube contact conductance, its dependence on the contact area, geometrical characteristics of the contact, as well as proximity of other contacts. A number of research questions related to the inter-tube conductance, including the ones that have been subjects of contradictory claims and controversial discussion in literature, are critically reviewed in Sect. 5.3. The results of mesoscopic modeling of the collective heat transfer by thousands of CNTs arranged into interconnected networks of bundles are reviewed in Sect. 5.4. While the development of mesoscopic models is still at the very early stage and many challenging technical issues still remain to be resolved, the first results of mesoscopic calculations are promising and demonstrate the ability of mesoscopic models to provide a bridge between the insights obtained in the atomistic simulations and the effective thermal conductance of CNT materials. Moreover, some of the predictions of the mesoscopic simulations, such as the strong (more than an order of magnitude) enhancement of the heat transfer in CNT materials due to the self-organization of CNTs into continuous networks of bundles or the dominant contribution of the intrinsic thermal resistance of CNTs to the effective thermal resistance of CNT materials composed of nanotubes that are longer than several hundreds of nanometers, are rather unexpected and make a strong impact on interpretation of experimental observations.

An apparent conclusion emerging from the overview of the computational results presented in this chapter is the one on the critical importance of combining the results obtained by different methods and dealing with processes occurring at different time and length scales into a well-integrated multiscale physical model of the heat transfer in CNT materials. In particular, the results of atomistic simulations can be used to design a reliable description of the inter-tube conductance suitable for incorporation into mesoscopic models which, in turn, can yield the information on the structural dependence of the thermal transport properties and stimulate the development of advanced theoretical models. Further insights obtained in atomistic studies of the effects of cross-links, polymer wrapping, or metal coating on the intrinsic conductivity of CNTs and the inter-tube conductance can be incorporated into the mesoscopic models and used for the exploration of the space of the material design parameters and optimization of the thermal transport properties of CNT network materials. Moreover, the dynamic nature of the mesoscopic models opens up opportunities for investigation of the effect of mechanical deformation on thermal conductivity of CNT materials. Mesoscopic simulations can be used for evaluation of the changing pathways of the heat flow in materials undergoing mechanical deformation as well as the thermal transport signatures of critical events, such as the onset of coordinated buckling and local failure, leading to an improved understanding of the performance of CNT-based thermal interface materials under conditions when large deformations are introduced by the requirements of compliance with complex geometrical shapes of heat sinks and sources in microelectronic devices.

On the side of the theoretical analysis of the general structure—thermal transport properties relationships, one critical challenge is to go beyond the assumption of randomly dispersed straight nanotubes and to design analytical descriptions capable of connecting the structural characteristics of real CNT materials (material density, CNT type and length, density of cross-links, pore and bungle size distributions) to the thermal transport properties. The ability of the mesoscopic simulations to reproduce realistic structures of various CNT materials (aerogels, films, forests, and fibers) is likely to play an important role in the design and verification of the advanced theoretical models.

Acknowledgments The authors acknowledge financial support provided by the National Aeronautics and Space Administration (NASA) through an Early Stage Innovations grant from NASA's Space Technology Research Grants Program (NNX16AD99G) and by the Air Force Office of Scientific Research (AFOSR) through the AFOSR's Thermal Sciences program (FA9550-10-10545). Computational support is provided by the National Science Foundation (NSF) through the Extreme Science and Engineering Discovery Environment (Projects TG-DMR110090 and TG-DMR130010).

References

1. Zou J, Liu J, Karakoti AS, Kumar A, Joung D, Li Q, Khondaker SI, Seal S, Zhai L (2010) Ultralight multiwalled carbon nanotube aerogel. ACS Nano 4:7293–7302
2. Gui X, Cao A, Wei J, Li H, Jia Y, Li Z, Fan L, Wang K, Zhu H, Dehai C (2010) Soft, highly conductive nanotube sponges and composites with controlled compressibility. ACS Nano 4:2320–2326
3. Sreekumar TV, Liu T, Kumar S, Ericson LM, Hauge RH, Smalley RE (2003) Single-wall carbon nanotube films. Chem Mater 15:175–178
4. Hennrich F, Lebedkin S, Malik S, Tracy J, Barczewski M, Rösner H, Kappes M (2002) Preparation, characterization and applications of free-standing single walled carbon nanotube thin films. Phys Chem Chem Phys 4:2273–2277
5. Muramatsu H, Hayashi T, Kim YA, Shimamoto D, Kim YJ, Tantrakarn K, Endo M, Terrones M, Dresselhaus MS (2005) Pore structure and oxidation stability of double-walled carbon nanotube-derived buckypaper. Chem Phys Lett 414:444–448
6. Ma W, Song L, Yang R, Zhang T, Zhao Y, Sun L, Ren Y, Liu D, Liu L, Shen J, Zhang Z, Xiang Y, Zhou W, Xie S (2007) Directly synthesized strong, highly conducting, transparent single-walled carbon nanotube films. Nano Lett 7:2307–2311
7. Xu M, Futaba DN, Yumura M, Hata K (2012) Alignment control of carbon nanotube forest from random to nearly perfectly aligned by utilizing the crowding effect. ASC Nano 6:5837–5844
8. Futaba DN, Hata K, Yamada T, Hiraoka T, Hayamizu Y, Kakudate Y, Tanaike O, Hatori H, Yumura M, Iijima S (2006) Shape-engineerable and highly densely packed single-walled carbon nanotubes and their application as super-capacitor electrodes. Nat Mater 5:987–994
9. Wang Z, Liang Z, Wang B, Zhang C, Kramer L (2004) Processing and property investigation of single-walled carbon nanotube (SWNT) buckypaper/epoxy resin matrix nanocomposites. Compos Part A Appl Sci Manuf 35:1225–1232
10. Whitten PG, Spinks GM, Wallace GG (2007) Mechanical properties of carbon nanotube paper in ionic liquid and aqueous electrolytes. Carbon 43:1891–1896
11. Whitby RLD, Fukuda T, Maekawa T, James SL, Mikhalovsky SV (2008) Geometric control and tuneable pore size distribution of buckypaper and buckydisks. Carbon 46:949–956

12. Poulin P, Vigolo B, Launois P (2002) Films and fibers of oriented single wall nanotubes. Carbon 40:1741–1749
13. Zhang X, Jiang K, Feng C, Liu P, Zhang L, Kong J, Zhang T, Li Q, Fan S (2006) Spinning and processing continuous yarn from 4-inch wafer scale super-aligned carbon nanotube arrays. Adv Mater 18:1505–1510
14. Behabtu N, Young CC, Tsentalovich DE, Kleinerman O, Wang X, Ma AWK, Bengio A, ter Waarbek RF, de Jong JJ, Hoogerwerf RE, Fairchild SB, Ferguson JB, Maruyama B, Kono J, Talmon Y, Cohen Y, Otto MJ, Pasquali M (2013) Strong, light, multifunctional fibers of carbon nanotubes with ultrahigh conductivity. Science 339:182–186
15. Zhang L, Zhang G, Liu C, Fan S (2012) High-density carbon nanotube buckypapers with superior transport and mechanical properties. Nano Lett 12:4848–4852
16. Girifalco LA, Hodak M, Lee RS (2000) Carbon nanotubes, buckyballs, ropes, and a universal graphitic potential. Phys Rev B 62:13104–13110
17. Thess A, Lee R, Nikolaev P, Dai H, Petit P, Robert J, Xu C, Lee YH, Kim SG, Rinzler AG, Colbert DT, Scuseria GE, Tománek D, Fischer JE, Smalley RE (1996) Crystalline ropes of metallic carbon nanotubes. Science 273:483–487
18. Rinzler AG, Liu J, Dai H, Nikolaev P, Huffman CB, Rodríguez-Macías FJ, Boul PJ, Lu AH, Heymann D, Colbert DT, Lee RS, Fischer JE, Rao AM, Eklund PC, Smalley RE (1998) Large-scale purification of single-wall carbon nanotubes: process, product, and characterization. Appl Phys A Mater Sci Process 67:29–37
19. Dettlaff-Weglikowska U, Skákalová V, Graupner R, Jhang SH, Kim BH, Lee HJ, Ley L, Park YW, Berber S, Tománek D, Roth S (2005) Effect of $SOCl_2$ treatment on electrical and mechanical properties of single-wall carbon nanotube networks. J Am Chem Soc 127: 5125–5131
20. Eom K, Nam K, Jung H, Kim P, Strano MS, Han J-H, Kwon T (2013) Controllable viscoelastic behavior of vertically aligned carbon nanotube arrays. Carbon 65:305–314
21. Prasher RS, Hu XJ, Chalopin Y, Mingo N, Lofgreen K, Volz S, Cleri F, Keblinski P (2009) Turning carbon nanotubes from exceptional heat conductors into insulators. Phys Rev Lett 102:105901
22. Hone J, Whitney M, Piskoti C, Zettl A (1999) Thermal conductivity of single-walled carbon nanotubes. Phys Rev B 59:R2514–R2516
23. Hone J, Whitney M, Zettl A (1999) Thermal conductivity of single-walled carbon nanotubes. Synth Met 103:2498–2499
24. Hone J, Llaguno MC, Nemes NM, Johnson AT, Fischer JE, Walters DA, Casavant MJ, Schmidt J, Smalley RE (2000) Electrical and thermal transport properties of magnetically aligned single wall carbon nanotube films. Appl Phys Lett 77:666–668
25. Hone J, Llaguno MC, Biercuk MJ, Johnson AT, Batlogg B, Benes Z, Fischer JE (2002) Thermal properties of carbon nanotubes and nanotube-based materials. Appl Phys A 74: 339–343
26. Gonnet P, Liang SY, Choi ES, Kadambala RS, Zhang C, Brooks JS, Wang B, Kramer L (2006) Thermal conductivity of magnetically aligned carbon nanotube buckypapers and nanocomposites. Curr Appl Phys 6:119–122
27. Yang DJ, Zhang Q, Chen G, Yoon SF, Ahn J, Wang SG, Zhou Q, Wang Q, Li JQ (2002) Thermal conductivity of multiwalled carbon nanotubes. Phys Rev B 66:165440
28. Yang DJ, Wang SG, Zhang Q, Sellin PJ, Chen G (2004) Thermal and electrical transport in multi-walled carbon nanotubes. Phys Lett A 329:207–213
29. Duong HM, Nguyen ST (2011) Limiting mechanisms of thermal transport in carbon nanotube-based heterogeneous media. Recent Pat Eng 5:209–232
30. Ivanov I, Puretzky A, Eres G, Wang H, Pan Z, Cui H, Jin R, Howe J, Geohegan DB (2006) Fast and highly anisotropic thermal transport through vertically aligned carbon nanotube arrays. Appl Phys Lett 89:223110
31. Aliev AE, Lima MH, Silverman EM, Baughman RH (2010) Thermal conductivity of multi-walled carbon nanotube sheets: radiation losses and quenching of phonon modes. Nanotechnology 21:035709

32. Zhang G, Liu C, Fan S (2013) Directly measuring of thermal pulse transfer in one-dimensional highly aligned carbon nanotubes. Sci Rep 3:2549
33. Marconnet AM, Panzer MA, Goodson KE (2013) Thermal conduction phenomena in carbon nanotubes and related nanostructured materials. Rev Mod Phys 85:1295–1326
34. Zhang KJ, Yadav A, Kim KH, Oh Y, Islam MF, Uher C, Pipe KP (2013) Thermal and electrical transport in ultralow density single-walled carbon nanotube networks. Adv Mater 25: 2926–2931
35. Berber S, Kwon Y-K, Tomanek D (2000) Unusually high thermal conductivity of carbon nanotubes. Phys Rev Lett 84:4613–4616
36. Che J, Çağın T, Goddard WA III (2000) Thermal conductivity of carbon nanotubes. Nanotechnology 11:65–69
37. Osman MA, Srivastava D (2001) Temperature dependence of the thermal conductivity of single-wall carbon nanotubes. Nanotechnology 12:21–24
38. Maruyama S (2002) A molecular dynamics simulation of heat conduction in finite length SWNTs. Phys B 323:193–195
39. Maruyama S (2003) A molecular dynamics simulation of heat conduction of a finite length single-walled carbon nanotube. Microscale Thermophys Eng 7:41–50
40. Padgett CW, Brenner DW (2004) Influence of chemisorption on the thermal conductivity of single-wall carbon nanotubes. Nano Lett 4:1051–1053
41. Moreland JF, Freund JB, Chen G (2004) The disparate thermal conductivity of carbon nanotubes and diamond nanowires studied by atomistic simulation. Microscale Thermophys Eng 8:61–69
42. Grujicic M, Cao G, Gersten B (2004) Atomic-scale computations of the lattice contribution to thermal conductivity of single-walled carbon nanotubes. Mater Sci Eng B 107:204–216
43. Grujicic M, Cao G, Roy WN (2005) Computational analysis of the lattice contribution to thermal conductivity of single-walled carbon nanotubes. J Mater Sci 40:1943–1952
44. Zhang G, Li B (2005) Thermal conductivity of nanotubes revisited: effects of chirality, isotope impurity, tube length, and temperature. J Chem Phys 123:114714
45. Bi K, Chen Y, Yang J, Wang Y, Chen M (2006) Molecular dynamics simulation of thermal conductivity of single-wall carbon nanotubes. Phys Lett A 350:150–153
46. Lukes JR, Zhong H (2007) Thermal conductivity of individual single-wall carbon nanotubes. J Heat Transf 129:705–716
47. Pan R-Q, Xu Z-J, Zhu Z-Y (2007) Length dependence of thermal conductivity of single-walled carbon nanotubes. Chin Phys Lett 24:1321–1323
48. Shiomi J, Maruyama S (2008) Molecular dynamics of diffusive-ballistic heat conduction in single-walled carbon nanotubes. Jpn J Appl Phys 47:2005–2009
49. Alaghemandi M, Algaer E, Bohm MC, Muller-Plathe F (2009) The thermal conductivity and thermal rectification of carbon nanotubes studied using reverse non-equilibrium molecular dynamics simulations. Nanotechnology 20:115704
50. Wu MCH, Hsu J-Y (2009) Thermal conductivity of carbon nanotubes with quantum correction via heat capacity. Nanotechnology 20:145401
51. Xu Z, Buehler MJ (2009) Strain controlled thermomutability of single-walled carbon nanotubes. Nanotechnology 20:185701
52. Nishimura F, Takahashi T, Watanabe K, Yamamoto T (2009) Bending robustness of thermal conductance of carbon nanotubes: nonequilibrium molecular dynamics simulation. Appl Phys Express 2:035003
53. Ren C, Zhang W, Xu Z, Zhu Z, Huai P (2010) Thermal conductivity of single-walled carbon nanotubes under axial stress. J Phys Chem C 114:5786–5791
54. Thomas JA, Iutzi RM, McGaughey AJH (2010) Thermal conductivity and phonon transport in empty and water-filled carbon nanotubes. Phys Rev B 81:045413
55. Shelly RA, Toprak K, Bayazitoglu Y (2010) Nose–Hoover thermostat length effect on thermal conductivity of single wall carbon nanotubes. Int J Heat Mass Transf 53:5884–5887
56. Lin C, Wang H, Yang W (2010) The thermomutability of single-walled carbon nanotubes by constrained mechanical folding. Nanotechnology 21:365708

57. Pan R-Q (2011) Diameter and temperature dependence of thermal conductivity of single-walled carbon nanotubes. Chin Phys Lett 28:066104
58. Wei N, Xu L, Wang H-Q, Zheng J-C (2011) Strain engineering of thermal conductivity in graphene sheets and nanoribbons: a demonstration of magic flexibility. Nanotechnology 22:105705
59. Qiu B, Wang Y, Zhao Q, Ruan X (2012) The effects of diameter and chirality on the thermal transport in free-standing and supported carbon-nanotubes. Appl Phys Lett 100:233105
60. Volkov AN, Shiga T, Nicholson D, Shiomi J, Zhigilei LV (2012) Effect of bending buckling of carbon nanotubes on thermal conductivity of carbon nanotube materials. J Appl Phys 111:053501
61. Nishimura F, Shiga T, Maruyama S, Watanabe K, Shiomi J (2012) Thermal conductance of buckled carbon nanotubes. Jpn J Appl Phys 51:015102
62. Cao A, Qu J (2012) Size dependent thermal conductivity of single-walled carbon nanotubes. J Appl Phys 112:013503
63. Imtani AN (2013) Thermal conductivity for single-walled carbon nanotubes from Einstein relation in molecular dynamics. J Phys Chem Solids 74:1599–1603
64. Feng D-L, Feng Y-H, Chen Y, Li W, Zhang X-X (2013) Effects of doping, Stone-Wales and vacancy defects on thermal conductivity of single-wall carbon nanotubes. Chin Phys B 22:016501
65. Zhu L, Li B (2014) Low thermal conductivity in ultrathin carbon nanotube (2,1). Sci Rep 4:4917
66. Salaway RN, Zhigilei LV (2014) Molecular dynamics simulations of thermal conductivity of carbon nanotube: resolving the effects of computational parameters. Int J Heat Mass Transf 70:954–964
67. Ma J, Ni Y, Volz S, Dumitrică T (2015) Thermal transport in single-walled carbon nanotubes under pure bending. Phys Rev Appl 3:024014
68. Sääskilahti K, Oksanen J, Volz S, Tulkki J (2015) Frequency-dependent phonon mean free path in carbon nanotubes from nonequilibrium molecular dynamics. Phys Rev B 91:115426
69. Mehri A, Jamaati M, Moradi M (2015) The effect of imposed temperature difference on thermal conductivity in armchair single-walled carbon nanotube. Int J Mod Phys C 26:1550105
70. Zhong H, Lukes J (2006) Interfacial thermal resistance between carbon nanotubes: molecular dynamics simulations and analytical thermal modeling. Phys Rev B 74:125403
71. Maruyama S, Igarashi Y, Taniguchi Y, Shiomi J (2006) Anisotropic heat transfer of single-walled carbon nanotubes. J Therm Sci Technol 1:138–148
72. Chalopin Y, Volz S, Mingo N (2009) Upper bound to the thermal conductivity of carbon nanotube pellets. J Appl Phys 105:084301
73. Kumar S, Murthy JY (2009) Interfacial thermal transport between nanotubes. J Appl Phys 106:084302
74. Xu Z, Buehler MJ (2009) Nanoengineering heat transfer performance at carbon nanotube interfaces. ACS Nano 3:2767–2775
75. Varshney V, Patnaik SS, Roy AK, Farmer BL (2010) Modeling of thermal conductance at transverse CNT-CNT interfaces. J Phys Chem C 114:16223–16228
76. Evans WJ, Keblinski P (2010) Thermal conductivity of carbon nanotube cross-bar structures. Nanotechnology 21:475704
77. Evans WJ, Shen M, Keblinski P (2012) Inter-tube thermal conductance in carbon nanotubes arrays and bundles: effects of contact area and pressure. Appl Phys Lett 100:261908
78. Hu G-J, Cao B-Y (2013) Thermal resistance between crossed carbon nanotubes: molecular dynamics simulations and analytical modeling. J Appl Phys 114:224308
79. Volkov AN, Salaway RN, Zhigilei LV (2013) Atomistic simulations, mesoscopic modeling, and theoretical analysis of thermal conductivity of bundles composed of carbon nanotubes. J Appl Phys 114:104301
80. Hu L, McGaughey AJH (2014) Thermal conductance of the junction between single-walled carbon nanotubes. Appl Phys Lett 105:193104

81. Salaway RN, Zhigilei LV (2016) Thermal conductance of carbon nanotube contacts: molecular dynamics simulations and general description of the contact conductance. Submitted
82. Zhigilei LV, Wei C, Srivastava D (2005) Mesoscopic model for dynamic simulations of carbon nanotubes. Phys Rev B 71:165417
83. Buehler M (2006) Mesoscale modeling of mechanics of carbon nanotubes: self-assembly, self-folding, and fracture. J Mater Res 21:2855–2869
84. Volkov AN, Zhigilei LV (2010) Mesoscopic interaction potential for carbon nanotubes of arbitrary length and orientation. J Phys Chem C 114:5513–5531
85. Cranford SW, Buehler MJ (2010) *In silico* assembly and nanomechanical characterization of carbon nanotube buckypaper. Nanotechnology 21:265706
86. Anderson T, Akatyeva E, Nikiforov I, Potyondy D, Ballarini R, Dumitrică T (2010) Towards distinct element method simulations of carbon nanotube systems. J Nanotechnol Eng Med 1:041009
87. Volkov AN, Zhigilei LV (2010) Structural stability of carbon nanotube films: the role of bending buckling. ACS Nano 4:6187–6195
88. Zhigilei LV, Volkov AN, Leveugle E, Tabetah M (2011) The effect of the target structure and composition on the ejection and transport of polymer molecules and carbon nanotubes in matrix-assisted pulsed laser evaporation. Appl Phys A 105:529–546
89. Xie B, Liu Y, Ding Y, Zheng Q, Xu Z (2011) Mechanics of carbon nanotube networks: microstructural evolution and optimal design. Soft Matter 7:10039–10047
90. Jacobs WM, Nicholson DA, Zemer H, Volkov AN, Zhigilei LV (2012) Acoustic energy dissipation and thermalization in carbon nanotubes: atomistic modeling and mesoscopic description. Phys Rev B 86:165414
91. Wang C, Xie B, Liu Y, Xu Z (2012) Mechanotunable microstructures of carbon nanotube networks. ACS Macro Lett 1:1176–1179
92. Li Y, Kröger M (2012) Viscoelasticity of carbon nanotube buckypaper: zipping-unzipping mechanism and entanglement effects. Soft Matter 8:7822–7830
93. Li Y, Kröger M (2012) A theoretical evaluation of the effects of carbon nanotube entanglement and bundling on the structural and mechanical properties of buckypaper. Carbon 50:1793–1806
94. Wang Y, Gaidău C, Ostanin I, Dumitrică T (2013) Ring windings from single-wall carbon nanotubes: a distinct element method study. Appl Phys Lett 103:183902
95. Ostanin I, Ballarini R, Potyondy D, Dumitrică T (2013) A distinct element method for large scale simulations of carbon nanotube assemblies. J Mech Phys Solids 61:762–782
96. Won Y, Gao Y, Panzer MA, Xiang R, Maruyama S, Kenny TW, Cai W, Goodson KE (2013) Zipping, entanglement, and the elastic modulus of aligned single-walled carbon nanotube films. Proc Natl Acad Sci USA 110:20426–20430
97. Ostanin I, Ballarini R, Dumitrică T (2014) Distinct element method modeling of carbon nanotube bundles with intertube sliding and dissipation. J Appl Mech 81:061004
98. Zhao J, Jiang J-W, Wang L, Guo W, Rabczuk T (2014) Coarse-grained potentials of single-walled carbon nanotubes. J Mech Phys Solids 71:197–218
99. Maschmann MR (2015) Integrated simulation of active CNT forest growth and mechanical compression. Carbon 86:26–37
100. Volkov AN, Zhigilei LV (2010) Scaling laws and mesoscopic modeling of thermal conductivity in carbon nanotube materials. Phys Rev Lett 104:215902
101. Volkov AN, Zhigilei LV (2012) Heat conduction in carbon nanotube materials: strong effect of intrinsic thermal conductivity of carbon nanotubes. Appl Phys Lett 101:043113
102. Keblinski P, Cleri F (2004) Contact resistance in percolating networks. Phys Rev B 69:184201
103. Foygel M, Morris RD, Anez D, French S, Sobolev VL (2005) Theoretical and computational studies of carbon nanotube composites and suspensions: electrical and thermal conductivity. Phys Rev B 71:104201
104. Duong HM, Papavassiliou DV, Lee LL, Mullen KJ (2005) Random walks in nanotube composites: improved algorithms and the role of thermal boundary resistance. Appl Phys Lett 87:013101

105. Kumar S, Alam MA, Murthy J Y (2007) Effect of percolation on thermal transport in nanotube composites. Appl Phys Lett 90:104105

106. Li J, Ma PC, Chow WS, To CK, Tang BZ, Kim J-K (2007) Correlations between percolation threshold, dispersion state, and aspect ratio of carbon nanotubes. Adv Funct Mater 17:3207–3215

107. Vassal J-P, Orgéas L, Favier D, Auriault J-L, Le Corre S (2008) Upscaling the diffusion equations in particulate media made of highly conductive particles. I. Theoretical aspects. Phys Rev E 77:011302

108. Vassal J-P, Orgéas L, Favier D, Auriault J-L, Le Corre S (2008) Upscaling the diffusion equations in particulate media made of highly conductive particles. II. Application to fibrous materials. Phys Rev E 77:011303

109. Duong HM, Papavassiliou DV, Mullen KJ, Maruyama S (2008) Computational modeling of the thermal conductivity of single-walled carbon nanotube–polymer composites. Nanotechnology 19:065702

110. Ashtekar NA, Jack DA (2009) Stochastic modeling of the bulk thermal conductivity for dense 1027 carbon nanotube networks. In: Proceedings of ASME IMECE2009, Paper IMECE2009-11282, Orlando, FL

111. Chalopin Y, Volz S, Mingo N (2010) Erratum: "Upper bound to the thermal conductivity of carbon nanotube pellets" [J Appl Phys 105:084301 (2009)]. J Appl Phys 108:039902

112. Bui KND, Grady BP, Papavassiliou DV (2011) Heat transfer in high volume fraction CNT nanocomposites: effects of inter-nanotube thermal resistance. Chem Phys Let 508:248–251

113. Yamada Y, Nishiyama T, Yasuhara T, Takahashi K (2012) Thermal boundary conductance between multi-walled carbon nanotubes. J Therm Sci Technol 7:190–198

114. Žeželj M, Stanković I (2012) From percolating to dense random stick networks: conductivity model investigation. Phys Rev E 86:134202

115. Frenkel D, Smit B (1996) Understanding molecular simulation: from algorithms to applications. Academic, San Diego

116. Matsui M (1989) Molecular dynamics study of the structural and thermodynamic properties of MgO crystal with quantum correction. J Chem Phys 91:489–494

117. Levashov VA, Billinge SJL, Thorpe MF (2007) Quantum correction to the pair distribution function. J Comput Chem 28:1865–1882

118. Turney JE, McGaughey AJH, Amon CH (2009) Assessing the applicability of quantum corrections to classical thermal conductivity predictions. Phys Rev B 79:224305

119. Yonetani Y, Kinugawa K (2003) Transport properties of liquid para-hydrogen: the path integral centroid molecular dynamics approach. J Chem Phys 119:9651–9660

120. Wang J-S (2007) Quantum thermal transport from classical molecular dynamics. Phys Rev Lett 99:160601

121. Wang J-S, Ni X, Jiang J-W (2009) Molecular dynamics with quantum heat baths: application to nanoribbons and nanotubes. Phys Rev B 80:224302

122. Dammak H, Chalopin Y, Laroche M, Hayoun M, Greffet J-J (2009) Quantum thermal bath for molecular dynamics simulation. Phys Rev Lett 103:190601

123. Savin AV, Kosevich YA, Cantarero A (2012) Semiquantum molecular dynamics simulation of thermal properties and heat transport in low-dimensional nanostructures. Phys Rev B 86:064305

124. Bedoya-Martínez ON, Barrat J-L, Rodney D (2014) Computation of the thermal conductivity using methods based on classical and quantum molecular dynamics. Phys Rev B 89:014303

125. Hernández-Rojas J, Calvo F, Gonzalez Noya E (2015) Applicability of quantum thermal baths to complex many-body systems with various degrees of anharmonicity. J Chem Theory Comput 11:861–870

126. Evans DJ (1982) Homogeneous NEMD algorithm for thermal conductivity—application of non-canonical linear response theory. Phys Lett A 91:457–460

127. Tersoff J (1988) New empirical approach for the structure and energy of covalent systems. Phys Rev B 37:6991–7000

128. Tersoff J (1988) Empirical interatomic potential for carbon, with applications to amorphous carbon. Phys Rev Lett 61:2879–2882

129. Lindsay L, Broido DA (2010) Optimized Tersoff and Brenner empirical potential parameters for lattice dynamics and phonon thermal transport in carbon nanotubes and graphene. Phys Rev B 81:205441

130. Brenner DW (1990) Empirical potential for hydrocarbons for use in simulating the chemical vapor deposition of diamond films. Phys Rev B 42:9458–9471

131. Brenner DW, Shenderova OA, Harrison JA, Stuart SJ, Ni B, Sinnott SB (2002) A second-generation reactive empirical bond order (REBO) potential energy expression for hydrocarbons. J Phys Condens Matter 14:783–802

132. Stuart SJ, Tutein AB, Harrison JA (2000) A reactive potential for hydrocarbons with intermolecular interactions. J Chem Phys 112:6472–6486

133. Yu C, Shi L, Yao Z, Li D, Majumdar A (2005) Thermal conductance and thermopower of an individual single-wall carbon nanotube. Nano Lett 5:1842–1846

134. Pop E, Mann D, Wang Q, Goodson K, Dai H (2006) Thermal conductance of an individual single-wall carbon nanotube above room temperature. Nano Lett 6:96–100

135. Wang J, Wang J-S (2006) Carbon nanotube thermal transport: ballistic to diffusive. Appl Phys Lett 88:111909

136. Donadio D, Galli G (2007) Thermal conductivity of isolated and interacting carbon nanotubes: comparing results from molecular dynamics and the Boltzmann transport equation. Phys Rev Lett 99:255502

137. Mingo N, Broido DA (2005) Length dependence of carbon nanotube thermal conductivity and the "problem of long waves,". Nano Lett 5:1221–1225

138. Mingo N, Broido DA (2005) Carbon nanotube ballistic thermal conductance and its limits. Phys Rev Lett 95:096105

139. Huxtable ST, Cahill DG, Shenogin S, Xue L, Ozisik R, Barone P, Usrey M, Strano MS, Siddons G, Shim M, Keblinski P (2003) Interfacial heat flow in carbon nanotube suspensions. Nat Mater 2:731–734

140. Yang J, Waltermire S, Chen Y, Zinn AA, Xu TT, Li D (2010) Contact thermal resistance between individual multiwall carbon nanotubes. Appl Phys Lett 96:023109

141. Yang J, Shen M, Yang Y, Evans WJ, Wei Z, Chen W, Zinn AA, Chen Y, Prasher R, Xu TT, Keblinski P, Li D (2014) Phonon transport through point contacts between graphitic nanomaterials. Phys Rev Lett 112:205901

142. Hsu I-K, Pettes MT, Aykol M, Chang C-C, Hung W-H, Theiss J, Shi L, Cronin SB (2011) Direct observation of heat dissipation in individual suspended carbon nanotubes using a two-laser technique. J Appl Phys 110:044328

143. Lin W, Shang J, Gu W, Wong CP (2012) Parametric study of intrinsic thermal transport in vertically aligned multi-walled carbon nanotubes using a laser flash technique. Carbon 50:1591–1603

144. Yue Y, Huang X, Wang X (2010) Thermal transport in multiwall carbon nanotube buckypapers. Phys Lett A 374:4144–4151

145. Shi L, Li DY, Yu CH, Jang WY, Kim D, Yao Z, Kim P, Majumdar A (2003) Measuring thermal and thermoelectric properties of one-dimensional nanostructures using a microfabricated device. J Heat Transfer 125:881–888

146. Yang J, Yang Y, Waltermire SW, Wu X, Zhang H, Gutu T, Jiang Y, Chen Y, Zinn AA, Prasher R, Xu TT, Li D (2012) Enhanced and switchable nanoscale thermal conduction due to van der Waals interfaces. Nat Nanotechnol 7:91–95

147. Colbourn EA (ed) (1994) Computer simulation of polymers. Longman, Harlow

148. Lee SW, Kim B-S, Chen S, Shao-Horn Y, Hammond PT (2009) Layer-by-layer assembly of all carbon nanotube ultrathin films for electrochemical applications. J Am Chem Soc 131:671–679

149. Yan XH, Xiao Y, Li ZM (2006) Effects of intertube coupling and tube chirality on thermal transport of carbon nanotubes. J Appl Phys 99:124305

150. Dresselhaus MS, Eklund PC (2000) Phonons in carbon nanotubes. Adv Phys 49:705–814

151. Zbib AA, Mesarovic SD, Lilleodden ET, McClain D, Jiao J, Bahr DF (2008) The coordinate buckling of carbon nanotube turfs under uniform compression. Nanotechnology 19:175704

152. Kim P, Shi L, Majumdar A, McEuen PL (2001) Thermal transport measurements of individual multiwalled nanotubes. Phys Rev Lett 87:215502
153. Wang S, Liang Z, Pham G, Park Y-B, Wang B, Zhang C, Kramer L, Funchess P (2007) Controlled nanostructure and high loading of single-walled carbon nanotubes reinforced polycarbonate composite. Nanotechnology 18:095708
154. Cao A, Dickrell PL, Sawyer WG, Ghasemi-Nejhad MN, Ajayan PM (2005) Super-compressible foamlike carbon nanotube films. Science 310:1307–1310

Part II
Applications

Chapter 6
Overview of Carbon Nanotubes for Horizontal On-Chip Interconnects

Jean Dijon

6.1 Introduction

The extraordinary development of micro- and nano-electronics is based on the brilliant idea of Gordon Moore, Robert Noyce, and others who proposed in the early 1970s a model development based on the shrinking of the integrated structures (transistors, connections) in the chips. This provides a long-term road map for technological development as well as a very efficient economic model. The size reduction, all other aspects being equal, results in performance improvements related to the possibility of making faster and more complex devices on the same area of silicon. Each node, typical scale length of the components, follows the same development cycle with massive investments for production and a return on investment at the end of the cycle related to the fact that better-performing, cheaper devices flood the market. The idea was also that the performance improvement was mostly a continuous process and not based on technological breakthrough at each node. Indeed it is reasonable to anticipate that such breakthroughs take a considerable amount of time to be fully realized and implemented. Initially the performances of the chips were largely limited by the active components which are the transistors. Since the mid-1990s this situation is completely reversed and now the chips are limited by interconnects. These limitations are so serious that they contribute to slowing down the microelectronic road map. A first revolution in the field of interconnects was the replacement of aluminum wires by copper wires and the introduction of low K dielectric materials instead of more conventional ones.

J. Dijon (✉)
CEA-LITEN/DTNM, Commissariat à l'énergie atomique et aux énergies alternatives,
17 avenue des martyrs, 38054 Grenoble cedex 9, France
e-mail: jean.dijon@cea.fr

© Springer International Publishing Switzerland 2017
A. Todri-Sanial et al. (eds.), *Carbon Nanotubes for Interconnects*,
DOI 10.1007/978-3-319-29746-0_6

To overcome current limitations a new material revolution is probably mandatory. Carbon materials such as carbon nanotubes, thanks to their superlative physical properties, can be the future material of choice.

In this chapter, the current state of the art of CNT line technology will be described, and the following pertinent questions will be addressed: why up to now is it not widely developed considering that the problem of interconnects scaling is a more and more serious one? What are the roadblocks not initially anticipated and what could be the future developments to bring this technology from the status of a promising to a useful one?

6.2   Brief Theoretical Reminder

A perfect single wall carbon nanotube is a one-dimensional conductor which has a conductance GoN_{ch}, where $Go = 4e^2/h$ is the basic unit of quantum conductance (i.e., around 77.5 µS) and N_{ch} is the number of conduction channel in the tube (exactly two for a metallic armchair SWCNT). Such non-dispersive conduction where the resistance is not dependent on the length is the so-called ballistic transport. This regime is valid up to the ballistic length L_{bm}. For tubes longer than L_{bm} electrons are scattered like in classical conductors and the resistance Rcnt of a carbon nanotube versus length L can be described by the following simple equation:

$$R_{CNT}(L) = Rq\left(1 + \frac{L}{Lm}\right) \tag{6.1}$$

L_m is the mean free path of the electrons and is equal to L_{bm} for perfect CNT.
Rq   is the quantum resistance (6.5 kΩ) inverse of the quantum conductance.

When the contact between the CNT and the electron reservoir is not ideal (inducing either electrons reflections or scattering at the contact level) a non-quantic contact resistance R_{nq} must be added and the resistance of one tube connected to electrodes becomes

$$R(L) = R_{nq} + R_{CNT} = Rc + \frac{Rq}{Lm}L \tag{6.2}$$

Rc   is the sum of the quantum and non-quantum resistances.

As we are interested in making electrical connections with various diameters and in decreasing the resistance, several tubes (n) are paralleled to form a bundle. The interconnection resistance can simply be represented by

$$R_{connect}(L) = \frac{R(L)}{n} = \frac{Rc}{NA} + \frac{Rq}{L_mN}\frac{L}{A} = \left(\frac{Rc}{NL} + \frac{Rq}{L_mN}\right)\frac{L}{A} = \rho_{connect}\frac{L}{A} \tag{6.3}$$

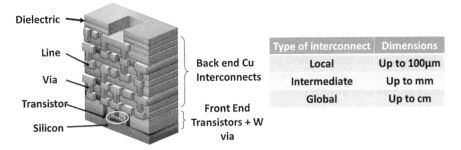

Fig. 6.1 Schematic representation of a silicon chip with interconnects levels and active components level. The typical dimensions of interconnects are given in the table

with A the cross-sectional area of the connection and N the CNT density (cm^{-2}) $(n = NA)$. If the tubes are not SWCNT, this kind of expression is considered to be still valid, ballistic transport has also been measured in MWCNTs [1]. The number, w, of conducting walls per tube must be taken into account and N is replaced by the product Nw in Eq. (6.3) considering that all the walls are connected. We will discuss the validity of this assumption in the paragraph devoted to contact. This simple equation will allow us to understand the main technology and material properties that must be optimized in order to achieve high performance CNT connections.

In a chip, interconnects are a combination of lines which are parallel to the chip surface (silicon) and via that are perpendicular to the surface and connects two levels of lines (see Fig. 6.1). Interconnects are usually categorized in three types: local (shortest and narrowest, closest to the transistors), intermediate and global (longest and widest, furthest away from the transistors). The typical lengths of these interconnects are also given in Fig. 6.1. Typically in a chip there is about 1.6 km of connection per square centimeter of silicon. Silicon components are largely copper components! The first level of via directly connected to the transistors is made of tungsten to avoid copper contamination of the transistors. Indeed copper induces deep levels in silicon which kill the carrier lifetime. Copper is a material which diffuses very easily with temperature making short circuits in the chip so that copper interconnects must be completely encapsulated with highly efficient barriers to contain the material. Process temperature above 450 °C is prohibited to avoid Cu diffusion in the dielectric materials during fabrication.

Now going back to Eq. (6.3) two situations arise. When we are dealing with short connections like via, their typical length L is shorter than Lm, thus via resistances are dominated by the first term of the equation, i.e. the contact resistance of the CNTs. When we are dealing with lines, L is typically much longer than the mean free path, so for reasonable tube contact resistance the contact contribution to the line resistance can be neglected. The performances of the lines are thus mostly dominated by the CNT quality, i.e. by the electron mean free path, physical parameter that qualifies the CNT defect content. For both interconnect elements (via and lines) the CNT density is a key factor. The achievable density is directly related to the technology of integration.

6.3   CNT Density in Interconnections

CNT growth is essentially a catalytic process on a metal [2], thus CNTs are growing selectively in the areas covered by the catalyst. Thus the integration and localization of the CNT to build interconnects need to precisely localize the metal catalyst on the chip. The simplest way to deposit the catalyst is to form a continuous film by physical vapor deposition (PVD) either sputtering or electron beam evaporation. The thin catalyst film deposited at room temperature on the adapted under-layer material (that can be considered part of the catalyst) will de-wet during the CNT growth process performed at higher temperature. This film de-wetting leads to formation of small droplets that seed the CNT growth and can be considered as a template for tube formation. The droplet size controls the CNT diameter [3] and the droplet density controls the CNT density. The maximum density is achieved when one CNT is growing on each droplet (due to the droplet size only one tube per droplet is possible). To maximize CNT density droplet density must be maximized. Thermodynamic considerations [4, 5] allow the maximum droplet density for a de-wetting film to be fixed. The Young Dupré equation gives the wetting angle theta between the catalyst droplet and the substrates when the surface energy γ of the materials and the interface are known:

$$\cos \theta = \frac{\gamma_{substrate} - \gamma_{int\,erface}}{\gamma_{catalyst}} \tag{6.4}$$

Generally this angle is close to or larger than $90°$ as the metal catalyst (iron, cobalt, nickel, and alloys) are high energy materials as compared to commonly used under-layer materials (oxide or nitride materials).

By minimization of the energy of a catalyst film with a thickness t deposited on a surface, the droplet diameter φ and their density ρ verify for $\theta > 90°$ [6].

$$\varphi_{droplet} >= 6t\frac{1}{(1 - \cos \theta)} \tag{6.5}$$

$$\rho_{droplet} <= \frac{1}{9\pi t^2}\frac{(1 - \cos \theta)}{2 + \cos \theta} \tag{6.6}$$

As the diameter of the growing CNTs is close to the droplet diameter, Eq. (6.5) gives a good rule of a good rule of thumb for the film thickness needed to grow a given tube diameter, typically a 0.5 nm film gives tubes around 3 nm in diameter.

Unfortunately the droplet density is not independent of their size and Eq. (6.6) provides the upper limit (the so-called de-wetting limit) to evaluate the achievable CNT density with this technique. Thermodynamics provides only extremal values which cannot be overcome but may be hard to reach experimentally. In particular the hypothesis of catalyst volume conservation is made in this calculation which ignores the possibility of catalyst loss by diffusion in the substrate during the process. Such diffusion occurs at the grain boundaries for polycrystalline materials like TiN [7] used as barrier layer for copper and even in oxide materials like alumina [8, 9]. In these cases, an equivalent catalyst thickness t* smaller than the deposited one has to be used in the above expressions. This explains the discrepancies observed in the literature between the deposited catalyst and the obtained tube diameters. To specify CNT diameters, the above droplet density must be compared to the requested CNT density for interconnects. The ITRS (International Technological Road map for Silicon) sets the requested conductive wall density that is needed to achieve a CNT material conductivity equivalent to copper one for the future nodes, see Table 6.1. The minimum performance entry in terms of conductivity for CNT interconnects is around 10 $\mu\Omega$ cm for future advanced nodes.

The density of a hexagonal array of tubes with a diameter d_{CNT} is given by

$$N = \frac{2}{(d_{CNT} + \delta)^2 \sqrt{3}} \tag{6.7}$$

where δ is the tube spacing which is at minimum the CNT wall distance (0.34 nm) for a compact array. This expression represents the absolute close pack limit possible in a bundle of CNT.

On average, the wall number is linked with the tube diameter by a linear relation [20, 21].

$$w = \alpha d_{CNT} - \beta \tag{6.8}$$

Thus the maximum density of conducting wall is given by

$$Nw = \frac{2 (\alpha d_{CNT} - \beta) F}{(d_{CNT} + \delta)^2 \sqrt{3}} \tag{6.9}$$

F is the fraction of metallic wall in the tube 1/3 for SWCNT with a tendency to be closer to 1 for MWCNT. Due to their larger diameter semiconducting shells have a small vanishing band gap and become conductors at room temperature.

α, β varies according to the authors and the used growth process typically a rule of thumb is that they are close to 1.

For the de-wetting limit by combining Eqs. (6.5), (6.6), and (6.8) it is

$$Nw = \frac{4 (\alpha d_{CNT} - \beta) F}{\pi\, d_{CNT}^2\, (1 - \cos\theta)\, (2 + \cos\theta)} \tag{6.10}$$

The plot of Eqs. (6.9) and (6.10) with $F = 1$ is given in Fig. 6.2 and compared with experimental results. Clearly it is extremely difficult to achieve the minimum conducting wall density around 5×10^{12} cm^{-2} as requested by ITRS analysis. Taking into account the fraction of metallic tubes, two strategies are possible: either to use SW with a diameter smaller than 2 nm or to use larger MWCNTs with a diameter smaller than 6 nm.

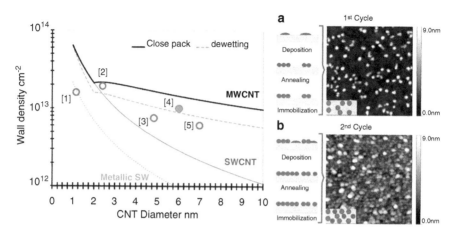

Fig. 6.2 *Left*: state of the art of high density CNT films with the limits given by Eqs. (6.9) and (6.10). The data [1], [2], [3], [5] are, respectively, from Zhong et al. [6]), Esconjauregui et al. [22], Dijon et al. [23], and Yamazaki et al. [24]. The point [4] is from our last not yet published results. *Right*: principle of cyclic catalyst deposition reprinted with permission from [22] Copyright © 2010 American Chemical Society

Table 6.1 Requested metallic SWCNT density to be competitive with Cu resistivity according to the year of production (ITRS 2013)

Year of production	2015	2017	2019	2021	2023	2025	2026
MPU/ASIC metal 1½ pitch (nm)	31.8	25.3	20	15.9	12.6	10.0	8.9
Cu effective resistivity ($\mu\Omega$ cm)	4.51	5.08	5.85	6.84	8.07	9.38	10.1
SWCNT minimum density (10^{13} cm^{-2})	1.08	0.96	0.81	0.66	0.56	0.48	0.45

Various CNT growth modes exist, nevertheless to be compatible with a microelectronic process flow one requirement is to be wafer compatible and made at a reasonable temperature. This excludes techniques like Arc discharge [25] which produces very high crystalline quality tubes mostly due to high temperature process but are not able to produce aligned structures. As a consequence the most suitable techniques for CNT integration in electronic components are chemical vapor deposition (CVD) process or Plasma Enhanced CVD process (PECVD). Initially PECVD was made in RF systems with diode structure which produces vertically aligned large diameter CNT [26] or fiber like materials. Point-arc microwave plasma technique [27], for example, is able to produce densely aligned SWCNT at low temperature (600 °C). Other CVD assisted techniques like hot filament (HFCVD) are also able to produce dense materials [28]. Works have been done to increase as much as possible the CNT density, trying to overcome the de-wetting limit. Instead of starting with catalyst films it is possible to use already formed nanoparticles. One system uses laser ablation then classification of particle sizes by a mass spectrometer to produce typically 4 nm Co particles [29]. With particle density exceeding the thermodynamic limit adapted growth processes have to be developed otherwise the particles sinter by Ostwald ripening mechanism creating larger tubes but not denser CNT material. Another approach is to design specific multilayered catalyst system either with enhanced anti-diffusion barrier [6] or to make a catalyst by cyclic deposition and annealing [22] (Fig. 6.2). The catalyst diffusion must be limited towards the substrate but also it has to be controlled on the surface. Thus in between each catalyst deposition cycle an immobilization step is introduced after the nanoparticle formation (Fig. 6.2). This step is either an oxidation of the particles or a carbon deposit [24]. With both techniques, tube densities above 10^{13} cm^{-2} have been achieved either with SWCNT close to 1 nm in diameter or DWCNT close to 2.4 nm, respectively. Up to now, these extremely promising materials have not been integrated in devices.

There are two strategies to realize pure carbon lines, either make ultra-dense CNT bundles as presented or to use individual MWCNTs for the local connections with small width, typically smaller than 10 nm this second option will be presented after. We have seen the kind of material specifications that are needed, the technologies that allow the integration of these materials in a device to make lines are the subject of the next part.

6.4 CNT Integration in Horizontal Lines

Horizontal CNT lines are much more difficult to realize than vertical via because high density CNT forest materials naturally grow perpendicular to the substrate. This point is easy to understand, the space between tubes is so small in a forest that the only available way for the tubes to escape from the substrate is on top of the growth front where there is free space. There are two main classes of process to realize lines: direct growth on the substrate of interest and transfer or assembly after the CNT growth.

Concerning direct growth, the CNTs can be grown either vertically then redirected horizontally or grown horizontally. The redirection of CNTs from dense vertically aligned carpet film has been achieved by a rolling method [30]. A roller is rotated and translated on the surface with a downward force to compress the film. To leave the film anchored on the surface a thin sheet of foil is located between the CNTs and the roller to avoid nanotube sticking to the roller surface. The foil also must be sheared along the direction of rolling in order to retain alignment in the film. After rolling a second step of densification can be used. This densification process exploits a liquid-induced collapse of SW or MW CNT forests [31]. When liquids are introduced into the forest and dried, the surface tension of the liquids and the strong Van der Waals (VdW) interactions stick the CNTs together to near-ideal graphitic spacing when the tubes are initially quite straight. Such liquid wetting and surface tension effects have also been used to flip CNTs over to a horizontal direction. A problem with flipping is that the direction of flipping needs to be accurately controlled to define the line direction. One possibility (Fig. 6.3) is to grow thin long CNT stripe structures and to flip them down perpendicularly to the stripe

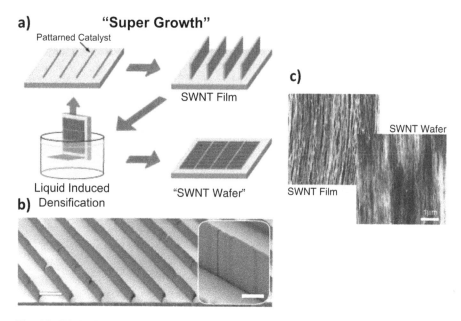

Fig. 6.3 Fabrication of an SWCNT wafer with CNT horizontally aligned on the surface. (**a**) Principle of the process. (**b**) SEM picture of the SWCNT arranged in long parallel stripes on the surface. (**c**) Atomic Force Microscopy images of the film and the wafer. Adapted from Hayamizu et al. [32]. Reprinted by permission from Macmillan Publishers Ltd: Nature Nanotechnology, copyright (2008)

direction by dipping the sample in isopropanol alcohol and pulling up at constant speed [32]. The achieved packing density in that case was 58 % of the maximum density. The macroscopic CNT surface direction is well defined and lines parallel to that direction can be patterned from the achieved CNT film. One other possibility to help alignment is to use a surface composed by different materials and surface energies to give preferred landing position. This option is not yet fully investigated.

As lines will be normally connected to via in order to make real connections, an integration flow that allows the growth of dense CNT lines starting from via has been developed [13]. Instead of making dense growth using classical Al_2O_3 Fe as catalyst system, CNTs are grown on the metal 1 of via made of aluminum. The interest of doing this is multiple. First CNTs are directly grown on a metal, then a thin Al_2O_3 layer always exists on the metal surface even after chemical de-oxidation which plays the role of an anti-diffusion barrier for the iron catalyst (this is mandatory due to the extremely high diffusion coefficient of iron in aluminum) and finally concave etching profiles of aluminum are conceivable as can be seen in Fig. 6.4. Such a profile allows to focus the CNT during their growth making a self-densification of the CNT bundles. This densification avoids detrimental interactions between the CNTs and the via side wall and allows the achievement of long, dense, and rather stiff structures such as those shown in Fig. 6.4. Typically tens of μm long structures

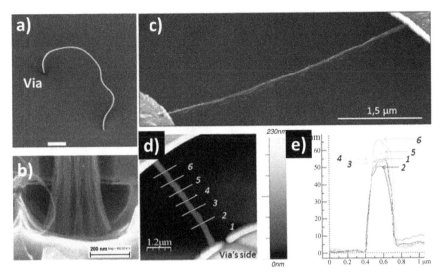

Fig. 6.4 CNT line made from a bundle grown in an individual via: (**a**) bundle structure grown from a 300 nm via (**b**) concave bottom shape of the via which allows the densification of the bundle during growth (**c**) 60 nm diameter CNT line formed between contact electrodes with a bundle grown in a 200 nm via (**d, e**) AFM profile of the line. Images (**a, b**) (Courtesy J. Dijon) (**c–e**) from Chiodarelli et al. [13] Copyright (2013), with permission from Elsevier

with 100 nm diameter are grown from 300 nm diameter via. The redirection of the lines is more complex than the redirection of stripes described previously due to the very low mechanical force induced by the liquid flow on the bundle as compared with its stiffness. A gentle mechanical action must thus be added using a dedicated tool that produces a shear flow to control the direction of single lines.

Instead of growing vertical structures and flipping them horizontally it is possible to grow well-oriented horizontal membranes starting from the edge of vertical patterned structures. One obvious technological difficulty is to localize the catalyst on this edge. A clever integration process [33] consists in making a first narrow trench in the substrate to localize the catalyst which is deposited at a given tilt angle (70° from normal incidence) to cover just one side of the trench. The trench shadowing effect avoids catalyst deposition on bottom surface. The resist used to define this first trench is also used to lift off the catalyst on the other surfaces. Then in a second step the full structure is completed allowing to make, for example, long narrow trenches needed to make lines (Fig. 6.5). The process is compatible with atomic layer deposition of the required catalyst under layer.

To accurately control the thickness of the growth zone an alumina layer can be sandwiched in between two TiN layers [34]. After dry etching of the sandwich and catalyst deposition the CNT growth area will be just located on the alumina layer and thus accurately controlled by the layer thickness. Indeed the catalyst diffuses in the TiN layer and selective growth on Al_2O_3 is performed without catalyst

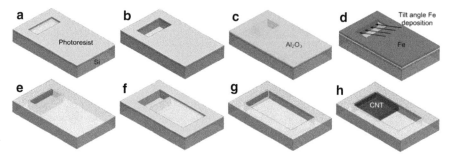

Fig. 6.5 Integration process of horizontal CNT from Lu et al. [33]. (**a**) Photolithography pattern-ing. (**b**) DRIE etching of the first trench. (**c**) ALD deposition of Al_2O_3 layer. (**d**) Tilt angle electron evaporation of the Fe catalyst layer. (**e**) Removal of photoresist. (**f**) Patterning of the second trench with photolithography. (**g**) Removal of photoresist after DRIE etching of the second trench. (**h**) The CVD growth of horizontal CNT. © IOP Publishing 2011. Reproduced with permission. All rights reserved

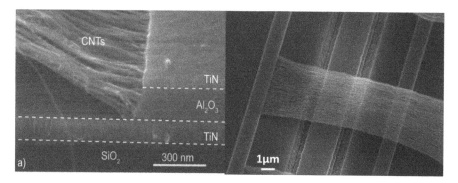

Fig. 6.6 *Left*: dense CNT membrane growing on the edge of Al_2O_3 layer sandwiched in between two TiN layer reprinted from [34] copyright (2014) with permission from Elsevier *Right*: CNT membranes before IPA dipping (courtesy H. Le Poche)

patterning. Dense DWCNT membranes were horizontally grown on the surface (Fig. 6.6) and IPA dipping was used to stick them on the surface.

CNT lines can also be made by wet processes assembly on the surface, which present the advantages of being a low temperature process compatible with Back End of the Line and of decoupling the growth conditions from the integration. Such a process is difficult for via. Sorted high quality metallic SWCNT can be used as a starting material which normally gives a conductivity benefit over direct growth. The drawbacks are the limited length of such CNTs (typically a few microns) and the difficulty of making dense oriented capillary assemblies. The short tube length is directly related to the chemical and sonication treatments needed to prepare the CNT solution which means that on long connections tubes-tubes contacts will play a significant role on the achieved resistivity values. The wet assembly methods demonstrated so far include: chemical functionalization of the substrate [35], self-assembly into a liquid crystal phase [36], di-electrophoresis (DEP) [37], Langmuir Blodgett (LB), and Langmuir Schaefer (LS) techniques [37].

DEP occurs when polarizable nano-objects are suspended in a non-uniform electric field. The object is polarized by the electric field and the induced dipole aligns along the field lines. Since the field is non-uniform, the pole experiencing the greatest electric field will dominate over the other, and the object will move. Since the direction of the force is dependent on field gradient rather than field direction, DEP will occur in continuous or alternative electric fields. High frequency DEP is thus one way to assemble individual tubes in between electrodes. One advantage of this process is its self-limitation [38] which means that when the connection is made between the electrodes no other tube will deposit, a decisive advantage to make devices or connection with individual tubes. Since the early demonstration of SW assembly in between parallel gold electrodes by DEP [37], the methodology has largely improved to yield control over the density and orientation of the CNTs which now reach 2.5×10^{11} cm^{-2} [39] but is still too low for interconnects. To make denser layers of oriented DWCNT, DEP coupled with capillary assembly has been developed [40]. The alignment of the CNTs is mostly controlled through the electrical parameters of the DEP (bias, frequency, etc.) while the density of the CNT trapped in the cavity where they are assembled is quasi-independently adjusted by tuning the parameters of the capillary assembly (temperature, scanning speed, contact angle, etc.). The DEP buried electrodes are electrically insulated from the CNT suspension which avoids damaging of the tubes by electrochemical reactions or by current flow, the achieved tube density is close to 10^{12} cm^{-2}.

LB or LS films are monolayers of particles or objects assembled on a liquid surface and then transferred on a substrate. To form a monolayer of single walled carbon nanotubes the CNTs are dispersed in dichloroethane (DCE) and then poured onto water. After evaporation of the DCE the CNTs form an isotropic phase at the water interface and mobile bars compress the film to form a well-aligned smectic phase. Very high density arrays of SWCNTs (500 SW per μm, 2.5×10^{13} cm^{-2}) with very low defect density were achieved by using purified solutions of SWCNTs, composed of 99 % semiconducting CNTs [41]. This technique has not yet been used to produce conducting films or interconnects.

6.5   CNT Contacts

The connection of 1D objects to the 3D world is not a simple matter. For CNTs there are both metallurgical problems at the carbon interface with other materials and the current injection in the nano-objects can be different from injection in bulk material. While the weight of the contacts is less important in the resistance budget of lines than via, it is nevertheless needed to understand their limitations, to optimize and stabilize their performances. Particularly the carrier injection into the tubes largely depends on the contact geometry and ultimately limits the performances of the lines.

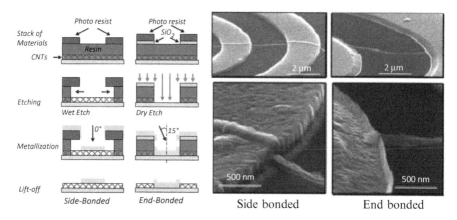

Fig. 6.7 Example of side-bonded and end-bonded contacts made on 100 nm diameter horizontal CNT lines [43] reproduced with permission copyright Cambridge University Press 2013

There are at least two contact geometries for bundles of CNTs or individual tubes, the so-called End-Bonded (EB) and Side-Bonded (SB) contacts. In the EB contact geometry, the tube is contacted at the tip. This is the case for tubes growing on catalyst metal particles or from contact formed with a reaction between the CNT and the metal. The objective is to make a strong bounding between the tube and the metal. In the SB geometry, the contact is made laterally, this is the most common contact made by deposition of metal on top of the structures. The nanotubes are embedded in the metal and the bonding is most often a weak VdW bonding. Other mixed contacts have also been proposed [42], Fig. 6.7 presents such contacts made on CNT bundles. For metallic CNTs the contact resistance is dominated by the barrier in between the metal and the CNT. The carriers must tunnel through this barrier. For semiconducting CNTs the formation of a Schottky barrier is the key issue. CNT contact with metals is difficult, particularly due to the low surface energy of graphene sheets which means that high surface energy metals will not easily wet the carbon in the SB configuration. It is also due to the difficulty in making bonds between metal and carbon in the EB geometry. Indeed stable carbides are made at rather high temperature typically above 550 °C for titanium carbide (TiC) [44] which, however, has one of the lowest temperatures of formation.

6.5.1 End-Bonded Contacts

Theoretical calculations have been made for EB metallic CNT contacted with Al, Cu, Pd, Pt, Ag, and Au [45]. Various metals have been ranked according to their electronic contact quality, thus their ability to form low resistance Ohmic contacts with the nanotube. The metal ranking is the following: $Ag \lesssim Au < Cu \ll Pt \approx Pd \ll Al$. The calculations have been performed for a 10 nm long (6, 0) CNT. To achieve a good metal CNT contact two parameters must be considered. The binding energy controls the distance between the metal and the carbon, thus the barrier thickness and the metal ability to wet the CNT surface. High binding energy is needed to decrease the contact distance and good wettability allows the formation of a continuous metal film. Palladium, titanium, and nickel are the most suitable metals, thanks to their ability to make continuous films on CNTs [46]. The bad ranking of Ag, Au, and Cu comes from their low binding energy; furthermore, they will be probably prohibited as highly contaminant metals for microelectronic applications. Pd has been known for a long time to give good Ohmic contact even on semiconducting CNTs with diameters larger than 2 nm [47]. Titanium is widely used due to its common availability as an adhesion layer even if contact performances are not as good as those achieved with Pd. Aluminum, while ranked high, is not widely used due to technological problems related with fast formation of Al_2O_3.

One technological problem with EB contact is to properly connect the internal shells of a MWCNT particularly with standard metal contact processes using lithography which restrict the control of the direction of the metal flux (Fig. 6.6). Also, low temperature metal deposition using sputtering or evaporation only yields a physical contact to the MWCNT, which results in a weak electronic coupling at the Fermi surface even for contamination-free contact interface. In a bundle either via or line the contact with the different tubes must be considered as statistical [43]. This may explain the worst contact resistances achieved with bundles as compared with individual tubes (best values instead of averaged values are generally published) as can be seen in Table 6.2.

The best EB contact with an MWCNT was made with carbon material deposited by electron beam induced deposition [17]. After annealing at 350 °C the 2.2 μm long connection made of one individual MWCNT presents a resistance of 116 Ω. All the CNT shells are connected. Other examples of good multiple shells contact

Table 6.2 CNT contact resistance achieved with different contact structures and contact materials

Contact material	Rc (Ω cm^2)	L$_T$ (nm)	Comments	References
Silver	2E-2	8000	SWCNT film SB	Jackson and Grahama [10]
Pt	1.1E-6	380	SWCNT film SB	Koechlin et al. [11]
Pd	3.46E-9	200	Individual SWCNT SB	Franklin and Chen [12]
Pd/Au	3.9E-8		CNT line EB	Chiodarelli et al. [13]
Ti/Au	5E-7		Via EB (sum bottom + top)	Chiodarelli et al. [14]
Ti/Al	1.4E-8		Via EB (sum bottom + top)	Lee et al. [15]
CNT	1.4E-7		MW/MW SB Rc	Santini et al. [16]
Carbon EBID	4.1E-10		Individual MW EB Rc	Kim et al. [17]
Carbon EBID	9E-9		Individual MW SB Rc	Kim et al. [17]
NiSi	1.0E-8		Planar contact on Si	Zhang et al. [18]
Cu-Cu	1.2E-9		Bonding	Chen et al. [19]

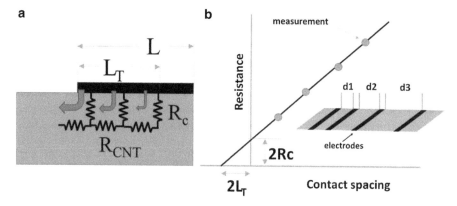

Fig. 6.8 Principle of TLM measurement. (**a**) Schematic of geometry and resistance distribution in electrode/contact interface. (**b**) TLM test structure with a representative plot of R vs d illustrating extraction of Rc and L_T

are obtained with tungsten [48], thanks to the formation of carbide, in that case a high temperature process is needed.

6.5.2 Side-Bonded Contacts

The technology of Side-Bonded contact is the easiest to realize for CNTs lying on a surface; unfortunately, the carrier injection is more complex than in EB contact. Classically the charge injection that applies to side contacts is related to the length over which injection occurs from the contact edge (Fig. 6.8). This length is the contact transfer length L_T. For contacts to bulk and thin-film materials the expression of the contact resistance R_c is [49]

$$Rc = \frac{\sqrt{\rho_C R_F}}{W} coth\left(\frac{L}{L_T}\right), \quad L_T = \sqrt{\frac{\rho_C}{R_F}} \tag{6.11}$$

where ρ_c is the intrinsic contact resistance (Ω cm^2), R_F is the film resistance under the contact W and L are, respectively, the width and length of the contact.

For highly anisotropic materials like CNT bundles and for cylindrical geometry this expression certainly has to be reconsidered. The case of SB contact on a ballistic conductor has been treated [50]. The transfer length measurement method (TLM) which consists of plotting the resistance for devices made with different gap between similar contact electrodes allows the extraction of both R_c and L_T (Fig. 6.8). The classical metric for contacts is the intrinsic contact resistance, product of the contact area by the contact resistance which allows to compare contact performances regardless of their area.

The SB contact geometry leads to injection problems in the different shells of MWCNTs and in the different tubes for CNT bundles. Indeed the inter-shell coupling or the tube-tube coupling is rather weak which means that in spite of the intimate contact the transverse CNT resistance is several orders of magnitude higher than the longitudinal one. Contrary to in-plane conduction, inter-shell conduction is governed by hopping and depends on the π orbital overlap of nearby shells. A rough estimate of the inter-shell resistance is the interlayer resistivity of pure graphite which is very sensitive to the material quality and the interlayer separation. Typically this resistivity is between 10^{-3} and 10^{-1} Ω cm [51, 52].

Theoretically inter-shell transport is forbidden for an infinitely long and perfect tube, since both the energy and the Bloch wave vector have to be conserved [53]. In fact the radial resistivity of MWCNTs has been measured to be as high as 10^2 Ω cm [54], much larger than that of bulk graphite. The inter-shell conductance is largely independent of temperature and is consistent with tunneling through the orbitals of nearby shells. As a result the current flows mainly along individual shells that are rather efficiently insulated from each other. It is accepted that only two or three shells at the tube surface are effectively participating in the MW conduction when SB contacted. The presence of semiconducting shells will also increase the electrical decoupling of the inner shells. SB contacted MWCNTs essentially behave electrically like large diameter SW or DW CNTs.

To reduce the CNT-metal contact resistance in the SB geometry graphitic interfaces have been proposed [55]. The idea is to make a carbon based interface that will wet more correctly both the CNT and the metal decreasing the tunneling barrier in between the electrode and the CNT. Amorphous carbon (a-C) is deposited on the tube but it does not improve the contact as it is a disordered material with sp^3 bounding. Thus a catalyzed annealing step is needed to promote graphitization. A 5 nm Ni layer is deposited on top of the a-C and annealed at 750 °C under H_2 flow [56]. Contact resistance improvement by a factor 4 has been obtained as compared with Pd SB contacts. Finally CNT-CNT contact resistance has been experimentally determined to be around 10^{-7} Ω cm^2 [16]. Table 6.2 summarizes the above discussion, adding some comparisons with other contact resistances measured in the microelectronic domain. Clearly the CNT contacts, although not perfect, are not drastically different from other technologies but there is still room for improvement. One of the most critical points is the rather long (200 nm) transfer length with planar SB contact which severely limits the integration density of carbon based electronics. For interconnect lines the resistivity is degraded due to the difficulty in connecting all the tubes and all the shells within each tube. Thus for SW as well as for MW, EB or mixed contacts are probably mandatory for ultimate performances. Carbon based contacts are promising but need a lot of developments to become the "standard" contact technology.

6.6 Performances of CNT Lines

Table 6.3 summarizes the published performances as well as the technology used to make integrated horizontal lines. It is clear that up to now the conductivity achieved is far from expectations as predicted by theoretical considerations based on results on individual tubes and simulations. The resistivity is two decades too high to be useful as a replacement material for copper and up to now the resistivity of integrated lines seems to saturate around 1 mΩ cm, the best result (i.e., 200 $\mu\Omega$ cm) being achieved with aligned SWCNT films obtained at Rice University [62].

It is, however, instructive to analyze the performances of other types of CNT structures, such as macroscopic CNT cable assemblies. A non-exhaustive list of the technology and the electrical performances obtained in this context is presented in Table 6.4. First the lowest resistivity achieved so far, 15 $\mu\Omega$ cm [64], is more than one decade better than the resistivity achieved with integrated lines or films. While these macroscopic tube assemblies have rather large diameters compared to integrated lines (typically a few tens of micrometers), they almost fulfill the requested 10 $\mu\Omega$ cm resistivity value for interconnects. Thus it proves that at least for ultimate interconnects real CNT assemblies are already close to the specifications.

Careful inspection of Table 6.4 raises a lot of interesting points and helps us to understand to date technological and material limitations of integrated lines. Small tube diameters, typically DWCNTs provide better wire performances than MWCNTs. In these assemblies tubes are connected together by their sides. So this result is consistent with our previous discussion on contacts where we have shown that only few walls are electrically connected with SB contacts. Cables made by spinning CNT forests [66] which are grown with a process similar to the one used to make lines (low temperature CVD at 700 °C with iron catalyst) have performances around 1 mΩ cm, i.e. similar to integrated lines and far below the best results obtained with tubes grown at higher temperature (typically close to 1150 °C with CH$_4$ or alcohol as precursors). These high temperature processes lead to highly crystallized CNT with Raman peak ratio IG/ID extremely high 25.2 [67] for the used DWCNT in record resistivity assembly [64]. The relatively low temperature used to make these lines as compared with the elaboration temperatures of the tubes assembled in wires partly explains the lower electrical performances of CNT lines.

Another very interesting and important point is the extremely high CNT density achieved in the cables as can be assessed by the density (g cm^{-3}) and by the cross section of the assembled materials (Fig. 6.8). The high tensile strength achieved with these cables is also a good indication of their density. One example of the conductivity improvement induced by densification is illustrated by the rolling process developed by Wang et al. [65]. Initially after spinning the cables have a rather high resistivity of 0.8 mΩ cm. After rolling the cable in between cylinders with a high pressure (Fig. 6.9), the resistivity dramatically decreases down to 50 $\mu\Omega$ cm. This impressive gain shows that in the less dense assembly, many of the tubes inside the structure are not electrically well connected. Part of the

Table 6.3 CNT lines technology and achieved electrical performances

CNT	CNT growth	Line technology	Line geometry W, T, L (μm)	CNT density (cm^{-2})	Resistivity (m Ω cm)	References
SWCNT 2.8 nm	Super growth C$_2$H$_2$ Fe on Al$_2$O$_3$	Flipped with alcohol	Films	1.00E+13	8	Hayamizu et al. [32]
DWCNT 2–3 nm	CCVD of CH4 at 1000 °C	Assembly by capillarity + dielectrophoresis	2, 0.009, 12	80–100 CNT μm^{-1} 1E+12	4.4	Seichepine et al. [40]
TWCNT 7 nm	CCVD C$_2$H$_2$ 750 °C	Alcohol flow	30, 7, 100	1.6E+12	1.7	Li et al. [57]
DWCNT 4.5 nm	CVD C$_2$H$_2$ 580 °C	Horizontal growth		1.00E+12	1.75	Guerin et al. [34]
MWCNT 10 nm	CCVD C2H4 775 °C Fe on Al203	Rolling + Alcohol densification		4.00E+11	1	Tawfick et al. [58]
TWCNT 5 nm	CCVD C$_2$H$_2$	Flipped with alcohol from via		1.00E+12	1.1	Chiodarelli et al. [13, 42]
SWCNT 1.9 nm	NA	Capillary assembly (doped with Pt)		NA	0.8	Kim et al. [59]
TWCNT 5 nm	CCVD C$_2$H$_2$ 580 °C	Flipped from via (doped with iodine)	0.1, 0.1, 10	3.00E+12	0.45	Dijon et al. [60]
MWCNT 50–100 nm		Individual CNT dielectrophoresis	0.07, 0.07, 3	NA	0.5	Close and Wong [61]
MWCNT 50–100 nm		Individual CNT dielectrophoresis	0.07, 0.07, 3	NA	2.5	Close and Wong [61]
SWCNT 3 nm	HFCVD 750 °C	Rolling	Films	1.00E+13	0.2	Pint et al. [62]

Table 6.4 CNT cables assembly with growth technology and electrical performances

CNT	CNT growth	Wire technology	Doping	CNT density (cm^{-2})	Resistivity (m Ω cm)	References
DWCNT 2–6 nm (3.2 nm)	?	Coagulation from chlorosulfonic acid	Iodine	8E+12 (1.3 g cm^{-3})	0.018	Behabtu et al. [63]
			Undoped		**0.035**	
DWCNT 2–3 nm	CVD CH4 1150 °C floating catalyst ferrocene + sulfur		Iodine	0.28 g cm^{-3}	0.015	Zhao et al. [64]
			Undoped		**0.05**	
DWCNT 2–7 nm	CVD 1150–1300 °C ethanol + ferrocene	Spinning	Undoped	1.3–1.8 g cm^{-3}	0.8	Wang et al. [65]
		Spinning + rolling			**0.05**	
MWCNT 15 nm	CVD C$_2$H$_2$ 620–700 °C iron on SiO$_2$	Spinning from CNT forest	Undoped		**1.1**	Liu et al. [66]

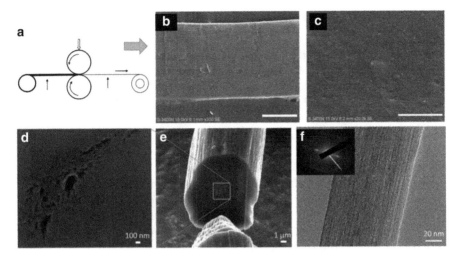

Fig. 6.9 Highly densified CNT cables that present record conductivity by dry (*top images*) and wet processes (*bottom images*). (**a**) Principle of rolling. (**b**) and (**c**) Highly densified CNT material Reprinted by permission from Macmillan Publishers Ltd: Nature Communications Wang et al. [65], copyright (2014) (**d-f**) [63] Reprinted with permission from AAAS copyright 2013

relatively poor results obtained after line integration comes from the low CNT density in the lines. Indeed growth of CNT forest leads to tubes with a lot of tortuosity, an indication of structural CNT defects. Tubes with different diameters are growing at different rates (the smaller growing faster). The Van Der Walls interactions between neighboring tubes keep the tubes in contact with each other preventing slippage between them [68]. Thus forest growth rate is smaller than tube growth rate and the tubes manage their length differences by creating individual CNT waviness, source of structural defects. The forest structure is also complex, a multi-scale structure not only made of individual tubes, but composed of small bundles that are mechanically stiffer than individual tubes. This intrinsic waviness limits densification and the transport properties of the tubes are certainly degraded as compared with individually grown tubes.

Finally the conductivity is improved by a factor between 2 and 4 by iodine doping. Doping has not been largely investigated for interconnects, this is certainly a way to improve interconnect performances.

As a summary, the combination of a rather low tube quality with a density far from ideal coupled with the difficulty in properly contacting all the tubes is certainly the best explanation of the results achieved so far in the context of CNT interconnects. Without a drastic change in the technology, these results probably rule out large CNT bundles made of pure carbon as a solution for global interconnects. On the contrary, for the end of the road map it is the very high current carrying capacity of CNTs [69, 70] which becomes the most appealing property. The paradigm of big bundle SWCNT as interconnect probably has to be changed and local interconnects made of individual CNTs are the most promising.

6.7   Local Interconnects Made of Individual CNTs

CNTs as a local interconnect use the bottom-up approach as compared with the well-established top-down approach based on lithography and etching. When dimensions shrink very high resolution lithography becomes more and more costly and challenging due to the intrinsic limitation of optics which is able to define structures with a size close to the used wavelength. The limitation of available wavelength and transparent materials below 154 nm calls for the challenging development of extreme UV 13 nm lithography. The use of bottom-up CNTs in this context is perfectly viable as compared with other contenders like graphene based interconnects. Such technology will use ultra-high resolution lithography and up to now it is impeded by performance degradation related to the open edge effects of the graphene ribbon when line width shrinks [71] (Fig. 6.10). For CNTs which are closed structures this is the opposite and performances are improving when dimension shrinks as can be seen in Fig. 6.11.

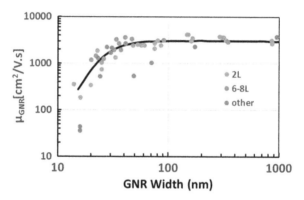

Fig. 6.10 Loss of mobility (thus of conductivity) versus line width for a graphene nanoribbon. Adapted from Yang and Murali [71]

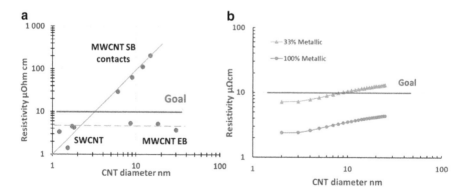

Fig. 6.11 Resistivity of individual CNTs versus their diameter. (**a**) Experimental results listed in Table 6.5. (**b**) Theoretical values according to the metallic wall content of the tubes, see text. The *red Goal line* is the ITRS specification

If performances are achieved, the other points to fulfill are to place the CNTs in between the points to connect and to make electrical connections between the CNTs and the other structures. Finally uniform properties between the tubes (diameter, wall number, etc.) will have to be achieved to minimize the fluctuations from one connection to the other. Concerning the resistivity performances, the consensus is that tubes are highly conductive. We have summarized some literature results in Table 6.5 used to draw Fig. 6.11

This figure provides a number of points worthy of discussion. As regards the experiments, MWCNT resistivity versus diameter aligns quite well on a continuous line with a slope of 2 in this log-log plot. This is what is expected for CNTs contacted with side contacts if a limited constant number of shells are conducting as explained in the contact paragraph. Thus the side-contacted MWCNTs exhibit a constant resistance per unit length, regardless of their diameter. Otherwise for metallic SW or MW with all the shells well contacted, a rather constant resistivity close to 5 $\mu\Omega$ cm is observed (horizontal dotted line). This value seems to be the limit of what is experimentally possible at room temperature considering that the CNTs are mostly grown at high temperature by CVD or Arc Discharge technique. The good news is that the performances are better than the resistivity target for advanced interconnects as given in Table 6.1. These results are not in contradiction with our discussion concerning the density. Indeed the packing density is achieved for individual tubes and for large tube diameter the number of conducting channels increases. Theoretically for MW the number of conduction channel per shell is expressed by Naeemi and Meindl [77].

$$N_{ch/shell}(d) = (ad + b)\, r \qquad (6.12)$$

where a $= 0.1836$ nm^{-1} and b $= 1.275$ and r is the metallic shell ratio.

The tube resistivity is plotted in Fig. 6.11b with two hypotheses concerning the metallicity of the tube and a ballistic length of 1 μm. The experiments agree with the theory. The possibility to obtain 100 % of purely metallic CNT or walls is not yet demonstrated with very few published results due to the more attractive need to develop 100 % semiconducting SW for carbon transistor applications. For MW interconnects, the situation is probably less critical in the sense that large wall diameter means small band gap. Typically the gap is inversely proportional to the shell diameter. The difficulty will be to achieve uniform performances from tube to tube due to the metallic wall statistic. To improve this aspect and the overall conductivity doping of the CNTs is certainly the best option.

The main road block for individual tube connection is the placement or local-ization of tubes on the substrate. Extensive developments have been made with SWCNT grown on quartz. This technique provides very good alignment and paves the way towards some answer for interconnects. The angular misalignment is lower than 0.01° [78]. The idea is to use crystal substrates having prominent interactions with CNTs in one specific direction. Both substrate lattice and miscut surfaces are

Table 6.5 Resistivity of individuals CNTs with different diameters and contact geometry

CNT	Growth process	Diameter (nm)	Resistivity ($\mu\Omega$ cm)	Contact geometry	References
m SWCNT	CVD on quartz 900 °C Fe	1.2	3.39	SB	Franklin and Chen [12]
m SWCNT	CH4 CVD 900 °C	1.5	1.41	SB	Li et al. [72]
m SWCNT	CVD 950 °C	1.8	4.24	SB	
m SWCNT	CVD 1000 °C	1.7	4.54	SB, EB	Mann et al. [73]
MWCNT	Arc	6	29.7	SB	Ebessen et al. [74]
MWCNT	CVD 700 °C	12	113	SB	Kreupl et al. [75]
MWCNT	Arc	9	63.6	SB	Bachtold et al. [76]
MWCNT	Arc	14.8	206	SB	Ebessen et al. [74]
MWCNT	Arc	18.2	5.2	SB	Ebessen et al. [74]
MWCNT	Arc	8.6	5.4	NA	[70]
MWCNT	NA	30	3.7	EB	[17]

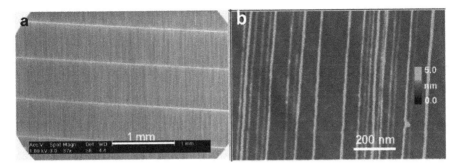

Fig. 6.12 (**a**) SEM image of high density array of long SWCNTs aligned along the [100] direction on ST-cut quartz substrate, the bright stripes correspond to patterned copper catalyst. (**b**) AFM image of 1 μm × 0.75 μm area, in which 22 SWNTs have been found. Adapted from Ding et al. [79] copyright 2008 American Chemical Society

used which provides either preferential directions for anisotropic VdW interaction or aligned terraces with regularly spaced edges where CNTs may align. The basic process consists of the deposition of stripes of catalyst perpendicular to the preferential growth direction, as illustrated in Fig. 6.12.

The process generally starts with a rather long annealing step of the quartz under oxygen at a temperature around 900 °C which smooths the surface and improves the alignment of the CNTs [80]. Then stripes of a very thin layer of catalyst (typically few tenths of nanometer of iron) are deposited on the surface and patterned using a lift off process. The resist residues are calcined at 450 °C before growth. The growth is made at high temperature (900 °C) in hydrogen with methane or ethanol as precursor. One of the road blocks of this technique is the low density of tubes that is growing on the surface. Typically it is around 10 CNT/μm while density higher than 50 CNT/μm has been reported with Cu as catalyst [79]. More than one growth cycle have also been used to increase the density [81]. The actual record density is 130 CNT/μm [82] using a new Trojan catalyst concept. The remaining difficulties are the variable pitch in between the tubes and the exclusive use of this technique for semiconducting SW. Up to now very few reports exist [83, 84] on the growth of enriched metallic SWCNTs and no demonstration of small multiwall CNTs growth was performed. One reason could be the lack of strong needs, there is an open field of research here. After growth the array of tubes must be transferred to the substrate of interest. Since there are schemes of interconnections with XY directions being made on two different layers, the fact that CNTs are growing in one direction is not an intrinsic problem but a technological one. The main advantage of transfer is the decoupling of the growth from the useful substrate, thus high temperature required for high quality CNTs is no more a problem for the back end compatibility. The drawback is the need to transfer and make registration of the tubes on the landing substrate. Adapted CNT transfer processes exist [85] and with the current graphene development that also need transfer, this step will probably become a routine part of the nanotechnology toolbox. The open question with this technique is

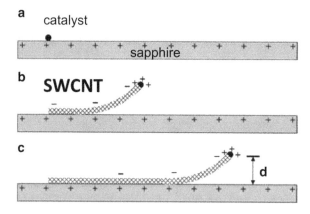

Fig. 6.13 Lattice oriented growth of SWCNT according to the "raised-head" model. The growth is a tip growth mode. In (**a**) the sapphire is positively charged due to oxygen depletion induced by the growth environment. In (**b**) due to charge transfer the SWCNT is negatively charged while the catalyst is positively charged, which prevents by electrostatic repulsion the contact between the catalyst and the surface (**c**). Reproduced from Yu et al. [87] copyright 2006 American chemical society

its applicability to the aligned growth of small MWCNTs. Other techniques of CNT alignment have been demonstrated using either scratches on the surface of silica [86] or patterned structures.

The understanding of the conditions which lead to CNT alignment on a surface may also trigger the development of new interconnects. There are basically two CNT growth modes one which is called the tip growth mode where the catalyst is lifted from the substrate and placed on top of the CNT, the other one is the root or base growth with the catalyst staying on the substrate surface. CNT Carpets or forests grow according to the base growth mode. The proposed mechanism of lattice oriented aligned SWCNT growth is a tip growth mode called the "raised-head" growth mode [87] (Fig. 6.13). The tube stays on the surface without sliding thus the orientation is built locally during the growth. The explanation of the flying catalyst lies on the positively charged sapphire surface related to Al-O dipoles and to the positively charged catalyst particle which is repulsed from the substrate by electrostatic force. Similar tip growth is also considered for the flow oriented growth of very long tubes (SW, DW, or TW) in this "Kite–mechanism" [88] the whole CNT floats above the substrate and is oriented by the gas flow. The base growth mechanism is much less favorable to induce good alignment since the whole nanotube needs to slide on the surface, once they rest on the surface. The nanotube/substrate interaction would increase as a function of the length leading either to growth arrest or to instability that will induce detachment of the tube from the substrate and loss of alignment.

6.8 CNT Doping

As CNTs are a mix of conducting and semiconducting shells a straightforward approach to improve the conductivity is to dope them. The success and growth of the microelectronics industry have been based on the precise control over the electrical conduction of semiconductors via ion implantation. Unfortunately direct doping by carbon substitution with nitrogen atoms induces a lot of defects and the formation of bamboo like structures [89]. In graphite a lot of work has been done on the intercalation of species in between the graphene sheets leading to materials with higher electrical conductivity than copper using, for example, $FeCl_3$ [90]. In MWCNT intercalation in between the shells is difficult due to the restricted possible change in the shell spacing. Stable doping within a limited temperature range is possible without degrading the sp^2 bonding by charge transfer between the doping species and the CNT network. The dopants can be physisorbed on the CNT surface or in its hollow part. Theoretical and experimental works have shown that modification of the structure of carbon nanotubes can drastically affect their electronic properties. Density functional theory (DFT) calculations show that decoration of SWNTs with platinum nanoclusters adds a number of electronic levels and enhances their density of states near the Fermi level for both metallic and semiconducting carbon nanotubes independently of the tube chirality. Chemical doping or decoration of carbon nanotubes by metal clusters represents a general approach to convert semiconducting walls to metallic ones and thus improves the overall conductivity. This process does not require either selective growth or post-growth selection of metallic CNTs for interconnects applications. Decrease in resistivity of the order of approximately 75 % [59] has been obtained with platinum clusters on lines made of aligned SWCNTs. Other very efficient and stable doping is obtained with iodine. Improvement in conductivity by a factor 4 is reported (Zao 2011; [60]). Extremely efficient conductivity enhancement has been recently reported [91], with an enhancement factor close to two decades on carpet of CNTs using molybdenum oxide MoO_3. A resistivity of 5×10^{-5} Ω cm is obtained. DFT calculations show that the CNTs are p doped while the surrounding oxide becomes conductive, thanks to the electrons injected by the tubes. Such results reinforce the idea that doping CNTs is one of the keys to achieve the performances and new ideas like this may trigger a revival of interest for CNT based interconnects.

6.9 Conclusion

While up to now only limited performances have been achieved with CNT interconnects, it is felt that CNTs can play a significant role in the future if the paradigm of the big CNT bundles made of pure carbon is changed and if the future developments at least for local interconnects are focused on controlling the placement of individual small MWCNT, doping them efficiently while all their shells are properly contacted. The electronics of the twenty-first century will certainly explore and develop circuits with active components in three dimensions as we are approaching the end of the classical road map and the limit of component size in two dimensions. In this context, stackable technology where more than one layer of active components is mandatory and non-contaminant interconnections following the same paradigm are probably needed. Carbon electronics using bi-dimensional materials and CNTs are possible contenders. CNT interconnects at least for local connections have to be developed in that direction.

References

1. Berger C, Yi Y, Wang ZL, de Heer WA (2002) Multiwalled carbon nanotubes are ballistic conductors at room temperature. Appl Phys A 74:363
2. Jourdain V, Bichara C (2013) Current understanding of the growth of carbon nanotubes in catalytic chemical vapour deposition. Carbon 58:2–39
3. Yamada T, Namai T, Hata K, Futaba DN, Mizuno K, Fan J, Yudasaka M, Yumura M, Iijima S (2006) Size-selective growth of double-walled carbon nanotube forests from engineered iron catalysts. Nat Nanotechnol 1:131–136
4. Dijon J, Fournier A, Szkutnik PD, Okuno H, Jayet C, Fayolle M (2010) Carbon nanotubes for interconnects in future integrated circuits: the challenge of the density. Diam Relat Mater 19(5–6):382–288
5. Nessim GD, Hart AJ, Kim JS, Acquaviva D, Oh J, Morgan CD, Seita M, Leib JS, Thompson CV (2008) Tuning of vertically-aligned carbon nanotube diameter and areal density through catalyst pre-treatment. Nano Lett 8(11):3587–3593
6. Zhong G, Warner JH, Fouquet M, Robertson AW, Chen B, Robertson J (2012) Growth of ultrahigh density single-walled carbon nanotube forests by improved catalyst design. ACS Nano 6(4):2893–2903
7. Yang J, Esconjauregui S, Robertson AW, Guo Y, Hallam T, Sugime H, Zhong G, Duesberg GS, Robertson J (2015) Growth of high-density carbon nanotube forests on conductive TiSiN supports. Appl Phys Lett 106:083108
8. Kim SM, Pint CL, Amama PB, Zakharov DN, Hauge RH, Maruyama B, Stach EA (2010) Evolution in catalyst morphology leads to carbon nanotube growth termination. J Phys Chem Lett 1:918–922
9. Robertson J, Zhong G, Esconjauregui CS, Bayer BC, Zhang C, Fouquet M, Hofmann S (2012) Applications of carbon nanotubes grown by chemical vapor deposition. Jpn J Appl Phys 51:01AH01
10. Jackson R, Grahama S (2009) Specific contact resistance at metal/carbon nanotube interfaces. Appl Phys Lett 94:012109
11. Koechlin C, Maine S, Haidar R, Trétout B, Loiseau A, Pelouard JL (2010) Electrical characterization of devices based on carbon nanotube films. Appl Phys Lett 96:103501
12. Franklin AD, Chen Z (2010) Length scaling of carbon nanotube transistors. Nat Nanotechnol 5:858–862
13. Chiodarelli N, Fournier A, Okuno H, Dijon J (2013) Carbon nanotubes horizontal interconnects with end-bonded contacts, diameters down to 50 nm and lengths up to 20 μm. Carbon 60: 139–145
14. Chiodarelli N, Delabie A, Masahito S et al (2011) ALD of Al_2O_3 for carbon nanotube vertical interconnect and its impact on the electrical properties. MRS Proc 1283:46–54
15. Lee S et al (2011) Integration of carbon nanotube interconnects for full compatibility with semiconductor technologies. J Electrochem Soc 158(11):K193–K196
16. Santini CA, Volodin A, Van Haesendonck C, De Gendt S, Groeseneken G, Vereecken PM (2011) Carbon nanotube–carbon nanotube contacts as an alternative towards low resistance horizontal interconnects. Carbon 4(9):4004–4012

17. Kim S, Kulkarni DD, Rykaczewski K, Henry M, Tsukruk VV, Fedorov AG (2012) Fabrication of an ultra-low resistance ohmic contact to MWCNT–metal interconnect using graphitic carbon by electron beam-induced deposition (EBID). IEEE Trans Nanotechnol 11(6):1223–1230

18. Zhang Z et al (2010) Sharp reduction in contact resistivities by effective Schottky barrier lowering with silicides as diffusion sources. IEEE Electron Device Lett 31:731–733

19. Chen KN, Fan A, Tan CS, Reif R (2004) Contact resistance measurement of bonded copper interconnects for three-dimensional integration technology. IEEE Electron Device Lett 25(1):10–12

20. Chiodarelli N, Li Y, Cott DJ, Mertens S, Peys N, Heyns M, De Gendt S, Groeseneken G, Vereecken PM (2011) Integration and electrical characterization of carbon nanotube via interconnects. Microelectron Eng 88:837–843

21. Chiodarelli N, Richard O, Bender H, Heyns M, De Gendt S, Groeseneken G, Vereecken PM (2012) Correlation between number of walls and diameter in multiwall carbon nanotubes grown by chemical vapor deposition. Carbon 50:1748–1752

22. Esconjauregui S, Fouquet M, Bayer BC, Ducati C, Smajda R, Hofmann S, Robertson J (2010) Growth of ultrahigh density vertically aligned carbon nanotube forests for interconnects. ACS Nano 4(12):7431–7436

23. Dijon J, Okuno H, Fayolle M, Vo T, Pontcharra J, Acquaviva D, Bouvet D, Ionescu AM, Esconjauregui CS, Capraro B, Quesnel E, Robertson J (2010) Ultra-high density carbon nanotubes on Al-Cu for advanced vias. IEDM 2010 IEEE international electron devices meeting, pp 33.4.1–33.4.4

24. Yamazaki Y, Saluma N, Katagiri M, Suzuki M, Sakai T, Sato S, Nihei M, Awano Y (2010) Synthesis of a closely packed carbon nanotube forest by a multi-step growth method using plasma-based chemical vapor deposition. Appl Phys Express 3:55002–55004

25. Journet C, Maser WK, Bernier P, Loiseau A, Lamyde la Chapelle M, Lefrant S, Deniard P, Leek R, Fischerk JE (1997) Large-scale production of single-walled carbon nanotubes by the electric-arc technique. Nature 388:756–758

26. Chhowalla M, Teo KBK, Ducati C, Rupesinghe NL, Amaratunga GAJ, Ferrari AC, Roy D, Robertson J, Milne WI (2001) Growth process conditions of vertically aligned carbon nanotubes using plasma enhanced chemical vapor deposition. J Appl Phys 90(10):5308–5317

27. Zhong G, Iwasaki T, Honda K, Furukawa Y, Ohdomari I, Kawarada H (2005) Low temperature synthesis of extremely dense and vertically aligned single-walled carbon nanotubes. Jpn J Appl Phys 44(4A):1558–1561

28. Pint CL, Pheasant ST, Parra-Vasquez ANG, Horton CC, Xu Y, Hauge RH (2009) Investigation of optimal parameters for oxide-assisted growth of vertically aligned single-walled carbon nanotubes. J Phys Chem C 113:4125

29. Sato S, Nihei M, Mimura A, Kawabata A, Kondo D, Shioya H, Iwai T, Mishima M, Ohfuti M, Awano Y (2006) Novel approach to fabricating carbon nanotube via interconnects using size-controlled catalyst nanoparticles. In: Proceedings second ITC conference, in Fukuoka, pp 230–232

30. Pint CL, Xu Y-Q, Pasquali M, Hauge RH (2008) Formation of highly dense aligned ribbons and transparent films of single-walled carbon nanotubes directly from carpets. ACS Nano 2(9):1871–1878

31. Futaba DN, Hata K, Yamada T, Hiraoka T, Hayamizu Y, Kakudate Y, Tanaike O, Hatori H, Yumura M, Iijima S (2006) Shape-engineerable and highly densely packed single-walled carbon nanotubes and their application as super-capacitor electrodes. Nat Mater 5:987–994

32. Hayamizu Y, Yamada T, Mizuno K, Davis RC, Futaba DN, Yumura M, Hata K (2008) Integrated three-dimensional microelectromechanical devices from processable carbon nanotube wafers. Nat Nanotechnol 3:289

33. Lu J, Miao J, Xu T, Yan B, Yu T, Shen Z (2011) Growth of horizontally aligned dense carbon nanotubes from trench sidewalls. Nanotechnology 22:265614

34. Guerin H, Le Poche H, Pohle R, Bernard LS, Buitrago E, Ramos R, Dijon J, Ionescu AM (2014) High-yield, in-situ fabrication and integration of horizontal carbon nanotube arrays at the wafer scale for robust ammonia sensors. Carbon 78:326–338

35. Wang Y, Maspoch D, Zou S, Schatz G, Smalley R, Mirkin C (2006) Controlling the shape, orientation, and linkage of carbon nanotube features with nano affinity templates. Proc Natl Acad Sci USA 103:2026–2031
36. Ko H, Tsukruk V (2006) Liquid-crystalline processing of highly oriented carbon nanotube arrays for thin-film transistors. Nano Lett 6:1443–1448
37. Chen XQ, Saito T, Yamada H, Matsushige K (2001) Aligning single-wall carbon nanotubes with an alternating-current electric field. Appl Phys Lett 78:3714
38. Vijayaraghavan A, Blatt S, Weissenberger D, Oron-Carl M, Hennrich F, Gerthsen D, Hahn H, Krupke R (2007) Ultra-large-scale directed assembly of single-walled carbon nanotube devices. Nano Lett 7(6):1556–1560
39. Steiner M, Engel M, Lin Y-M, Wu Y, Jenkins K, Farmer DB, Humes JJ, Yoder NL, Seo J-WT, Green AA et al (2012) High-frequency performance of scaled carbon nanotube array field-effect transistors. Appl Phys Lett 101:053123
40. Seichepine F, Salomon S, Collet M, Guillon S, Nicu L, Larrieu G, Flahaut E, Vieu C (2012) A combination of capillary and dielectrophoresis-driven assembly methods for wafer scale integration of carbon-nanotube-based nanocarpets. Nanotechnology 23:095303
41. Cao Q, Han S-J, Tulevski GS, Zhu Y, Lu DD, Haensch W (2013) Arrays of single-walled carbon nanotubes with full surface coverage for high-performance electronics. Nat Nanotechnol 8:180–186
42. Chiodarelli N, Fournier A, Dijon J (2013) Impact of the contact's geometry on the line resistivity of carbon nanotubes bundles for applications as horizontal interconnects. Appl Phys Lett 103:053115. doi:10.1063/1.4817648
43. Dijon J, Chiodarelli N, Fournier A, Okuno H, Ramos R (2013) Horizontal carbon nanotube interconnects for advanced integrated circuits. Mater Res Soc Symp Proc 1559: © 2013 Materials Research Society. doi:10.1557/opl.2013
44. Leroy WP, Detavernier C, Van Meirhaeghe RL, Kellock AJ, Lavoie C (2006) Solid-state formation of titanium carbide and molybdenum carbide as contacts for carbon-containing semiconductors. J Appl Phys 99:063704
45. Zienert A, Schuster J, Gessner T (2014) Metallic carbon nanotubes with metal contacts: electronic structure and transport. Nanotechnology 25:425203. doi:10.1088/0957-4484/25/42/425203
46. Zhang Y, Franklin NW, Chen RJ, Dai H (2000) Metal coating on suspended carbon nanotubes and its implication to metal-tube interaction. Chem PhysLett 331:35–41
47. Kim W et al (2005) Electrical contacts to carbon nanotubes down to 1 nm in diameter. Appl Phys Lett 87:173101
48. Wang M-S, Golberg D, Bando Y (2010) Superstrong low-resistant carbon nanotube–carbide–metal nanocontacts. Adv Mater 22:5350–5355
49. Reeves GK, Harrison HB (1982) Obtaining the specific contact resistance from transmission line model measurements. IEEE Electron Device Lett 3:111–113
50. Solomon PM (2011) Contact resistance to a one-dimensional quasi-ballistic nanotube/wire. IEEE Electron Device Lett 32:246–248
51. Casparis L (2010) Conductance anisotropy in natural and HOPG graphite. Master Thesis, University of Basel
52. Primak W (1956) C-axis electrical conductivity of graphite. Phys Rev 103:544
53. Yoon YG, Delaney P, Louie SG (2002) Quantum conductance of multiwall carbon nanotubes. Phys Rev B 66:073407
54. Bourlon B, Miko C, Forro L, Glattli DC, Bachtold A (2004) Determination of the intershell conductance in multiwalled carbon nanotubes. Phys Rev Lett 93(17):176806
55. Chai Y, Hazeghi A, Takei K, Chen H-Y, Chan PCH, Javey A, Wong HSP (2010) Graphitic interfacial layer to carbon nanotube for low electrical contact resistance. IEDM, San Francisco, pp 210–213
56. Lin A (2010) Carbon nanotube synthesis device fabrication, and circuit design for digital logic applications. PhD Thesis, Stanford University

57. Li H, Liu W, Cassell AM, Kreupl F, Banerjee K (2013) Low-resistivity long-length horizontal carbon nanotube bundles for interconnect applications—part II: characterization. IEEE Trans Electron Devices 60(9):2870
58. Tawfick S, O'Brien K, Hart AJ (2009) Flexible high-conductivity carbon-nanotube interconnects made by rolling and printing. Small 5(21):2467–2473
59. Kim YL, Li B, An X, Hahm MG, Chen L, Washington M, Ajayan PM, Nayak SK, Busnaina A, Kar S, Jung YJ (2009) Highly aligned scalable platinum-decorated single-wall carbon nanotube arrays for nanoscale electrical interconnects. ACS Nano 3:2818–2826
60. Dijon J, Ramos R, Fournier A, Le Poche H, Fournier H, Okuno H, Simonato JP (2014) Record resistivity of in-situ grown horizontal carbon nanotube interconnect. In: Technical proceedings of the 2014 NSTI nanotechnology conference and expo, NSTI-Nanotech 2014, vol 3, pp 17–20
61. Close GF, Wong H-SP (2008) Assembly and electrical characterization of multiwall carbon nanotube interconnects. IEEE Trans Nanotechnol 7(5):596–600
62. Pint CL, Xu Y-Q, Morosan E, Hauge RH (2009) Alignment dependence of one-dimensional electronic hopping transport observed in films of highly aligned, ultralong single-walled carbon nanotubes. Appl Phys Lett 94:182107
63. Behabtu N, Young CC, Tsentalovich DE, Kleinerman O et al (2013) Strong, light, multifunctional fibers of carbon nanotubes with ultrahigh conductivity. Science 339:182–185
64. Zhao Y, Wei J, Vajtai R, Ajayan PM, Barrera EV (2011) Iodine doped carbon nanotube cables exceeding specific electrical conductivity of metals. Sci Rep 83:1–5. doi:10.1038/srep00083
65. Wang JN et al (2014) High-strength carbon nanotube fibre-like ribbon with high ductility and high electrical conductivity. Nat Commun 5:3848. doi:10.1038/ncomms4848
66. Liu K, Sun Y, Zhou R, Zhu H, Wang J, Liu L, Fan S, Jiang K (2010) Carbon nanotube yarns with high tensiles strength made by a twisting and shrinking method. Nanotechnology 21:045708. doi:10.1088/0957-4484/21/4/045708
67. Wei J, Ci L, Jiang B, Li Y, Zhang X, Zhu H, Xua C, Wua D (2003) Preparation of highly pure double-walled carbon nanotubes. J Mater Chem 13:1340–1344
68. Bedewy M, Meshot ER, John Hart A (2012) Diameter-dependent kinetics of activation and deactivation in carbon nanotube population growth. Carbon 50:5106–5116
69. Collins PG, Hersam M, Arnold M, Martel R, Avouris P (2001) Current saturation and electrical breakdown in multiwalled carbon nanotubes. Phys Rev Lett 86(14):3128–3131
70. Wei BQ et al (2001) Reliability and current carrying capacity of carbon nanotubes. Appl Phys Lett 79:1172–1174
71. Yang Y, Murali R (2010) Impact of size effect on graphene nanoribbon transport. IEEE Electron Device Lett 31:237–239
72. Li S, Yu Z, Yen SF, Tang WC, Burke PJ (2004) Carbon nanotube transistor operation at 2.6 GHz. Nano Lett 4(4):753–756
73. Mann D, Javey A, Kong J, Wang Q, Dai H (2003) Ballistic transport in metallic nanotubes with reliable Pd ohmic contacts. Nano Lett 3(11):1541–1544
74. Ebbessen TW, Lezec HJ, Hiura H, Bennet JW, Ghaemi HF, Thio T (1996) Electrical conductivity of individual carbon nanotubes. Nature 382:54–56
75. Kreupl F, Graham AP, Duesberg GS, Steinhögl W, Liebau M, Unger E, Hönlein W (2002) Carbon nanotubes in interconnects applications. Microelectron Eng 64:399–408
76. Bachtold A, Fuhrer MS, Plyasunov S, Forero M, Anderson EH, Zettl A, McEuen PL (2000) Scanned probe microscopy of electronic transport in carbon nanotubes. Phys Rev Lett 84:6082
77. Naeemi A, Meindl JD (2006) Compact physical models for multiwall carbon-nanotube interconnects. IEEE Electron Device Lett 27(5):338–340
78. Rutherglen C, Jain D, Burke P (2009) Nanotube electronics for radiofrequency applications. Nat Nanotechnol 4:811
79. Ding L, Yuan D, Liu J (2008) Growth of high-density parallel arrays of long single-walled carbon nanotubes on quartz substrates. J Am Chem Soc 130:5428

80. Ibrahim I, Bachmatiuk A, Börrnert F, Blüher J, Zhang S, Wolff U, Büchner B, Cuniberti G, Rümmeli MH (2011) Optimizing substrate surface and catalyst conditions for high yield chemical vapor deposition grown epitaxially aligned single-walled carbon nanotubes. Carbon 49:5029

81. Zhou W, Ding L, Yang S, Liu J (2011) Synthesis of high-density, large-diameter, and aligned single-walled carbon nanotubes by multiple-cycle growth methods. ACS Nano 5(5): 3849–3857

82. Hu Y et al (2015) Growth of high-density horizontally aligned SWNT arrays using Trojan catalysts. Nat Commun 6:6099. doi:10.1038/ncomms7099

83. Hou P-X, Li W-S, Zhao S-Y, Li G-X, Shi C, Liu C, Cheng H-M (2014) Preparation of metallic single-wall carbon nanotubes by selective etching. ACS Nano 8(7):7156–7162

84. Wang Y, Liu Y, Li X, Cao L, Wei D, Zhang H, Shi D, Yu G, Kajiura H, Li Y (2007) Direct enrichment of metallic single-walled carbon nanotubes induced by the different molecular composition of monohydroxy alcohol homologues. Small 3:1486

85. Patil N, Lin A, Myers ER, Ryu K, Badmeav A, Zhu C, Wong H-SP, Mitra S (2009) Wafer-scale growth and transfer of aligned single-walled carbon nanotubes. IEEE Trans Nanotechnol 8(4):498–504

86. Choi WJ, Chung YJ, Kim YH, Han J, Lee Y-K, Kong K, Chang H, Lee YK, Kim BG, Lee J-O (2014) Drawing circuits with carbon nanotubes: scratch-induced graphoepitaxial growth of carbon nanotubes on amorphous silicon oxide substrates. Sci Report 4:5289. doi:10.1038/srep05289

87. Yu Q, Qin G, Li H et al (2006) Mechanism of horizontally aligned growth of single-wall carbon nanotubes on R-plane sapphire. J Phys Chem B 110:22676–22680

88. Huang S, Woodson M, Smalley R, Liu J (2004) Growth mechanism of oriented long single walled carbon nanotubes using fast-heating chemical vapor deposition process. Nano Lett 4(6):1025–1028

89. Terrones M, Ajayan PM et al (2002) N-doping and coalescence of carbon nanotubes: synthesis and electronic properties. Appl Phys A 74:355–361

90. Dresselhaus MS, Dresselhaus G (2002) Intercalation compounds of graphite. Adv Phys 51(1):1–186. doi:10.1080/00018730110113644

91. Esconjauregui S, D'Arsie L, Guo Y, Yang J, Sugime H, Caneva S, Cepek C, Robertson J (2015) Efficient transfer doping of carbon nanotube forests by MoO3. ACS Nano 9(10):10422–10430

Chapter 7
Carbon Nanotubes as Vertical Interconnects for 3D Integrated Circuits

Sten Vollebregt and Ryoichi Ishihara

7.1 Introduction

The continued downscaling of transistor dimensions, as described by Moore's law, has been the driving force of the semiconductor industry for many decades. Each new generation of transistors is roughly 1.4 times smaller, which doubles the number of transistors which can be put in the same area. While downscaling the transistors generally improves delay, power consumption, and most importantly lowers the price per transistor, the same is not true for the downscaling of the interconnects. Already in the mid-1990s it was predicted that interconnect RC-delay, the time it takes to charge up a line with a resistance R and capacitance C, would become a serious performance issue. Indeed, it surpassed gate delay at the end of the decade, as shown in Fig. 7.1. This forced the industry to shift to Cu as interconnect material, lowering R. Furthermore, the so-called low-k dielectrics (where k is the dielectric constant) were introduced instead of SiO_2 for the electrical and mechanical separation of the individual lines [1, 2]. However, in 1995 it was already predicted that this change of materials would "reach practical limits in just a few generations" [1].

Indeed, current interconnect technology suffers from many issues although the practical limits have not yet been reached due to much engineering effort. The further reduction in wire dimensions made grain and surface boundary scattering an important factor, which increases the resistivity of the Cu line above that of bulk resistivity value with a factor of 5–10 [3]. The line resistance is even further increased due to the poor scaling of the relative high resistive barrier layers that are required to prevent Cu diffusion. Furthermore, the pursue of materials with an even lower dielectric constant resulted in many mechanical stability issues in

S. Vollebregt • R. Ishihara (✉)
Department of Microelectronics, Delft University of Technology, Delft, The Netherlands
e-mail: s.vollebregt@tudelft.nl; r.ishihara@tudelft.nl

© Springer International Publishing Switzerland 2017
A. Todri-Sanial et al. (eds.), *Carbon Nanotubes for Interconnects*,
DOI 10.1007/978-3-319-29746-0_7

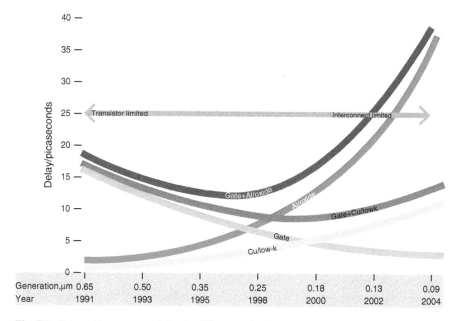

Fig. 7.1 Gate and interconnect delay for different technology nodes, data from Sun [2]

the resulting interconnects. Besides the electrical performance expressed by the interconnect delay and power dissipation, the electrical reliability is also becoming an issue. A decade ago the ITRS predicted that by now the current density in metal lines would have surpassed the maximum current density (J_{max}) of that of bulk Cu for high-end VLSI (very-large scale integration) circuits. A shift from aggressive clock-scaling to multi-core designs alleviated this issue, but still the most recent 2013 update of the roadmap predicts problems 8 years from now [4].

One way of reducing the RC-delay, and power consumption, of interconnects is simply by reducing their total length. In planar VLSI technology this is rather difficult to achieve, as the chip size remains large due to an increase in the number of transistors, while the transistors become smaller. However, a radical change in interconnect length could be obtained by stacking layers of transistors on top of each other and using many vertical interconnects (vias) to connect these individual transistors and the horizontal interconnects among them, a process called 3D integration which is illustrated by Fig. 7.2a, b [5]. Due to the added dimension for the wiring, the total interconnect length can be significantly reduced. Two approaches exist for 3D integration: chip or wafer stacking (Fig. 7.2b), and the direct fabrication of multiple transistor layers on a single wafer (Fig. 7.2d), the latter being dubbed monolithic integration [5–7].

Chip stacking is a technique which recently became mature enough for mass production, and is used in, for instance, high band-with memories. For this technique (thinned) chips are stacked on top of each other after testing. State-of-the-art chip stacking uses through-silicon vias (TSV) to connect the different device layers,

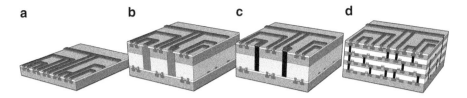

Fig. 7.2 Overview of current and future interconnect technology: (**a**) traditional planar technology, all metal layers are routed on top of the transistors; (**b**) current 3D IC technology using Cu TSV to connect chip or wafer stacked layers of transistors; (**c**) 3D IC TSV technology with CNT TSV; and (**d**) monolithic 3D IC with high-aspect ratio CNT vias

see also Fig. 7.2b. These vias run all the way through the active layer and the Si substrate, connecting the front-side with the back-side. As the maximum achievable aspect ratio for Cu plating is limited, the TSV take up considerable space, which otherwise would have been available for transistors. Moreover, this limits the density of vias, expressed as the number of vias per unit area. Increasing the via aspect ratio from 5–10 to 10–20 will reduce the wasted area, but in turn will give problems with Cu filling. Finally, the use of Cu for TSV introduces all kinds of reliability problems due to Cu diffusion into the transistor area, and the mismatch between the thermal expansion coefficients of Si and Cu.

The highest via density and performance can be obtained by using monolithic integration, in which the layers of transistors are just a few μm apart separated by a dielectric layer, as shown in Fig. 7.2d. Fabricating vias with a sufficiently high-aspect ratio (HAR) to achieve the high via density for monolithic integration, expected to be up to a billion vias per cm^2 [7], will be one of the challenges for this technology. Cu is likely unable to obtain high enough aspect ratios: already in planar IC technology it is reaching its practical limits. Besides that, contamination issues will become even more severe as the vias have to be placed much closer to the transistors.

Another issue related to 3D integration is thermal management. Current VLSI chips already operate many tens of degrees above room temperature, with power densities being reached of 100 W/cm^2. Stacking multiple layers of transistors on top of each other will increase the amount of transistors per unit area beyond that of traditional scaling, further increasing the power density. The heat of the transistors has to be transported efficiently to the surface, through all the interconnect layers. As the dielectrics surrounding the interconnects normally have a thermal conductivity of just a few W/mK, most heat will flow through the interconnect lines.

Much effort has been put in finding alternatives for Cu as interconnect material. One material which gained particular interest are carbon nanotubes (CNT), and especially for the application in vias as they can easily be fabricated vertically [8, 9]. CNT are an allotrope of sp^2 hybridized carbon, and can be visualized as rolled up sheets of graphene (in which graphene is a single atom thick layer of graphite). Nanotubes consisting of one sheet of graphene are called single-walled CNT (SWCNT), and tubes consisting of multiple sheets are dubbed multi-walled CNT (MWCNT).

One of the biggest advantages of CNT as interconnect material is their current carrying capacity due to the strong sp^2 bonds, which has been demonstrated to be in the order of 10^9 A/cm^2, even at elevated temperature [10]. In comparison, the maximum current density of Cu is in the order of 10^6 A/cm^2. Besides being able to withstand a large current density, CNT are in general good electrical conductors. Simulations have shown that potentially CNT are able to outperform Cu as interconnect material in terms of electrical resistance [11, 12].

Besides their electrical properties, CNT have two more characteristics which make them attractive as future via material. One of them is the high thermal conductivity, up to 3500 W/mK as was shown by measurements [13], which is almost nine times higher than that of Cu. This can aid in the anticipated thermal management issues of 3D IC. Finally, it has been demonstrated that CNT can be fabricated at HARs [14], which is caused by their bottom-up nature of fabrication. These features make CNT very attractive for use as CNT TSV, as shown in Fig. 7.2c, and future monolithic 3D IC as displayed in Fig. 7.2d.

In this chapter the application of CNT for vertical interconnects, and ultimately for the use as vias in 3D IC, is discussed. We start by discussing the requirements which are necessary for the utilization of CNT in 3D IC in Sect. 7.2. This is followed by a review of the research performed on the application of CNT as via material in VLSI technology in Sect. 7.3. In Sects. 7.4 and 7.5 we discuss results obtained on using CNT for TSV and in monolithic 3D IC, respectively. We conclude with the prospect of the use of CNT in vias in 3D IC.

7.2 Requirements for CNT Integration

In the previous section the potential advantages of CNT were discussed. To introduce a new material into existing semiconductor technology several boundary conditions will have to be met in order to successfully integrate the material, and make it an attractive candidate to replace current materials. Needless to say the performance of the new technology should be at least equal to that of the upcoming technology nodes with similar costs, but besides that some more practical limitations have to be kept in mind.

For the use as vertical interconnects in VLSI technology, or TSV, integration in the so-called back-end of line (BEOL) of semiconductor fabrication is required. The BEOL is basically everything which has to be done after the transistors have been fabricated and before wafer dicing, and consists mostly of the fabrication of the different interconnect levels and integrated components like capacitors.

As the processing in the BEOL takes place after the active devices have been finished, the material restrictions are less constraining. For instance, metals like Cu can be used for the interconnects. However, due to the use of advanced low-k dielectrics which are sensitive to temperature, and the thermal expansion coefficient mismatch between the many materials used, the maximum allowed temperature for processes in the BEOL is only 350–400 °C.

For the CNT deposition (or growth), which is performed by using chemical vapour deposition from a carbon feedstock on metal nanoparticles [15], this implies that most of the popular catalyst metals like Ni and Co can be used. Unfortunately, the catalyst which is used in perhaps the majority of publication in the literature, Fe, is regarded as a contaminant in most BEOL processes [16].

Furthermore, as will be seen in the next section, growing CNT at temperatures lower than 400 °C is not trivial. Generally, a lower deposition temperature results in a lower crystallinity of the CNT as was, for instance, shown by Raman spectroscopy [17]. As the electrical and thermal transport properties of the CNT depend on their crystallinity, this effect is unwanted and should be counteracted by optimizing the deposition conditions.

For monolithic 3D IC the frond-end and BEOL will be integrated with each other, as interconnect routing will be required between the different device layers. It is to be expected that it will put strict requirements on the materials which can be used in order to not introduce unwanted effects in the active devices. The advantage of Ni and Co is that they are already used as source/drain silicides in transistors in order to form low ohmic contacts [18], and should thus be compatible with 3D monolithic ICs. The Fe catalyst does not form a stable silicide, and is therefore likely not allowed in 3D IC processes.

7.3 CNT for Vertical Interconnects

CNT are relative straightforward to grow vertically. They either self-align by the crowding effect due to van der Waals interaction from other tubes [19] or can be aligned using an electric field generated by a plasma [20]. There are two major fabrication procedures for creating CNT vias, either by the traditional top-down approach or by using bottom-up integration [21], see also Fig. 7.3.

The top-down process closely resembles the procedure currently used to create metal vias. In this approach a contact opening is etched in the dielectric layer between the interconnect levels and subsequently filled with CNT followed by the optional filling of the holes with some kind of filler material (often a dielectric) combined with chemical mechanical polishing (CMP). Finally, the deposition of the next metal layer is performed.

In case of the bottom-up method CNT are first grown in vertical bundles, followed by dielectric deposition and bundle encapsulation, CMP to make the CNT resurface, and subsequently the top metallization. Distinct advantages of the bottom-up approach are that no small openings have to be etched, and that CNT do not have to be grown inside these small openings.

Due to their potential, it should come as no surprise that many different groups have been working on vias consisting of CNT. Already in 2002 Kreupl et al. [8, 22] from Infineon Technologies displayed interest in CNT as interconnect and demonstrated some first integration results. Being the first, their growth temperature of 700 °C and electrical resistance (see Fig. 7.6) were both high. Not much later

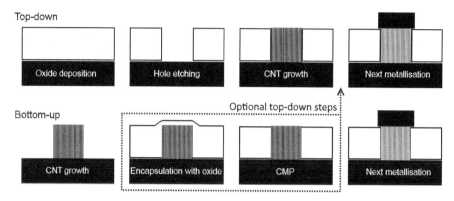

Fig. 7.3 Schematic overview of the top-down and bottom-up integration approaches, the *dashed box* represents optional steps which can be inserted into the top-down integration flow

Fujitsu Ltd. and related groups started working on fabricating CNT vias using the top-down approach [23–25]. Besides Infineon and Fujitsu also Samsung displayed interest in CNT vias [26, 27], but again the reported resistance and growth temperature (600 °C) are too high.

The research of Fujitsu (later continued as MIRAI-Selete) resulted in many break-throughs towards the high density low temperature integration of CNT vias with a low electrical resistance. For instance, they identified ways of decreasing the contact resistance by using Ti or TiN as contact material, which is thought to form a low resistance TiC layer [28]. The use of CMP was found to improve the electrical contact by removing the tips of the CNT and therefore allowing conduction through multiple MWCNT shells instead of just the outer shell [29–31].

In order to further increase the CNT bundle density and lower the resistivity, a technique was developed to directly deposit metal nanoparticles on top of the substrate. This resulted in a higher nanoparticle density than what was achieved before by using metal thin-films [32]. Later ballistic transport for 60 nm high vias was demonstrated [33]. In 2008 the current record holding CNT via in terms of electrical resistivity was fabricated, at a growth temperature of just 390 °C [34]. Still, its resistivity of 0.69 mΩ-cm is two orders of magnitude higher than that of bulk Cu. Unfortunately, some of the equipments MIRAI-Selete employed for catalyst deposition and growth are non-standard in semiconductor technology and can have potential scaling issues to full-wafer sizes.

Besides industry, several research groups at universities have also investigated the use of CNT for via application. The group of Robertson at Cambridge University investigated methods to obtain ultra high density growth (expressed in tubes/cm^2) [35, 36] and growth on $CoSi_2$ (an often used material for source/drain contacts in transistors) [37]. Unfortunately, most high density growth is obtained on non-conductive Al_2O_3 layers and with a growth temperature of 650 °C. Although electric tunnelling through a thin Al_2O_3 layer to the CNT was demonstrated [38]. Recently a reduction of growth temperature to 450 °C was demonstrated, but no electrical measurements were shown [39]. A cooperation between Cambridge and CEA-LETI resulted in the demonstration of vias with an ultra high density (2.5×10^{12} tubes/cm^2) grown at 590 °C [40]. Their measured resistance of 10 kΩ was very high, though, likely due to bad electrical contacts to the CNT.

KU Leuven/Imec investigated the use of CNT grown at 470 °C as vias in future sub-32 nm nodes on 8 in. wafers and provided full electrical characterization. While the resulting resistivity was initially still high, [41–43] by improving the integration flow a steady decline in the resistivity was obtained. Recently they have demonstrated growth at temperatures as low as 400 °C [44], and a 540 °C process with improved electrical contacts which resulted in a resistivity of 5 mΩ-cm [45, 46]. Very importantly, they were also the first to demonstrate integration on industrial wafer-scale and to provide statistics of their CNT vias.

Finally, in our group at Delft University of Technology we studied the fabrication of CNT test vias on 4 in. wafers using both the top-down and bottom-up approach [47]. Just like Fujitsu and Imec, TiN was used as support layer for the catalyst, as it is a well-known electrically conductive barrier layer in semiconductor fabrication. We did find, however, that this layer is very sensitive to plasma damage, which can even prevent vertically aligned CNT growth at relative low temperatures [48]. We observed that for our samples the contact resistance between the CNT and the metal contacts was significantly lower than the resistance of the CNT themselves, which was attributed to the embedding of the CNT tips in the contact metal as shown in the SEM cross-section in Fig. 7.4 [49]. In an attempt to lower the growth temperature

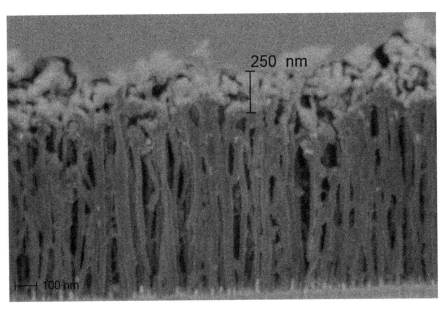

Fig. 7.4 Close up of the CNT tips as imaged by the back-scatter electron detector in the dual-beam FIB/SEM, displaying a clear contrast between Ti and CNT. Image courtesy of A.N. Chiaramonti from NIST, Boulder, CO

and improve material compatibility we investigated Co and CoAl as catalyst. It was found that with these catalysts it was possible to grow at record-low temperatures of 350 °C. The cause for this is the 0.1 eV lower activation energy of the Co catalyst on the TiN layer, compared to that of Fe on TiN [50].

Another aspect of CNT vias which was investigated by our group is the temperature coefficient of resistance (TCR) of CNT vias. As ICs are generally operated well above room temperature the resistance of the metal interconnects increases, which subsequently raises the RC-delay. For CNT it has been suggested that their TCR is negative [51]. For fabricated CNT vias, however, Yokoyama et al. [34] found a positive TCR. In our systematic studies of CNT vias fabricated using either Fe or Co at temperatures ranging from 350 to 500 °C, we did find a negative TCR of −300 to 400 ppm/K for Fe-grown CNT and −800 pmm/K for Co-grown CNT [50, 52], confirming the expected negative TCR behaviour. We did observe an interesting dependency of the TCR on CNT bundle length and width, as is shown in Fig. 7.5. The exact cause of this is still unclear, but might involve non-neglectable positive TCR contributions of the contact resistance.

The reported electrical measurements expressed in electrical resistivity from the groups mentioned above are shown in Fig. 7.6. Only results obtained from fully integrated vias are considered (i.e. not from the direct probing of the bundle with contact needles) in order to allow for a realistic use of the CNT bundles. A large

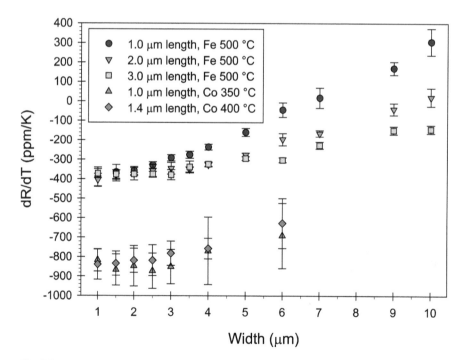

Fig. 7.5 Thermal coefficient of resistance values for the measured temperature range in ppm/K, plotted against via width and for different lengths

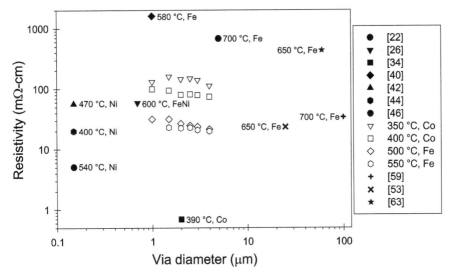

Fig. 7.6 Overview of the reported electrical performance in terms of complete via resistivity of fully integrated CNT vias and TSV obtained from the literature, versus the diameter of the reported via. The *open symbols* are data taken from Vollebregt et al. [50]

spread of values is observed, mainly due to the differences in quality of the CNT material, and the sometimes very high contact resistance. Overall, all values are considerably higher than that of bulk Cu which has a resistivity of $1.68\,\mu\Omega$-cm. This indicates that the CNT deposition conditions and integration flow should be studied in more detail in order to reduce the resistivity by improving the quality and contact resistance in order to enable the use of CNT in the BEOL.

7.4 Carbon Nanotube TSV

Current TSV technology uses electroplated Cu as metal filler, in combination with deep reactive ion etching (DRIE) in order to fabricate the through-silicon holes. In a conventional VLSI process the TSV are fabricated in the BEOL. This means that, in order to reach the back-side of the wafer or die, the holes have to be etched through the layer containing the active devices. Due to the limited aspect ratio of DRIE, and to ensure void-free Cu electroplating, a significant amount of device area is lost to reserve space for these Cu TSV. This issue is partly resolved by wafer thinning to allow for smaller via widths; however, this in turn results in fragile wafers which complicates handling. Another issue with Cu as filler metal are the mismatches in the thermal expansion coefficient with Si and the dielectrics, which gives rise to stress. This introduces additional failure mechanisms in the back-end metal stack and the stress influences the performance of nearby transistors, thus requiring a large exclusion zone in which no devices are placed.

CNT are an attractive candidate for replacing Cu for TSV. Due to their bottom-up nature in fabrication, much higher aspect ratios are envisioned which are only limited by the DRIE [53]. Moreover, CNT bundles have been shown to have sponge-like mechanical behaviour [54], and have a low thermal expansion coefficient [55]. Both of these should reduce mechanical failure mechanisms. Finally, according to recent theoretical models [56, 57], the electrical performance of CNT is comparable to that of Cu, while the thermal conductivity of CNT is even better than that of Cu.

Only a few integration results using CNT TSV have been demonstrated in the literature so far. The first attempt was performed by Xu et al. [58] on SiO_2 support layers. Although growth in deep silicon holes was demonstrated, no complete vias were fabricated. The first complete TSV were demonstrated by Wang et al. [59, 60], this time on Al_2O_3 support layers. The growth temperature used ($700\,°C$) and resistance are still too high for practical applications, though. Recently, also the group of Robertson published about CNT grown on Al_2O_3 at $650\,°C$ for TSV [53].

What all these publications have in common is that as support layer a non-conductive layer is used. While it has been shown that it is possible to form an electrical contact through $<2\,nm$ Al_2O_3 layers [38], this will result in a relative high ohmic contact to the CNT, as tunnelling will be the electrical transport mechanism. Although growth on metal support layers like TiN [48, 50] or $CoSi_2$ [37] has been demonstrated, to the best of our knowledge, no CNT growth over $100\,\mu m$ on these two conductive layers has been reported. Generally the growth stops after several tens of μm. The cause of this is either catalyst poisoning or the encapsulation of the catalyst by the CNT growth front, which in turn stops the catalytic reaction [61]. This is in contrast to Al_2O_3, on which lengths over several hundred μm have been reported.

Recently our group investigated ZrN as support layer [62, 63], which is commonly used as hard protective anti-corrosive coating but is hardly found in semiconductor fabrication. From a material perspective, Zr can be considered clean-room compatible. The advantage of this layer is that it is electrically conductive, with a minimum resistivity of $12\,\mu\Omega$-cm, although the layers used by our group have a much higher resistivity of $5.5\,m\Omega$-cm [63].

We found that this layer allowed similar CNT heights as could be obtained with Al_2O_3 as is shown in Fig. 7.7. It is suspected that the low surface energy of this layer is one of the reasons that CNT growth is made possible. As was shown by Zhang et al. [37] the difference of surface energy between the support layer and the metal catalyst is important in the formation of nanoparticles from which the CNT can nucleate. The surface energy of ZrN is comparable to that of other support layers which allows the growth of CNT: TiN, Al_2O_3, and SiO_2 [63]. Other conductive metal-oxides or nitrides which were tested (specifically MoO_3 and MoN) did not result in CNT growth and were found to have a much higher free surface energy. The exact reason why ZrN allows for longer CNT than TiN is unclear, but this could be related to the growth method of the catalyst: ZrN and Al_3O_3 result in base growth, while TiN enables tip growth. Using ZrN we were the first to fabricate CNT TSV which were contacted on both sides with metal thin-films [63].

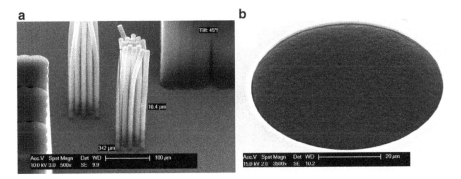

Fig. 7.7 SEM images of CNT grown on ZrN using 0.8 nm of Fe as catalyst at 650 °C: (**a**) high-aspect ratio bundles and (**b**) 60 μm wide CNT bundle grown inside a 300 μm deep through-silicon hole

The resistivities obtained from the few works in which the CNT bundles are electrically measured were added to Fig. 7.6. It is clear that the resistivities are currently still too high. However, research on CNT TSV has only recently started to attract attention. If CNT TSV are first introduced in interposer technology (the so-called 2.5D), the required minimum growth temperatures can be significantly higher, likely allowing growth temperatures up to 600–650 °C if the TSV are created before the rest of the metal interconnects. This would enable the use of CNT as TSV at an earlier stage.

7.5   Towards the Integration of CNT with Monolithic 3D IC

While the work cited in Sect. 7.3 provides a clear overview of the current state of CNT for via applications in VLSI and the remaining challenges, all of the results have a singular thing in common: the CNT are fabricated purely as test vias without any actual front-end electronics being present. In current VLSI technology the front-end and BEOL are separated, which implies that if CNT can meet all BEOL requirements, integration into the whole process cycle should be relatively straightforward. For monolithic 3D IC, on the other hand, no such clear division is going to be present. This implies that CNT vias will have to be fabricated directly alongside active devices, which poses additional challenges.

The only examples of CNT being integrated alongside electronics are by pick and place techniques after the CNT have been harvested from an external source and purified [64, 65], or by growing them locally on an IC by using on-chip heaters [66]. However, such techniques are regarded as cumbersome and inaccurate, resulting in bad industrial scalability [67]. Our group was the first to investigate the direct growth of CNT vias alongside active devices. We did this by combining a process which can be used to fabricate multiple layers of transistors with CNT vias grown using wafer-scale chemical vapour deposition [68].

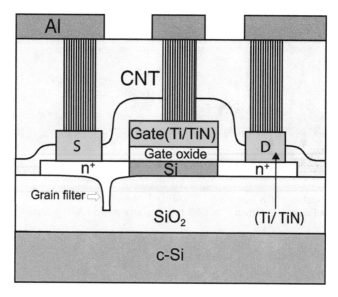

Fig. 7.8 Schematic cross-section of a transistor with CNT vias to source (S), drain (D), and gate

The process which was used to fabricate the transistors is the so-called μ-Czochralski process, which uses an excimer laser to locally melt an a-Si layer in order to form location controlled single-grain crystals in which high performance transistors can be fabricated. This technique was previously used by our group to fabricate various monolithic 3D ICs, including simple logic circuits [69, 70]. By combining this established process with our process flow for fabricating CNT test vias with Fe [48], a single layer of high mobility transistors with CNT vias directly grown on the source, drain, and gate of the transistors, as sketched in Fig. 7.8, could be fabricated [68].

We studied the impact of the μ-Czochralski process on the performance of the CNT vias, and vice versa the impact of the CNT vias and their growth process on the transistor performance. Using CNT via test structures available on the wafer alongside the single layer of transistors, the resistivity of the CNT vias was determined as is shown in Fig. 7.9. As can be seen the resistivity of the vias fabricated alongside transistors is higher, while the fabrication process for the CNT vias was kept constant. The cause for this seems to be the TiN layer, which is being affected by the excimer laser dopant activation process. This in turn has a negative impact on the density and alignment of the CNT bundles.

On the other hand, the transistors with CNT vias to the source, drain, and gate demonstrated equal characteristics as were obtained before with traditional metallization. This is shown by, for example, the I_D–V_G characteristics displayed in Fig. 7.10. All performance indicators like the mobility, on/off ratio, and subthreshold slope (SS) were found to be similar to that of transistors obtained before with Al contacts [70]. This indicates that for the 3D IC process employed by our group, the

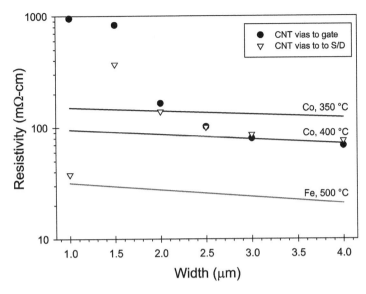

Fig. 7.9 Resistivities obtained from CNT vias integrated alongside the transistors, compared to values obtained before from test vias without transistors

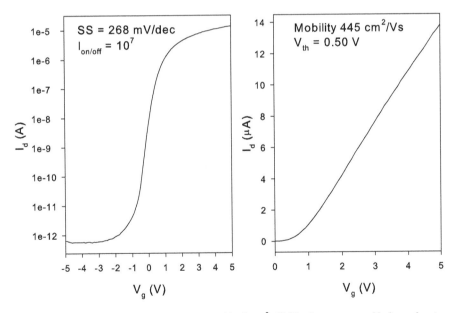

Fig. 7.10 I_D–V_G of an NMOS transistor with $1 \, \mu m^2$ CNT vias to source/drain and gate, demonstrating the high transistor performance. $V_D = 0.1$ V

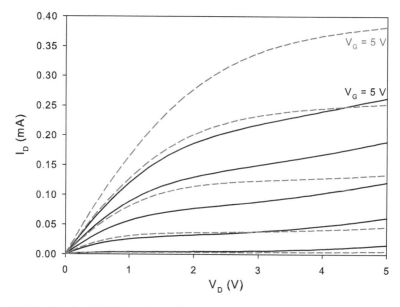

Fig. 7.11 I_D–V_D of an NMOS transistor with 1 μm (*solid line*) and 2 μm (*dashed line*) wide CNT vias, V_G ranges from 1 to 5 V in steps of 1 V. The mobilities and threshold voltages of both devices were similar. The impact of the CNT via resistance on the maximum on-current can clearly be observed

short CNT growth process at 500 °C has no significant impact on the performance of the transistors. We did find that the high CNT resistivity limits the maximum on-current of the transistors, as illustrated by Fig. 7.11 for two transistors with different via sizes but equal mobility and threshold voltage. This is not surprising, as the resistivity is still four orders in magnitude higher than that of Cu.

As Fe is not regarded as a cleanroom compatible material we recently investigated the use of Co as alternative catalyst in the 3D IC process, and fabricated two layers of transistors, 3D CMOS inverters, and 3D memory cells with CNT vias [71]. The combination of Co and TiN as support layers was found to be insensitive to the excimer laser process steps, resulting in CNT vias with an equal resistivity as that of test vias obtained before.

7.6 Conclusion and Future Prospects

Nowadays interconnects are regarded as the most performance limiting part of VLSI. 3D integration is suggested as the most promising method to alleviate this issue, by reducing the interconnect length as an additional dimension for routing is introduced. For this technology it is required to fabricate reliable, low resistance, and HAR vias. Two types of vias can be distinguished: TSV for chip and wafer-scale integration, and high-aspect ratio vias for monolithic integration. For both type of vias CNT are an interesting candidate to replace the current Cu interconnect technology because of their excellent electrical, thermal, and mechanical properties and bottom-up fabrication technique.

For successful integration of CNT into semiconductor technology the growth temperature of the CNT CVD growth process and the used materials, especially the transition metal catalyst, will have to be compatible with the limitations set by the BEOL. For monolithic integration these limitations will be even more strict, as the division between the front-end and back-end will fade. Specifically this implies that the growth temperature should be 350–400 °C, and Fe should be avoided as catalyst.

Much research effort has been made to demonstrate the potential of CNT as vias in standard VLSI technology, from both groups in industry and academia. Import break-through, like the lowering of the electrical contact resistance and deposition temperature, have been achieved and full-wafer integration has been demonstrated. Unfortunately, due to the temperature restrictions the minimum reported resistivity of 0.69 mΩ-cm is still more than two orders in magnitude higher than that of bulk Cu.

The use of CNT for TSV and monolithic 3D IC has been much less investigated, partly because 3D integration is a relative new concept. The same issue exists as for regular CNT vias: the resistivity is currently too high. For CNT TSV most published research has been obtained from CNT growth on non-conductive layers, but recently long CNT could be grown on electrically conductive ZrN layers which should enable lower contact resistances and easier integration. Finally, for monolithic 3D integration it was demonstrated that CNT can be directly grown on transistors, and that the CNT process does not damage the pre-fabricated transistors.

CNT remain a high potential replacement for current metallization schemes, although many technical challenges still have to be overcome. Specifically the crystallinity at low growth temperature should be improved by optimizing the CVD process. Furthermore, the contact resistance should be further reduced, for which embedding of the CNT tips appears to be a promising route. For TSV it is crucial that growth on conductive substrates is employed, as alternative new electrically conductive support layers like ZrN should be studied. Especially in future technology nodes, where the resistivity of metals increases sharply while reliability goes down, and for 3D application where a HAR is required, CNT can potentially have a big impact.

References

1. Bohr MT (1995) Interconnect scaling - the real limiter to high performance ULSI. In: IEEE international electron devices meeting, pp 241–244
2. Sun S (1997) Process technologies for advanced metallization and interconnect systems. In: IEEE international electron devices meeting, pp 765–768
3. Rossnagel SM, Wisnieff R, Edelstein D, Kuan TS (2005) Interconnect issues post 45nm. In: IEEE international electron devices meeting, pp 89–91

4. ITRS (2013) International technology roadmap for semiconductors. http://www.itrs2.net/
5. Koyanagi M, Kurino H, Lee KW, Sakuma K, Miyakawa N, Itani H (1998) Future system-on-silicon LSI chips. IEEE Micro 18(4):17
6. Chan VWC, Chan PCH, Chan V (2000) Three dimensional CMOS integrated circuits on large grain polysilicon films. In: IEEE international electron devices meeting, pp 161–164
7. Topol AW, La Tulipe JDC, Shi L, Frank DJ, Bernstein K, Steen SE, Kumar A, Singco GU, Young AM, Guarini KW, Ieong M (2006) Three-dimensional integrated circuits. IBM J Res Dev 50(4/5):491
8. Kreupl F, Graham AP, Lieba M, Duesber GS, Seide R, Unger E (2004) Carbon nanotubes for interconnect applications. In: IEEE international electron devices meeting, pp 683–686
9. Robertson J (2007) Growth of nanotubes for electronics. Mater Today 10(1–2):36
10. Wei BQ, Vajtai R, Ajayan PM (2001) Reliability and current carrying capacity of carbon nanotubes. Appl Phys Lett 79(8):1172
11. Naeemi A, Meindl JD (2008) Performance modeling for single- and multiwall carbon nanotubes as signal and power interconnects in gigascale systems. IEEE Trans Electron Devices 55(10):2574
12. Li H, Srivastava N, Mao JF, Yin WY, Banerjee K (2011) Carbon nanotube vias: does ballistic electron-phonon transport imply improved performance and reliability. IEEE Trans Nanotechnol 58(8):2689
13. Pop E, Mann D, Wang Q, Goodson K, Dai H (2006) Thermal conductance of an individual single-wall carbon nanotube above room temperature. Nano Lett 6(1):96
14. Hutchison DN, Morrill NB, Aten Q, Turner BW, Jensen BD, Howell LL, Vanfleet RR, Davis RC (2010) Carbon nanotubes as a framework for high-aspect-ratio MEMS cabrication. J Microelectromech Syst 19(1):75
15. Dai H (2001) Nanotube growth and characterization. In: Carbon nanotubes. Topics in applied physics, vol 80. Springer, Berlin, pp 29–53
16. Istratov A, Hieslmair H, Weber ER (2000) Iron contamination in silicon technology. Appl Phys A 70:489
17. Vollebregt S, Ishihara R, Tichelaar FD, Hou Y, Beenakker CIM (2012) Influence of the growth temperature on the first and second-order Raman band ratios and widths of carbon nanotubes and fibers. Carbon 50(10):3542
18. Kikkawa T, Inoue K, Imai K (2004) Self-Oriented Regular Arrays of Carbon Nanotubes and Their Field Emission Properties Cobalt silicide technology. In: Silicide technology for integrated circuits. The Institution of Engineering and Technology, London, pp 77–94
19. Fan S, Chapline MG, Franklin NR, Tombler TW, Cassell AM, Dai H (1999) Self-oriented regular arrays of carbon nanotubes and their field emission properties. Science 283(5401):512
20. Bower C, Zhu W, Jin S, Zhou O (2000) Plasma-induced alignment of carbon nanotubes. Appl Phys Lett 77(6):830
21. Li J, Ye Q, Cassell A, Ng HT, Stevens R, Han J, Meyyappan M (2003) Bottom-up approach for carbon nanotube interconnects. Appl Phys Lett 82(15):2491
22. Kreupl F, Graham AP, Duesberg GS, Steinhögl W, Liebau M, Unger E, Hönlein W (2002) Carbon nanotubes in interconnect applications. Microelectron Eng 64(1–4):399
23. Nihei M, Kawabata A, Awano Y (2003) Direct diameter-controlled growth of multiwall carbon nanotubes on nickel-silicide layer. Jpn J Appl Phys 42(6B):L721
24. Katagiri M, Yamazaki Y, Wada M, Kitamura M, Sakuma N, Suzuki M, Sato S, Nihei M, Kajita A, Sakai T, Awano Y (2011) Improvement in electrical properties of carbon nanotube via interconnects. Jpn J Appl Phys 50:05EF01
25. Katagiri M, Wada M, an Yuichi Yamazaki BI, Suzuki M, Kitamura M, Saito T, Isobayashi A, Sakata A, Sakuma N, Kajita A, Sakai T (2012) Fabrication and characterization of planarized carbon nanotube via interconnects. Jpn J Appl Phys 51:05ED02
26. Choi YM, Lee S, Yoon HS, Lee MS, Kim H, Han I, Son Y, Yeo IS, Chung UI, Moon JT (2006) Integration and electrical properties of carbon nanotube array for interconnect applications. In: Sixth IEEE conference on nanotechnology, pp 262–265

27. Lee S, Moon S, Yoon HS, Wang X, Kim DW, Yeo IS, Chung UI, Moon JT, Chung J (2008) Selective growth of carbon nanotube for via interconnects by oxidation and selective reduction of catalyst. Appl Phys Lett 93(18):182106

28. Nihei M, Horibe M, Kawabata A, Awano Y (2004) Carbon nanotube vias for future LSI interconnects. In: IEEE international interconnect technology conference, pp 251–253

29. Horibe M, Nihei M, Kondo D, Kawabata A, Awano Y (2004) Mechanical Polishing Technique for Carbon Nanotube Interconnects in ULSIs. Jpn J Appl Phys 43(9A):6499

30. Nihei M, Kondo D, Kawabata A, Sato S, Shioya H, Sakaue M, Iwai T, Ohfuti M, Awano Y (2005) Low-resistance multi-walled carbon nanotube vias with parallel channel conduction of inner shells. In: IEEE international interconnect technology conference, pp 234–236

31. Yokoyama D, Iwasaki T, Yoshida T, Kawarada H, Sato S, Hyakushima T, Nihei M, Awano Y (2007) Low temperature grown carbon nanotube interconnects using inner shells by chemical mechanical polishing. Appl Phys Lett 91(26):263101

32. Sato S, Nihei M, Mimura A, Kawabata A, Kondo D, Shioya H, Iwai T, Mishima M, Ohfuti M, Awano Y (2006) Novel approach to fabricating carbon nanotube via interconnects using size-controlled catalyst nanoparticles. In: IEEE international interconnect technology conference, pp 230–232

33. Nihei M, Hyakushima T, Sato S, Nozue T, Norimatsu M, Mishima M, Murakami T, Kondo D, Kawabata A, Ohfuti M, Awano Y (2007) Electrical properties of carbon nanotube via interconnects fabricated by novel damascene process. In: IEEE international interconnect technology conference, pp 204–206

34. Yokoyama D, Iwasaki T, Ishimaru K, Sato S, Hyakushima T, Nihei M, Awano Y, Kawarada H (2008) Electrical properties of carbon nanotubes grown at a low temperature for use as interconnects. Jpn J Appl Phys 47(4):1985

35. Esconjauregui S, Fouquet M, Bayer BC, Ducati C, Smajda R, Hofmann S, Robertson J (2010) Growth of ultrahigh density vertically aligned carbon nanotube forests for interconnects. ACS Nano 4(12):7431

36. Esconjauregui S, Fouquet M, Bayer BC, Eslava S, Khachadorian S, Hofmann S, Robertson J (2011) Manipulation of the catalyst-support interactions for inducing nanotube forest growth. J Appl Phys 109:044303

37. Zhang C, Yan F, Allen CS, Bayer BC, Hofmann S, Hickey BJ, Cott D, Zhong G, Robertson J (2010) Growth of vertically-aligned carbon nanotube forests on conductive cobalt disilicide support. J Appl Phys 108(2):024311

38. Esconjauregui S, Xie R, Guo Y, Pfaendler SML, Fouquet M, Gillen R, Cepek C, Castellarin-Cudia C, Eslava S, Robertson J (2013) Electrical conduction of carbon nanotube forests through sub-nanometric films of alumina. Appl Phys Lett 102(11):113109

39. Sugime H, Esconjauregui S, Yang J, D'Arsié L, Oliver RA, Bhardwaj S, Cepek C, Robertson J (2013) Low temperature growth of ultra-high mass density carbon nanotube forests on conductive supports. Appl Phys Lett 103:073116

40. Dijon J, Okuno H, Fayolle M, Vo T, Pontcharra J, Acquaviva D, Bouvet D, Ionescu AM, Esconjauregui CS, Capraro B, Quesnel E, Robertson J (2010) Ultra-high density Carbon Nanotubes on Al-Cu for advanced Vias. In: IEEE international electron devices meeting, pp 33.4.1–33.4.4

41. Chiodarelli N, Li Y, Cott DJ, Mertens S, Peys N, Heyns M, Gendt SD, Groeseneken G, Vereecken PM (2010) Integration and electrical characterization of carbon nanotube via interconnects. Microelectron Eng 88(5):837

42. Chiodarelli N, Masahito S, Kashiwagi Y, Li Y, Arstila K, Richard O, Cott DJ, Heyns M, Gendt SD, Groeseneken G, Vereecken PM (2011) Measuring the electrical resistivity and contact resistance of vertical carbon nanotube bundles for application as interconnects. Nanotechnology 22(8):085302

43. Chiodarelli N (2011) Integration of carbon nanotubes as future interconnections for sub-32nm technologies. Ph.D. thesis, Katholieke Universiteit Leuven

44. Vereecke B, van der Veen MH, Barbarin Y, Sugiura M, Kashiwagi Y, Cott DJ, Huyghebaert C, Tökei Z (2012) Characterization of carbon nanotube based vertical interconnects. In: Extended abstracts of the 2012 international conference on solid state devices and materials, pp 648–649

45. van der Veen MH, Vereecke B, Huyghebaert C, Cott DJ, Sugiura M, Kashiwagi Y, Teugels L, Caluwaerts R, Chiodarelli N, Vereecken PM, Beyer GP, Heyns MM, Gendt SD, Tökei Z (2012) Electrical characterization of CNT contacts with Cu Damascene top contact. Microelectron Eng 106:106

46. van der Veen MH, Barbarin Y, Vereecke B, Sugiura M, Kashiwagi Y, Cott DJ, Huyghebaert C, Tökei Z (2013) Electrical Improvement of CNT Contacts with Cu Damascene Top Metallization. In: IEEE international interconnect technology conference, pp 193–195

47. Vollebregt S, Ishihara R, van der Cingel J, Beenakker K (2012) Low-temperature bottom-up integration of carbon nanotubes for vertical interconnects in monolithic 3D integrated circuits. In: Proceedings of the 3rd IEEE international 3D system integration conference, pp 1–4

48. Vollebregt S, Ishihara R, Derakhshandeh J, van der Cingel J, Schellevis H, Beenakker CIM (2011) Integrating low temperature aligned carbon nanotubes as vertical interconnects in Si technology. In: 11th IEEE conference on nanotechnology, pp 985–990

49. Vollebregt S, Chiaramonti AN, Ishihara R, Schellevis H, Beenakker CIM (2012) Contact resistance of low-temperature carbon nanotube vertical interconnects. In 12th IEEE conference on nanotechnology, pp 424–428

50. Vollebregt S, Tichelaar FD, Schellevis H, Beenakker CIM, Ishihara R (2014) Carbon nanotube vertical interconnects fabricated at temperatures as low as 350°C. Carbon 71:249

51. Naeemi A, Meindl JD (2007) Physical modeling of temperature coefficient of resistance for single- and multi-wall carbon nanotube interconnects. IEEE Electron Device Lett 28(2):135

52. Vollebregt S, Banerjee S, Beenakker CIM, Ishihara R (2013) Size-dependent effects on the temperature coefficient of resistance of carbon nanotube vias. IEEE Trans Electron Devices 60(12):4085

53. Xie R, Zhang C, van der Veen MH, Arstila K, Hantschel T, Chen B, Zhong G, Robertson J (2013) Carbon nanotube growth for through silicon via application. Nanotechnology 24(12):125603

54. Poelma RH, Morana B, Vollebregt S, Schlangen E, van Zeijl HW, Fan X, Zhang GQ (2014) Tailoring the mechanical properties of high-aspect-ratio carbon nanotube arrays using amorphous silicon carbide coatings. Adv Funct Mater 24(36):5737–5744

55. Jiang H, Liu B, Huang Y, Hwang KC (2004) Thermal expansion of single wall carbon nanotubes. J Eng Mater Technol 126(3):265

56. Kim BC, Kannan S, Gupta A, Mohammed F, Ahn B (2010) Development of carbon nanotube based through-silicon vias. J Nanotechnol Eng Med 1(2):021012

57. Xu C, Li H, Suaya R, Banerjee K (2010) Compact AC modeling and performance analysis of through-silicon vias in 3-D ICs. IEEE Trans Electron Devices 57(12):3405

58. Xu T, Wang Z, Miao J, Chen X, Tan CM (2007) Aligned carbon nanotubes for through-wafer interconnects. Appl Phys Lett 91(4):042108

59. Wang T, Olofsson KJN, Campbell EEB, Johan Liu C (2009) Through silicon vias filled with planarized carbon nanotube bundles Nanotechnology 20(48):485203

60. Wang T, Jeppson K, Ye L, Liu J (2011) Carbon-nanotube through-silicon via interconnects for three-dimensional integration Small 7(16):2313

61. Wang X, Feng Y, Unalan HE, Zhong G, Li P, Yu H, Akinwande AI, Milne W (2011) The mechanism of the sudden termination of carbon nanotube supergrowth. Carbon 49:214

62. Banerjee S (2014) Super-growth of CNTs based on ZrN for TSV application. Master's thesis, Delft University of Technology

63. Vollebregt S, Banerjee S, Tichelaar FD, Ishihara R (2015) Carbon nanotubes TSV grown on an electrically conductive ZrN support layer. In: IEEE international interconnect technology conference, pp 281–283

64. Wei H, Patil N, Lin A, Wong HSP, Mitra S (2009) Monolithic three-dimensional integrated circuits using carbon nanotube FETs and interconnects. In: IEEE international electron devices meeting, pp 1–4

65. Patil N, Lin A, Myers ER, Ryu K, Badmaev A, Zhou C, Wong HSP, Mitra S (2009) Wafer-Scale growth and transfer of aligned single-walled carbon nanotubes. IEEE rans Nanotechnol 8(4):498

66. Santra S, Ali SZ, Guha PK, Zhong G, Robertson J, Covington JA, Milne WI, Gardner J, Udrea F (2010) Post-CMOS wafer level growth of carbon nanotubes for low-cost microsensors a proof of concept. Nanotechnology 21:485301
67. Duesberg GS, Graham AP, Kreupl F, Liebau M, Seidel R, Unger E, Hoenlein W (2004) Ways towards the scaleable integration of carbon nanotubes into silicon based technology Diam Relat Mater 13:354
68. Vollebregt S, Chiaramonti AN, van der Cingel J, Beenakker K, Ishihara R (2013) Towards the integration of carbon nanotubes as vias in monolithic three-dimensional integrated circuits. Jpn J Appl Phys 52(4):04CB02
69. Mofrad MRT, Derakhshandeh J, Ishihara R, Baiano A, van der Cingel J, Beenakker K (2009) Stacking of single-grain thin-film transistors. Jpn J Appl Phys 48:03B015
70. Derakhshandeh J, Golshani N, Ishihara R, Mofrad MRT, Robertson M, Morrison T, Beenakker CIM (2011) Monolithic 3-D Integration of SRAM and image sensor using two layers of single-grain silicon. IEEE Trans Electron Devices 58(11):3954
71. Vollebregt S, Ishihara R (2016) The direct growth of carbon nanotubes as vertical interconnects in 3D integrated circuits. Carbon 96:332–338

Chapter 8
Carbon Nanotubes as Microbumps
for 3D Integration

Dominique Baillargeat and E.B.K. Tay

8.1 Introduction

Radio frequency (RF) packaging in electronics combines all assembly techniques in order to build with the best accuracy, repeatability, and stable RF systems. The combination of active and passive devices forms the RF system. Interconnections connect all blocks (subsystems, chips) together in order to mesh the entire system. Thus, interconnections play a major role in the functioning of the entire RF system. Interconnections need to bring the power and the signal with the lowest losses possible and the shortest delay reachable. With the increase of the frequency, the interconnections need to be more and more compact with a constant degree of performance.

Moore's law predicts that transistor sizes will continue to linearly decrease in future electronics chips. Hence, smaller interconnections on the first and second level will be necessary (Fig. 8.1). Currently, metallic interconnections are used in conventional high-frequency electronics up to the microscale. However, bulk materials such as metals have serious limitations at both the nano and micron scales (such as electromigration and grain scattering issues). Indeed, with a low current density below 20 nm, metal behaviour is not suitable for using as interconnections

D. Baillargeat (✉)
XLIM UMR CNRS 7252, Université de Limoges/CNRS, 123 Avenue Albert Thomas,
Limoges 87060, France
e-mail: dominique.baillargeat@xlim.fr

E.B.K. Tay
CINTRA CNRS/NTU/THALES, UMI 3288, Research Techno Plaza, 50 Nanyang Drive,
Level 6, Singapore 637553, Singapore

NOVITAS, School of EEE, Nanyang Technological University, Block S1-B3A-01,
50 Nanyang Drive, Singapore 639798, Singapore
e-mail: ebktay@ntu.edu.sg

© Springer International Publishing Switzerland 2017
A. Todri-Sanial et al. (eds.), *Carbon Nanotubes for Interconnects*,
DOI 10.1007/978-3-319-29746-0_8

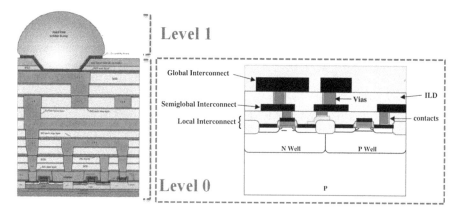

Fig. 8.1 Level 1 and level 0 of interconnections. Level 1 connects the entire chip while the level 0 interconnections are smaller and only bond unique components together such as transistors

in terms of performance and mechanical, thermal constraints. To overcome such limitations, CNTs could be a perfect candidate for interconnections because of their remarkable properties such as their long mean free path and their high current-carrying capability.

Two main levels of interconnection exist in integrated circuits: the first and second (also called level 0) levels regard their size, length, and purpose (Fig. 8.1). The first level lists all "chip-to-chip packaging" interconnects such as wire bonding, the flip chip, or a via-hole. These are used to connect the chip to its package and are the principal interest of the work presented in this chapter. On the other hand, we have the back-end-of-line interconnection (BEOL), which lists all smaller interconnections other than the first level. These are usually located in the integrated circuits themselves.

8.1.1 Level 0 of Interconnection Using CNTs

Interconnections bring both the signal and the power to electronics components. Interconnections at level 0 are split into two categories because the interconnections to bring a signal do not have the same technical requirements (size, material, etc.) as the interconnections for power. Moreover, these interconnections are indexed into three categories according to their length: local interconnects for the smallest length, semiglobal interconnects and global interconnects for the bigger lengths. Most simulations have been performed on interconnects to verify if the use of CNTs in interconnections is able to outperform the performance of current metallic interconnections. Most work (e.g., modeling and fabrication) so far with CNTs has been focused on level 0 [1, 2].

8.1.1.1   Local Interconnections Using CNTs [3]

(a) Local Interconnects: Properties and Issues

Because electronics are at the edge of the nanoscale, the size of the smallest interconnects in integrated circuits will necessarily shrink in the future. Indeed, they will be the most affected interconnections due to the size reduction of the electronic circuits themselves. This reduction in size will cause some problems, such as how to develop fabrication tools and equipment small enough to manipulate nanoscale components (e.g., nanoscale lithography, etc.) as well as electromigration issues (e.g., grain boundary scattering, surface scattering, and a high resistive diffusion barrier layer) that will occur with bulk materials at the nanoscale.

The length of the local interconnects is short (500–1500 nm). However, after they are aggregated, their length usually becomes larger. That is the reason why, today, 50 % of the interconnect power is dissipated at local interconnects. One of the main issues to solve with local interconnects is lateral capacitance. This high capacitance decreases the performance of interconnects by increasing the transmission signal delay. On the other hand, to carry the power to interconnects, the smallest resistance is necessary, which implies a thicker interconnect; this also avoids the electromigration effect. Thus, we will see how CNT bundles are able to solve these issues when approaching the nanoscale.

(b) Local Interconnects: Performance of CNTs (SW vs. MW) [2, 4]

Simulations have shown that controlled-diameter single-wall CNTs (SWCNTs) are able to lower the lateral capacitance by a factor of four and the power dissipation by a factor of two compared to a metallic local interconnect. This improves the speed of the interconnect by 50 %. Indeed, by controlling the number of CNTs in a given area, we are able to decrease the total lateral capacitance by decreasing the number of CNTs in the bundle. The SWCNT diameter can be also decreased (lower CNT surface) in order to slow down the lateral capacitance. But to bring the power to local interconnects, because of an important quantum resistance that exists in an individual CNT, it is necessary to keep a minimum of parallel CNTs. This will have the effect of decreasing the total interconnect resistance. Thus, a trade-off between the capacitance and resistance needs to be investigated regarding the CNTs' bundle density. Finally, because of the short lengths of the local interconnects (i.e., below the value of the CNT mean free path), CNT resistance remains high compared to a copper wire, for example, and may not be suitable for this level of interconnects.

8.1.1.2   Semi-Global Interconnections Using CNTs

The length of the semiglobal interconnects is logically located between the local and global interconnects, that is, between a few micro and hundreds of microns. These interconncts represent only a small proportion of interconnections length.

Regarding the state of these interconnects, densely packed double-wall CNT (DWCNT) bundles are preferred. Indeed, because the product R × C (where R is the interconnect resistance and C is the interconnect capacitance) determines the delay, an important number of CNTs with the lowest capacitance are necessary, and thus DWCNTs are preferred over SWCNTs due to the higher proportions of metallic CNTs that we are able to achieve with current fabrication processes.

8.1.1.3 Global Interconnections Using CNTs

Global interconnects have a length higher than a few hundred microns. They are mostly made from aluminum or copper. Using these materials, the resistance becomes proportional to the square root of the materials. With CNT bundles and the correct density of CNTs, we are able to decrease the width of the interconnections made by metal. However, a relatively large density of CNTs is necessary in order to reach a total interconnection resistance lower than a metal wire. If we use the SWCNTs instead of multiwall CNTs (MWCNTs), a perfect control of the CNT chirality during the growth needs to be done in order to get only metallic ones. An expected control of the CNTs chirality during the growth has so far not been achieved.

At the global level, for the same reasons as the local interconnects level, capacitance in CNT bundles can be easily lower than the metal interconnections, especially because the interconnect width at this level is larger. Finally, the inductance in CNT interconnections includes both kinetic and magnetic inductance effects. The kinetic inductance effect will be very low and can be neglected because of the large number of CNTs in parallel in the bundle. The magnetic inductance remains the same as the metal interconnection. So, at the global level of interconnects, physical models have shown that a bundle of large diameter (50 nm) MWCNTs could be used for power distribution because of their lower resistance, whereas dense SWCNT bundles (with at least 40 % of metallic SWCNT) can outperform the signal interconnect made by metal. The percentage of metallic versus semiconducting CNT is the major issue for the use of SWCNT in interconnects. A CNT density as high as $9.10^{15}/m^2$ has been obtained [5].

8.1.1.4 Conclusion

In conclusion, CNTs (both SW and MW) have pros and cons for use in these three levels of interconnection. According to the dimensions of the interconnections (length, diameter, etc.), CNT bundles with the correct density and type of CNTs may be able to outperform the performance of metal interconnections. Finally, CNTs would be an optimal material for a multilevel interconnect network by combining carbon-based structures. However, the state of the art shows us that the types of interconnections made by CNTs still need to be improved on the CNT growth control in order to integrate in future devices. So far, several CNTs bundles have been successfully fabricated for IC applications.

8.1.2 Level 1 of Interconnection Using CNTs [4, 6–8]

We now present the connections from the chip to it package. These interconncts play a major role in packaging because they need to deal with powering, cooling, and protecting electronic components in RF systems. Not surprisingly, the performance, size, cost, and reliability of the packaging—and more precisely the interconnections—will be strongly dependent on both the material used and the method employed (e.g., wire bonding, flip chip, via hole).

As with the local and global interconnections presented previously, because of the scaling down of transistor sizes in integrated circuits, further miniaturization will be necessary for interconnection. Three-dimensional packaging will increase the level of integration in electronics, and in this part of the discussion, we focus on using wire bonding, via hole and flip chip technologies. These three methods to interconnect a chip to an IC have a high degree of integration; however, their use involves a manufacturing process dealing with both thermal and reliability issues. Shrinking the interconnect size will even further complicate the fabrication process.

Thus, it becomes important to discover new materials with unique and excellent properties compatible with the nanoscale. CNTs may be a solution for off-chip interconnections used for signal and power transmission, in the same way as local and global interconnects. Moreover, because most of the time these interconnects "feed" an entire chip, they are more exposed to thermal and mechanical constraints. Again, the solution may come from CNTs and their unique mechanical, electrical, and thermal properties.

8.1.2.1 Wire Bonding

Wire bonding has been the most-used technology so far and the first way discovered to interconnect integrated circuits. The wire is usually made of gold (but can also be made of copper or aluminum) with a minimum diameter of 15 μm. We encounter two major problems with this type of interconnection. First, integration is not optimal because the wire has a length between 200 and 400 μm located around the chip. Secondly, the wire has a very low diameter (15 μm), which implies a high magnetic inductance. So, the use of wire bonding technology may be an issue above 50 GHz because of this high magnetic inductance.

8.1.2.2 Hot-Via [9–11]

This second interconnection configuration, called hot-via, is based on flip chip technology. Instead of flipping the chip in order to interconnect it from the topside, the backside is directly interconnected. Thus, a metallic bump links first the motherboard with the backside of the IC, and a via is then connected through the substrate to the IC. This technology maintains the advantages of allowing a glance on the IC as well as the possibility of realizing a transition between the coplanar accesses on the motherboard to a microstrip line on the chip. For the bumps, metals such as copper and gold are currently employed. Thus, size reduction of the chip will still affect them, and the use of CNTs as the via hole transition may be an answer of the downscaling issues.

8.1.2.3 Flip Chip

In recent years, much research has focused on compact designs, 3D integration, and high operation frequency in electronics. Flip chip is one of the technologies offering lower insertion loss [12], compact packages [13] and low-cost fabrication [14], but such requirements are limited at higher frequencies (i.e., higher than 100 GHz) because of the size limitation of the interconnection. As the name of the technology suggests, the flip chip consists of flipping the chip/IC and connecting the back face with the motherboard by metal bumps. In contrast with the hot-via presented previously, the back face of the chip, where the integrated circuit is, is connected. No vias through the substrate are necessary and only a planar technology is sufficient to make the flip chip compatible. This advantage of topology avoids the issues of via growth with CNTs and thus simplifies the steps of the fabrication process. IBM first introduced flip chip interconnections in 1970. At that time, flip chip technology was called C4 (controlled collapse chip connection).

Today, flip chip technology is widely used throughout the industry and is used daily in many electronics devices. To meet the demands of compact-size multifunctional designs at higher frequencies, metal bumps need to be downscaled. However, downscaling metal bumps in the microscale range is highly susceptible to electromigration issues, hence affecting adjacent devices or causes reliability issues [15] as already presented with local and global interconnections.

(a) Flip Chip Using CNTs (State of the Art)
   The use of vertical CNT bundles as flip chip bumps is an alternative approach for overcoming this problem and proposes interesting capabilities. Indeed, compared to gold, an ideal CNT is able to reach an electrical conductivity three times higher and a mean free path up to 25 orders of magnitude larger [4]. Its large kinetic inductance and a negligible magnetic inductance [4] also allow CNT to have a negligible skin depth effect. Another advantage of using CNTs for interconnection is their high current-carrying capacity (10^9 A/cm^2) and their

excellent thermal conductivity (3000 W m^{-1} K^{-1}) [16], potentially giving a better thermal management and power dissipation to a flip chip structure.

According to the International Technology Roadmap for Semiconductors, the forecasted requirement for flip chip bump pitches will shrink below 20 μm. However, traditional solder bumps encounter difficulties downscaling below 30 μm pitch due to the high diffusive and softening nature of solder [17], electromigration, and grain scattering issues [15]. CNTs are a suitable choice for future flip chip interconnections [4]. These advantages have motivated researchers and our group to evaluate the performance of CNT bumps for interconnect usage in both DC and high-frequency applications [4, 18]. Thus, a combination of both flip chip and CNT has the potential to make devices workable beyond 100 GHz with high performance. Indeed, in the literature for flip chip technology, we cannot find interconnections with a diameter lower than 20–25 μm [19, 20] working at a higher frequency than 100 GHz.

Much work on CNT bumps has been performed in recent years, and Soga et al. reported low DC resistance of 2.3 Ω for a 100-μm diameter bump [21]. But only a few works have presented a flip-chip-based technology. Flip chip using CNT bundles as bumps for a high-power amplifier was presented in [20] by Iway's group (Fig. 8.2). They showed a high-density (10^{11} cm^{-2}) CNT bundle as a bump for flip chip technology. Gold was plated on the surfaces of patterned metal and CNT bumps to allow thermal-assisted flip chip bonding bumps. A temperature of 345 °C was then necessary to fix the flip chip. The bundle diameter was 10 μm and the bundle resistance was found negligible because of the high number of CNTs.

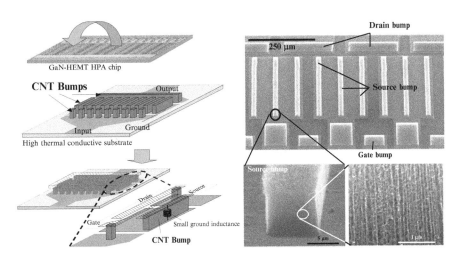

Fig. 8.2 On *left*, concept of a flip chip using carbon nanotube interconnections for thermal management. On *right*, SEM images of the corresponding CNT bundles. A high CNT density is observed. *Source*: [20]

As we can see in this work, flip chip technology coupled with CNTs is able to exhibit a good heat-removal technology. Indeed, the thermal management in interconnects requires new material and new approaches at the nanoscale. CNT-based flip chip technology could be a very efficient method. A flip chip interconnect was also presented by Hermann et al. using CNT pads on one side and metal contact pads on the other [18]. However, to realize a permanent mechanical contact, the bumps were filled with epoxy and heated to 200 °C to achieve 2.2 Ω bumps resistance. Hermann et al. demonstrated a reliable electrical flip chip interconnect using CNT bumps working over 2000 temperature cycles [18]. In all the works above, the CNT bumps were grown using the chemical vapor deposition (CVD) approach. The mechanism for vertical alignment during the thermal CVD approach is achieved mainly by the electrostatic and van der Waals forces between CNTs, resulting in tubes that are not exactly "aligned" [22]. Poor "alignment" forms bends, reduces the mean free path, and increases the resistance of CNT [23]. Plasma-enhanced chemical vapor deposition (PECVD) is able to solve this issue by introducing an electric field to achieve alignment as well as lower the growth temperature [16].

To summarize, a high aspect ratio (with CNT bundle diameter below 20 μm) is necessary to get good performance with the increase of frequency. High aspect ratio CNT bumps are reachable for better high-frequency performance. Lower process temperatures need to be reached, and beyond proof of the flip chip concept by 3D interconnected CNTs, future devices with a better control of density and a higher density will be suitable to increase the number of CNT contacts. Finally, another problem encountered in nanopackaging concerns the solder. Indeed, mechanical and thermal constraints become too high at a very low scale to support such a process. Thus, the idea of using a contact between CNTs and the electrostatic and van der Waals forces as the mechanical solder will be presented later.

8.2 CNT-Based Microbumps

In the flip chip scenario, the grown CNT bumps are usually pressed onto prepatterned conductive adhesive [21] or solder materials [24] to form the connections we saw in the previous part. This approach requires heating up the structure to a minimum of 200 °C in order to reflow the materials to obtain good contacts. Yung et al. [25] demonstrated a large-scale assembly process using vertically aligned CNT interconnection bundles showing that CNTs adhere well to each other by van der Waals forces and electrostatic interactions. However, no work using CNT interconnection bumps was reported for a CNT bump pitch below 150 μm, which is a requirement for future flip chip technology.

In the present work, we demonstrate the CNT interconnection bump-joining methodology for a pitch smaller than 150 μm. The fabrication methodology is divided into three parts:

1. Fabrication of the structure dedicated to RF application;
2. Growth of CNT bumps on both sides of the substrate using the PECVD approach;
3. Alignment and "insertion of the CNT bumps into each other using a flip chip bonder machine.

The first flip chip device working up to 40 GHz is fabricated and is characterized to set up and optimize the process and finally to demonstrate proof of the proposed innovative approach to interconnect CNTs. Moreover, we discuss the technological aspects of developing the test structure, and will present the DC and RF behaviour of small-scale CNT interconnection bumps. Finally, modelling using electromagnetic (EM) and hybrid EM/analytic approaches will be conducted in order to extract CNT parameters and understand the electrical behaviour of the test device. Then we will present a theoretical approach to flip chip working up to 110 GHz.

8.2.1 CNT Growth on Gold Metallization

As will be explained later, in order to fabricate CNT-based flip chip bonding, we need to develop CNTs directly on Au metallization. Consequently, one of the issues encountered during the process is the controllability of CNT growth density on Au as well as the understanding of interaction between different barrier layers and Au underlayers. Gold inhibits CNT growth, and thus a study for suitable barrier layers is required between the catalyst and Au metallization [26]. CNT growth has been reported to be influenced by different types of barrier layers: Titanium (Ti), titanium nitride (TiN), and chromium (Cr) were reported to support CNT growth, while aluminum (Al), Ni–Cr alloy, Cu, and Au inhibit growth [26–28]. The reason for using different barrier layers depends on the needs of applications; TiN is claimed to be the best barrier layer on Cu, whereas Ti and Cr are common barrier layers used to improve the adhesion of Au on Si and in the under-bump metallization for bumps [28, 29].

8.2.1.1 Test Structure

The test structure consists of 50-Ω coplanar waveguide (CPW) with a length of 0.7 mm for the signal (center) and 1.2 mm for the ground (Fig. 8.3). The line and bump patterns are obtained using photolithography techniques, as we will detail in the next part. First, 700 nm of thermal oxide is grown at 1200 °C on high-resistivity Si wafers (>10 kΩ cm). The high-resistivity wafer is necessary for the microwave applications. Next, 10 nm Ti is used as an adhesion promoter followed by depositing 1-μm thick Au metallization lines. These two layers are deposited using an e-beam process. Using the lift-off approach, 50-μm diameter bumps are formed followed by 50-nm barrier layer and 20-nm nickel (Ni) catalyst deposition.

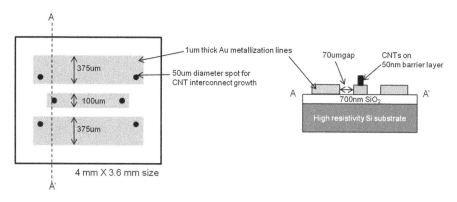

Fig. 8.3 Schematic of coplanar waveguides (CPW) test structure used (not drawn to scale). *Left* shows the top view of the CPW with a length of 0.7 mm for signal and 1.2 mm for ground. The *dark circles* indicate the locations of the CNT interconnects. *Right* shows the cross-section view indicating different layers of the test structure. Note that the 50 nm barrier layer is not shown in this figure

All metal depositions are carried out using the e-beam evaporation technique at room temperature, and TiN is sputter-deposited at 75 °C. The final CNT growth process is performed at 8 mbar inside a PECVD chamber. The growth process includes pretreatment of the Ni catalyst in NH_3 environment for 2 min at 800 °C. The growth is carried out for a duration of 15 min using C_2H_2 as the carbon feedstock gas at a 4:1 ratio. The cathode voltage is biased at −707 V providing a DC plasma power of 100 W. Ex situ characterizations are carried out using the LEO 1550 Gemini scanning electron microscope (SEM) for a CNT structure at different stages, and 532 Renishaw Visible Raman for Raman analysis. The RF measurements are done using a probe station connected to an Agilent HP 8510C network analyzer. The test environment is calibrated using short-open-load-thru scheme in order to remove the losses from the setup (cables, connectors, vector network analyzer (VNA), etc.) and S-parameters are extracted for the range of 1–20 GHz with a step of 0.1 GHz.

8.2.1.2 Results and Discussion

The influence of CNT growth parameters on Au metallization lines is observed at various stages of the growth process as shown in Fig. 8.4. Figure 8.4a shows the morphology of the deposited Au metallization, which is smooth with no cracks under the observation of SEM. As a reminder, the Au is deposited with e-beam equipment with a process temperature always below 70 °C. Figure 8.4b shows the morphology of Au metallization after being subjected to high-temperature processing at 800 °C in ambient NH_3, resulting in the formation of crack lines. Finally, Fig. 8.4c shows the surface roughening of the Au metallization after

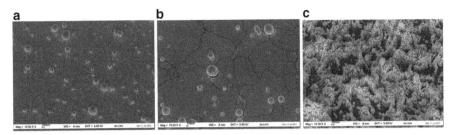

Fig. 8.4 SEM micrographs of the Au metallization at different stages of CNT growth. (**a**) Au-deposited film is smooth with no crack-lines. The circular islands are formed during the deposition of Au metallization by e-beam evaporation process. (**b**) After annealing for 2 min at 800 °C in ambient NH_3 and crack-lines are observed. (**c**) After subjecting to C_2H_2 and plasma treatment for 15 min in a PECVD chamber, the Au film formed a hill-like morphology

exposure to C_2H_2 feedstock gas and plasma treatment. The Au has agglomerated and formed a hill-like structure. De Los Santos et al. [30] suggested that the top layer of the Au melted, and that Au atoms diffused and nucleated at new sites. The addition of plasma at 800 °C raised the surface energy and likely induced the melting of the top layers of Au.

Using a two-point probe technique, the resistance of the Au coplanar strips lines registered an initial resistance of 2.03 and 7.73 Ω after the PECVD process. The increase in resistance is likely the result of film cracking and the formation of eutectic composition at the Au interface during annealing at 800 °C [31]. Thus, from the SEM micrographs and two-point probe testing, it is shown that the Au lines have degraded and suffered as a result of the CNT growth process.

Despite the structural degradation of the Au coplanar lines, experimental S-parameter properties shown in Fig. 8.5 are acceptable. To eliminate errors due to physical differences between test structures during the fabrication processes, the same coplanar structures (A and B) are used at every stage of measurements. Figure 8.5a, c show the return loss parameter (S_{11}) and insertion loss parameter (S_{21}) of test structure A, respectively, whereas Fig. 8.5b, d refer to measurements from test structure B. Each individual graph includes measurements at three different stages, namely, (1) the As-deposited film, (2) after 2 min annealing in ambient NH_3, and (3) after 15 min of CNT growth.

The results show that the cracks formed on the Au metallization after annealing caused the S-parameters measurements to worsen as compared to the As-deposited coplanar lines in Fig. 8.5. Insertion losses (S_{21}) decrease by 1.5–2 dB and a degradation equal to 5 dB is visible on the S_{11} reflection parameter. However, the line performances were still correct between 1 and 20 GHz. Interestingly, the same test structures show improvement as compared to the annealed case after being subjected to the plasma effect. After plasma treatment, the S_{11} reflection parameter had improved and behaved like the Au-deposited lines (highest peak at −17 dB for structure A and −19 dB for structure B).

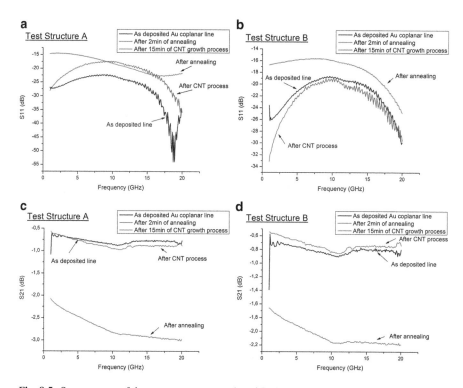

Fig. 8.5 S-parameters of the two test structures A and B: (**a**) return loss of structure A; (**b**) return loss of structure B; (**c**) insertion loss of structure A; and (**d**) insertion loss of structure B. Note that the peaks at 1 GHz in each S-parameter results above are due to the calibration process for the probe station. The calibration shows no successful results at this frequency

Moreover, the S_{21} transmission parameter of the coplanar line was also improved by an average of 1.87 dB for structure A and 1.32 dB for structure B as compared to the annealed lines. The observed results have two possible explanations: plasma could improve the quality of the line by cleaning off native oxide or amorphous carbon and/or plasma increases the surface energy of the Au lines leading to surface change at the Au metallization and forms new agglomerated structure, which presents new channels to carry the signal [32, 33].

8.2.2 RF Flip Chip Test Structure Based on CNT Bumps

8.2.2.1 Design and Fabrication

(a) Description of the Test Structure

In order to demonstrate proof of high-frequency CNT-based flip chip bonding, we designed a simple test structure composed of a 50-Ω CPW flip chip bonded on a substrate carrier. The topology of the signal line on the substrate carrier is optimized to compensate for the parasitic capacitance that occurs between the two substrates. All the dimensions are given in Fig. 8.6. An entire modelling (EM and hybrid EM/analytical) of the device was done including CNT interconnections in order to extract some CNT properties such as the contact resistance between CNTs. To demonstrate the feasibility of using the PECVD approach for achieving fine-pitch CNT bumps, three different sets of test structures comprised of (Structure 1) 170 μm by 150 μm, (Structure 2) 120 μm by 100 μm, and (Structure 3) 70 μm by 50 μm CNT bumps size are designed (Fig. 8.6)

In order to improve the mechanical support of the test structure, we introduced rows of dummy CNT bumps positioned at the sides of the chip and die to increase the densities of the CNT bumps. The dummy bumps provided additional mechanical support to hold the weight and levelled the attached die.

(b) Design: Hybrid (EM/Circuit) Modelling

The flip chip test structure is studied applying a hybrid approach based on 3D EM/circuit modelling between 1 and 40 GHz. This modelling combines full EM simulation and an analytical model of the CNT bundles. The analytical model depends on several parameters such as CNT diameter, CNT length, bundle CNT density and CNT components that will be described later. An optimization of the flip chip performance becomes possible as well. All the structures, except the CNT bumps, are considered by 3D EM simulations in order to define a generalized matrix [SG]. [SG] is defined between input/output ports (port 1 and port 2) and internal lumped accesses (3–6) for connecting circuit models of CNT bumps (Figs. 8.7 and 8.8).

Electromagnetic simulations are performed with the software Ansoft HFSS. In Fig. 8.8, [SG] represents the matrix extracted from EM simulation of the flip chip structure. The "RLC transmission line" box corresponds to the CNT bundle model that will be presented in the following discussion. $R_{CNT-CNT}$ is the contact resistance between interconnected CNTs.

The circuit model based on an RLC transmission line describing the electrical behaviour of the CNT structure is presented in Fig. 8.8. The CNT circuit models are inserted between the lumped accesses (ports 3–4 and ports 5–6) of the flip chip matrix [SG]. The RLC transmission line is composed of lumped and distributed components as described in [34]. Briefly, the contact capacitance C_C and the contact resistance R_C model the contact between the CNTs. The metal and values are found according to this other work [35]. The quantum resistance of R_{CNT} is considered equal to 20 kΩ/μm in comparison with previous experimental work [36]. The two other distributed components for nanowire effects are a kinetic inductance $L_k = L_{k-bundle} = \frac{h}{2v_F} \times l \times \frac{1}{N_{i-channel} \times N_{CNT}}$ and a quantum capacitance $C_Q = C_{Q-bundle} = \frac{2e^2}{hv_F} \times l \times N_{i-channel} \times N_C$ where ($v_F = 9.7 \times 10^5$ m/s: Fermi's velocity). According to the physical properties of the CNTs we consider in our test structure, there are $N_{i-channel} = 374$ in our model. The magnetic inductance L_M is considered negligible compared to the

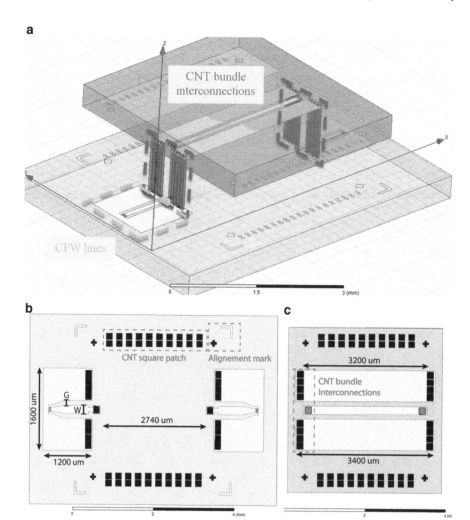

Fig. 8.6 Schematic of the flip chip structure fabricated to test for both DC and RF functionality. (**a**) The final flip chip after the chip is flipped and combined with the carrier; (**b**) carrier design that comprises an "open" coplanar line structure; (**c**) chip design that comprises of coplanar line; (**d**) shows the variation of the size of three flip chip test structure

Structure	1 (170 μm by 150 μm)	2 (120 μm by 100 μm)	3 (70 μm by 50 μm)
W /μm	200	150	100
G /μm	115	87	61
Number of CNT bundle	58	74	94

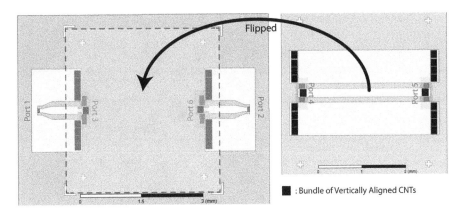

Fig. 8.7 Top view of the schematic of the flip chip structure

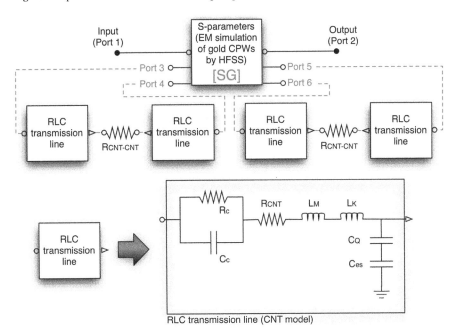

Fig. 8.8 Diagram of the hybrid flip chip structure model

very high kinetic inductance L_k in nanowires [34]. Electrostatic capacitance C_{ES} represents the coupling effect between the ground plane and the CNT. According to the experimental CNT we consider, C_{ES} and C_Q can be neglected. At last, in order to model a bundle of CNTs, we also consider the coupling between weak CNTs because of the large distance between them (about 200–500 nm). Then, the RLC CNT models are placed in parallel according to the estimated number of CNTs in the bundle, $N_{MWCNT} = 1.35 \times 10^5$. All components values are visible in Table 8.1

Table 8.1 CNT bundle components values

N_{MWCNT}	R_C	C_C	R_{CNT}	L_K	$R_{CNT-CNT}$
1	10 kΩ	4.1×10^{-6} aF	20 kΩ/μm	0.86 nH	324 kΩ
1.35×10^5	74 mΩ	55 pF	0.15 kΩ/μm	6.34 fH	2.4 Ω

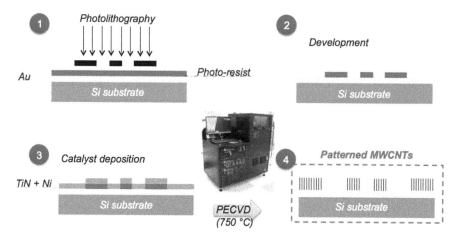

Fig. 8.9 Photolithography, e-beam, and PECVD steps to deposit pattern a catalyst material before CNT growth by PECVD. (*1*) Shows the UV illumination of the photo resist using a shadow mask. (*2*) Shows the development of the photo resist in order to remove nonwanted patterns. (*3*) Is the deposition of the barrier layer and catalyst on the all sample by e-beam. Finally, in (*4*) the CNTs are grown by PECVD and the entire photo resist is removed in order to keep only the remaining bundle of CNTs

As explained, all the lumped elements are defined according to the ideal technological properties of the CNT bundles, and will be used to design the whole test structure that will be fabricated. At this stage of modelling, the contact resistance between two CNTs, $R_{CNT-CNT}$, is considered equal to zero. It will be extracted after the flip chip process and will allow us to define an accurate hybrid (EM/analytical) model of the test structure. We will present later how the value of $R_{CNT-CNT}$ is extracted. Such a model will be used to understand and optimize the behaviour of the test structure.

(c) Fabrication Process of the Test Structure

The fabrication is carried out using standard lithography and a metal deposition system. A schematic of the process is presented in Fig. 8.9.

The substrate is a 4-in.-high resistivity (HR) <100> wafer with resistivity >10 kΩ. The use of an HR wafer is necessary in RF application to minimize attenuation losses. Next, the wafers are cleaned and placed into the tube furnace to perform wet oxidation of 600 nm of SiO$_2$. The use of the SiO$_2$ layer is not necessary for RF application, but from a DC point of view, it is necessary to reduce the leakage current. Next, lithography patterning is performed to create the opening for the depositing of Au or Cu metallization by e-beam evaporation.

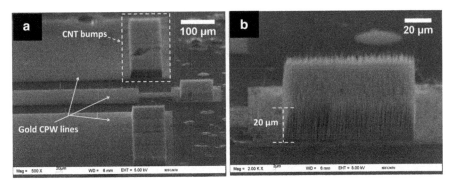

Fig. 8.10 (a) Low-magnification view of CNT bumps formed on the Au metal lines. (b) SEM image of a CNT bump grown using the PECVD approach. The grown CNTs were vertically aligned and their length was approximately 20 μm

After the first metallization is completed, a second lithography step is performed for the catalyst (used for CNT growth) and its barrier layer is required for CNT growth [37–39]. The final 4-in. wafer is then cut up using a dicing machine and is ready for the growth of CNT. The CNT growths are then carried out in the Aixtron Black Magic PECVD deposition system at 650 °C with a plasma power of 85 W for 30 min. The growth pressure is at 6 mbars with a gas ratio of gas of 1:5 (C_2H_2:NH_3).

From the literature point of view, there have been reports that using NH_3 during growth results in N-doped CNT. The NH_3 function is necessary to act as an etchant source and possibly as N-dopants according to various group [40, 41]. CNT bumps height of 20 μm is obtained as shown in Fig. 8.10. Each of these CNT bumps comprises of MWCNT with an average diameter about 100 nm.

Subsequently, a Panasonic flip chip bonder machine is used to perform the die alignment and attachment. The top part of the flip chip is flipped at 180° above the bottom part and a downward force is applied (Fig. 8.11). A load setting from 0.5 to 3 kg with a bonding time of 30 s is used as the bonding parameter.

The bonding load of 0.5 kg is sufficient to cause the CNT from the bottom carrier to "insert" and touch the top die. The load of 0.5 kg, which is equivalent to 4.5 N or 3.125 kg/cm^2, is much lower than that applied in previous reported flip chip experiments [18, 20, 21]. However, the load of 1.5 kg is found to provide the optimum electrical properties. Indeed, a higher electrical resistance is obtained when a low load (0.5 kg) is applied on it [39]. We believe that the CNTs start to be compressed when a larger load is used (greater than 1.5 kg), just as in the work of [21].

Indeed, with a load of 3 kg, we obtained the higher rate of successful devices, but the achieved resistances were all higher than the one measured with an applied load of 1.5 kg. We believe the compression increased the depth of

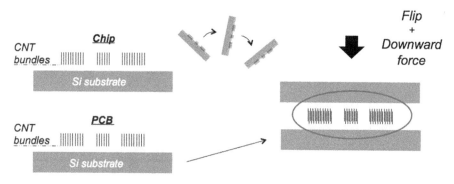

Fig. 8.11 Principle of the flip chip using interconnected bundle of carbon nanotubes. *Top part* (chip or die) is rotated at 180° above the bottom part (PCB or carrier) and a downward force is applied on it. CNTs will stick together by the electrostatic and van der Waals forces

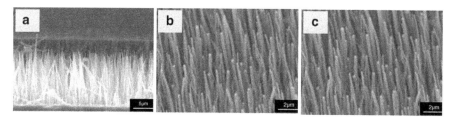

Fig. 8.12 Interconnection length regarding the force applied: (**a**) no load, (**b**) 0.03 g, and (**c**) 3 kg. At 3 kg, the interconnection length is maximal

penetration and the chances of CNTs contacting with one another (Fig. 8.12). After the bonding process is completed and the load is removed, CNTs return to their original vertical configuration. No bonding temperature is used in our experiment. Finally, we notice that the mechanical adhesion between the two parts is good enough to carry and measure our structures in DC/RF, but too weak to perform a stress test of the device. Thus, an improvement of the mechanical adhesion between the two flip chip parts needs to be investigated.

8.2.2.2 Fabrication Results

To demonstrate the feasibility of using PECVD approaches to achieve fine-pitch CNT bumps, three different sets of test structures (Fig. 8.13) are fabricated. The SEM images in Fig. 8.13 show the CNT bumps grown using the PECVD approach allows pitch sizes down to 80 μm (70 μm bump size + 10 μm distance between two bumps) in our experiment. Homogenous CNT bump heights can also be observed throughout the carrier and die (dummy and CNT bumps on electrodes). Larger catalyst pattern geometry would result in longer CNT length due to the differences in partial pressure of carbon feedstock gas. However, the effect of catalyst pattern

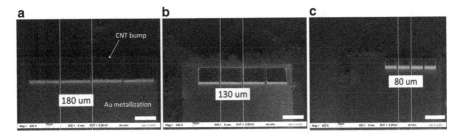

Fig. 8.13 The CNT bumps on an Au electrode with three different pitches. The smallest pitch designed is 80 μm. The dimension of each CNT bump in (**a**) structure 1 is 170 μm × 150 μm, (**b**) structure 2 is 120 μm × 100 μm, and (**c**) structure 3 is 70 μm × 50 μm. The scale bar at the bottom right of each image represents 100 μm

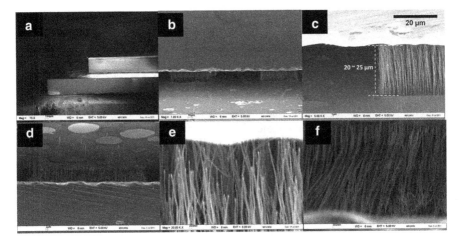

Fig. 8.14 SEM images of the CNT interconnection bumps demonstrated using a flip chip concept (**a–f**). The dimensions of CNT bumps were 100 μm × 100 μm for all images. (**a**) Die attached to a carrier at a tilted angle of 75°; (**b**) magnified view of two CNT interconnection bumps; (**c–d**) show the distance separating the two parts of the flip chip; (**e**) CNTs from bottom carrier are observed to be touching the die substrate indicative of the connections made; and (**f**) shows the CNTs touching each other

geometry is not significant in this experiment, and the height of all CNT bumps is assumed to be 20 μm regardless of the bumps dimensions.

Thus, for the first time, CNT interconnection bump joining methodology for fine-pitch bumps is achieved, as shown on Fig. 8.14. In Fig. 8.14a, the flip chip test structure is observed at an angle of 75° under the SEM. The assembled final structure, as shown in Fig. 8.14b–d, is separated by an average distance of 20–25 μm as observed in the SEM micrograph. It is also observed that some CNTs are able to reach the opposite part and be in contact with it, creating alternative paths to the CNT–CNT interconnections. A combination of both contacts could improve the device performance. Due to the equipment limitations, microphotographs of

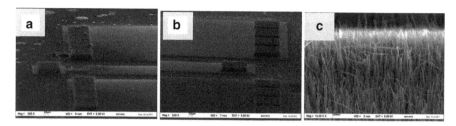

Fig. 8.15 SEM images of the CNT bumps morphology after removal of the top die. (**a**) CNT bumps on the carrier; (**b**) CNT bumps on the chip joined to the carrier bump in (**a**, **c**) with a magnified view of the bump. The remaining vertical alignments of the CNTs show the mechanical flexibility of CNTs

the CNT bumps during the bonding and release process could not be performed to demonstrate the mechanical flexibility of CNT bump as observed by [21]. However, based on the SEM images in Fig. 8.14c–f, the vertical alignment of CNT bump can still be observed, which is likely due to the mechanical flexibility of CNT bumps. In this experiment, structures 1 ($170 \times 150 \, \mu$m bump size) and 2 ($120 \times 100 \, \mu$m bump size) are tilted to 75° in the SEM to observe the gap between test structures. This is remarkable because no bonding temperature or adhesive is used during the bonding process to mechanically bond the die to the carrier.

The attached die is subsequently removed from the carrier using a tweezers to observe the effect of the CNT bumps after bonding. The carrier and die are then loaded into the SEM chamber with the same orientation to observe the condition of the CNT bumps that are in contact with one another, as shown in Fig. 8.15. A portion of the CNT bump appeared to be smeared as seen from Fig. 8.15a, but a high percentage of CNT bump retained their original structure. This is similar to [25] observations for a large-scale CNT-to-CNT interconnection structure, which demonstrates that the bonding process is reworkable.

8.2.2.3 DC Measurements: CNT Bump Resistance and Reworkability

In order to extract the resistance of a single CNT bump resistance, a four-point DC technique is used. The main advantage of using the four-point probe design is to eliminate the contact resistance of the probe pins and metal lines, thus removing unwanted resistance from the measurements. The resistance of the metallization is found to be in the same order as the CNT bumps resistance in our experiments. We have to consider it during the extraction of the CNT bump resistance. For improving the electrical performance of the CNT interconnections bumps on metal, the choice of barrier layer is crucial. Using Ti and TiN as the barrier layer in our study, Ti/TiN is found to be a better choice as compared to using TiN barrier layer. This is due to better conductivity of Ti and Ti is a preferred choice of adhesive layer for various metals in the industries.

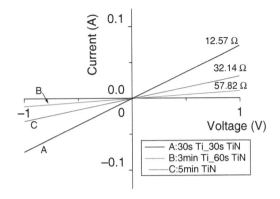

Fig. 8.16 I-V measurements—resistance of CNT bumps on three different barrier layers. The results on 30 s Ti_30 s TiN gives the lowest bump resistance equal to 6.25 Ω

In Fig. 8.16, it is also observed that the resistance decreases when the barrier layer thickness decreases. It further shows that electrons tunnel through the barrier layer and affect CNT conductivity. By comparing the resistivity values, the resistivity extracted are 0.164 Ω cm for the 30 s Ti/30 s TiN, 5.175 Ω cm for the 3 min Ti/60 s TiN and 0.695 Ω cm for 5 min TiN barrier layer. This clearly shows that thinner barrier layers help to improve the CNT conductivity. Due to the measurement technique, Fig. 8.16 shows the total resistance between the input and output accesses. Consequently, the two resistances of both input and output signal bumps are considered and are supposed to be the same. Later in this work, we will consider the case that gives the lowest bump resistance equal to 6.25 Ω.

In addition, we studied the effect of the bonding load. The relationship between the bonding load and resistance are plotted and shown in Fig. 8.17. The solid square and dots symbols represent the average resistance of one CNT interconnection and the standard errors are plotted as the error bars. Two observations can be made from the experiment. First, when the bonding load increases, the depth of penetration increases and the gap between the top chip and carrier decreases as shown in the inset. SEM image shown that the distances between the chip and carrier reduce from 35 μm (at 1 kg) to 30 μm (at 2 kg) to 28 μm (at 3 kg).

Second, an increase in the loading weight leads to a decrease of the resistance in an exponential manner. The variation of resistance also fluctuates less above 2 kg as indicated by the error bars. The decrease in measured resistance is mainly due to the increase in the number of parallel paths for electron transfer from one CNT to another. As CNT have large elastic modulus, a minimum load will be require to deform the CNT and to increase the number of conducting paths [42]. Below the optimum loading weight, CNT demonstrate resilient and flexibility which can return to its original position after bending [21]. The measured resistance is thus highly dependent on the probability to form conducting paths that result in the large error bar. Above the minimum load, the vertically aligned CNT starts to bend and form contacts between its neighboring CNT of the top chip, which creates more parallel paths for electrons transfer, thus reducing the measure resistance. When the CNT to CNT interspacing distances is decreased, the resistance also decreased [43]. The adhesion between CNT to CNT are solely by van der Waals forces interactions [25].

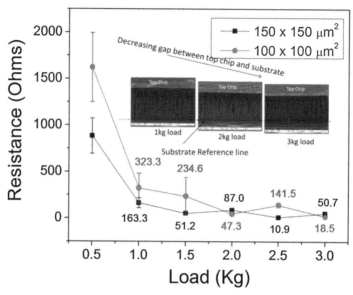

Fig. 8.17 Relationship between the measured resistances and bonding load. The SEM images as shown in the inset show the decreasing gap (distance) between the top chip and substrate with increasing bonding load

It is very important for the CNT interconnection bump to be reworkable in order to save cost. To verify our measurement repeatability, the top chip is removed and bonded again to the same carrier. The measurements are compared and shown in Fig. 8.18. The slight deviation of the second attempt could be caused by the differences in chip placement due to the limited alignment accuracy of the bonding machine. Figure 8.18b–d shows the SEM images of the CNT bumps after repeated bonding which are still intact.

8.2.2.4   RF Measurements: Discussion

High-frequency measurements from 1 to 40 GHz are also performed on the flip chip structures. Measurements are run with a VNA Agilent HP8510C and 150-μm pitch Cascade Microtech Infinity probes. A thru-reflect-line (TRL) calibration process is done before each measurement with a maximum error bar equal to 0.1 dB. TRL is a calibration process done before a high-frequency measurement by a VNA. This calibration process consists of replacing the device under test by a thru line, a short circuit, and finally a load circuit at 50 Ω.

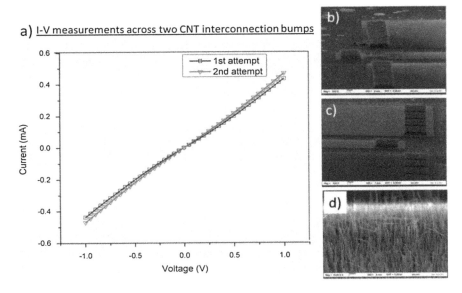

Fig. 8.18 (**a**) I-V characteristics after repeated bonding; (**b**) SEM image of the carrier after repeated bonding; (**c**) SEM image of the chip after repeated bonding; (**d**) the vertical alignment within the CNT bumps can still be observed in both structures

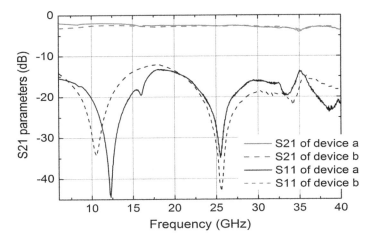

Fig. 8.19 S-parameters measurements of the flip chip device *a* (*solid lines*) and device *b* (*dash lines*) between carrier input and output

As shown in Fig. 8.19 two identical test structures (device a and b) with a 170×150 µm bump size exhibit a S_{21} parameters of -1.21 and -3.69 dB from 1 to 40 GHz. The return loss (S_{11}) stayed below -13 dB. The small variations above 30 GHz are due to the calibration process, which returns a measurement error between 0.1 and 0.2 dB on some frequency points.

At this experimental stage, proof of high-frequency CNT-based flip chips is demonstrated. The experimental results are encouraging.

In addition, to complete the experimental analysis on RF behaviour, we proceed to the estimation of the CNTs bump insertion losses. Thus, before we realize the flip chip process, we measure up to 40 GHz the S-parameters of the bonded CPW after the PECVD process. The length of the CPW top line is 3.2 mm. Insertion losses remain near −1 dB while the return losses remain below −20 dB. For instance, at 20 GHz, the top line insertion losses were −1.6 dB. Considering that the total length between the input and output accesses of the flip chip test structure is equal to 5 mm, we can estimate that the insertion losses due to the CPW line only are equal to −2.5 dB (−1.6 dB for a length of 3.2 mm). As shown in Fig. 8.19, the insertion losses of the test structure are near −2.5 dB. We can also deduce that the insertion losses due to CNT bumps only are negligible at higher frequencies. The same comparison can be made on all the measurement frequency bands.

Even if experimental results are encouraging, they present low performance compared to the classical approach with classical metal bumps. In the future, we need to fully understand and optimize not only their high-frequency behaviour, but also their mechanical properties.

To summarize, we have demonstrated the first measurement of a flip chip technology based on an interconnected bundle of CNTs from DC to 40 GHz. An innovative way to interconnect CNTs similar to the Velcro principle is used with successfully mechanical and electrical adhesions. No underfill or heat is necessary to interconnect the CNT bundles, which decreases the number of process steps and the cost of the device. The bumps size is 170×150 μm and only composed by 100-nm-diameter MWCNTs. This first measurement gives encouraging performance for using CNTs to replace metal bumps in flip chip technology. Moreover, a bonding repeatability is observed in DC, which opens the door to a new type of device: CNT interconnected bumps can be flipped several times and it becomes possible to replace a defective component or to verify components in a complex environment. However, the performances are still lower than metal bumps at this frequency range.

Future devices with a better control of density and a higher density will be suitable to increase the number of CNT–CNT contacts. A densification of the CNT bundle after bonding can also be a solution to increase the CNT density and so decrease the distance between CNTs. We can imagine changing the CNT bundle geometry in order to increase the surface in contact between the two flip chip parts. A better optimization of the access lines would also increase the performance of the device. The CNT growth temperature also needs to be reduced in the future. A temperature lower than 350 °C will be reached soon and will allow us to create a fully CMOS-compatible flip chip device based on CNT interconnections. Finally, we need to carry on mechanical and temperature stress tests in order to verify the resistance of our device in an aggressive environment.

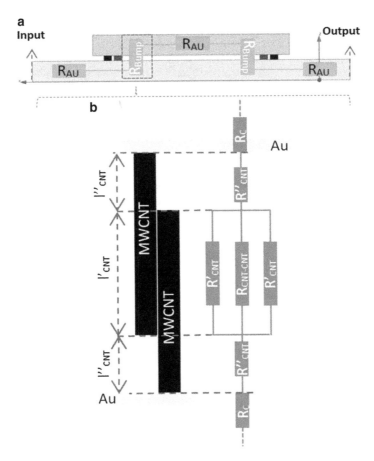

Fig. 8.20 (a) Side view of the flip chip structure and the corresponding equivalent electrical circuit with a breakdown of the bump resistances; (b) details on Rbump corresponding to the bump resistance for two interconnected MWCNTs according to the (a) schematic

8.2.2.5   Hybrid (EM/Analytical) Modelling

In order to understand the physical phenomena involved in the flip chip interconnect. and also to extract an accurate hybrid model, high-frequency simulations are done. We will first extract from measurements and theoretical definition all the parameters of the CNT bump equivalent model. This model is described in Fig. 8.20 which shows the model of the CNT bumps we measured in DC whevn considering the flip chip test structure (see Sect. 8.2.2.3). This model will permit us to extract the unknown contact resistance $R_{CNT-CNT}$ between two CNTs.

R_{bump} is the DC resistance of the bump. From measurements in Sect. 8.2.2.3 we found R_{bump} equal to 6.25 Ω. R' resistance represents the linked CNT parts, whereas R'' resistance represents the nonlinked part. Finally, $R_{CNT-CNT}$ is the contact

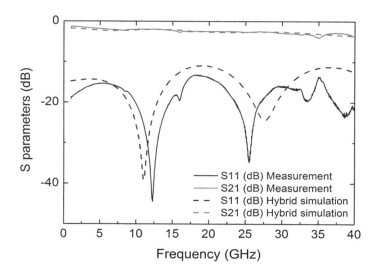

Fig. 8.21 S-parameters of the flip chip structure measured between port 1 and port 2. *Solid lines* correspond to the measurements while *dashed lines* show the hybrid EM/circuit simulation results

resistance between two MWCNTs. The contact resistance R_C is found according to the work of [35]. The resistances R'' and R' can be estimated from previous works done under similar conditions [36]. Consequently, knowing approximately the number of CNTs in a bundle (1.35×10^5), $R_{CNT-CNT}$ can be extracted. This contact resistance between two MWCNTs $R_{CNT-CNT}$ is equal to 324 kΩ. This value is similar to the one found in the work of other group [44].

This resistance will then be added to the analytical CNT model we presented previously in Sect. 8.2.2.1.b, in order to model the entire flip chip structure using the hybrid EM/analytical approach. All component values can be found in the Table 8.1. As shown in Fig. 8.21, good agreement is found between hybrid modeling and measurements. Consequently, the model is validated.

The hybrid model is used to perform parametric studies of the test structure by modifying the number of CNTs, their composition, their resistance, and other CNT parameters. Several tests are done to determine the effects of the bump resistance R_{CNT} and of the contact resistance $R_{CNT-CNT}$. By improving the CNT fabrication process, R_{CNT} will decrease and the performances will be slightly improved. For instance, considering R_{CNT} equal to 10 kΩ/μm (instead of 20 kΩ/μm in Table 8.1) the transmission level is only improved of 0.5 dB. We modify also $R_{CNT-CNT}$. Again, we observe very few effects on the behaviour of the test structure. For instance, by dividing by three $R_{CNT-CNT}$, we only improved the transmission level by 0.5 dB at low frequency (1 GHz) and by 0.25 dB at a higher frequency (40 GHz). With $R_{CNT-CNT} = 10$ kΩ, we improve S_{21} by 0.75 dB on all the bandwidth. Thus, we observed that decreasing the contact resistance between CNTs alone is not sufficient for improving more than 0.5 dB the S_{21} parameter.

In conclusion, even if an important work needs to be done on the CNT qualities by improving the fabrication process, it won't be sufficient to get a significant improvement of the electrical behaviour. So finally, we modify the number of CNTs in the bundle from 0.13×10^6 CNTs to 1.10^6 CNTs. In this case, the transmission level increases significantly of about 1.5 dB at low frequency (1 GHz) to 1 dB at higher frequency (40 GHz). Indeed, by increasing the number of CNTs in parallel, two resistances decrease at the same time: the contact resistance $R_{CNT-CNT}$ (more contacts in parallel) and the equivalent resistance due to the CNTs in parallel. The behaviour is very close to the optimum case we can expect for an ideal bump. In conclusion, focusing on the CNT density (or number of CNTs in a bundle) might be the best choice to improve the RF performances of the bump interconnect.

8.3 Conclusion and Future Work

The feasibility of using the PECVD approach to achieve CNT bumps pitch smaller than 150 μm was studied. The introduction of TiN barrier layer between Ni catalyst and Au metallization allows an efficient growth of CNT bumps directly on Au while maintaining good electrical connections between CNT and Au electrodes. The successful growth of CNT on Au metallization opens up opportunities to evaluate the performance of vertically aligned CNT bundles in very high frequency domain. The proof of concept of innovative RF CNTs bump flip chip is demonstrated. A test structure is built and tested. For the first time, very encouraging CNT-based flip chip measurements are obtained from DC up 40 GHz. In addition to the experimental work, a theoretical study based on 3D EM simulation and the hybrid EM/analytical approach was done. Electrical parameters of the CNT bumps were extracted for better understanding.

We conclude that even if the process and the current electrical performances need to be improved, many advantages exist such as the high aspect ratio, the possibility to flip several times, the lack of underfill, and so on. Suggestions to improve the process and electrical performances have been proposed.

In addition, we aim to extend this proof in the submillimeter wave domain. Indeed, at high frequencies, to realize small size bumps with metal (smaller than 20 μm in diameter) is challenging because of electro-migration effects and low current density capabilities [45]. Thus, we expect that above 100 GHz, CNT bumps should outperform metal bumps due to their excellent electrical conductivity for becoming an alternative technology compared to a current one. From the previous study, we consider a new test structure operating around 100 GHz. The new structure is identical to the one described in Fig. 8.6, with new dimensions suitable for submillimeter wave applications. The size of the bump is $20 \times 20 \times 20$ μm. As shown on Fig. 8.22, we apply parametric studies based on the hybrid EM/circuit model described previously.

From this parametric study, we observe that 2400 MWCNTs is not enough to have acceptable submillimeter performances (red dash curve). By increasing the

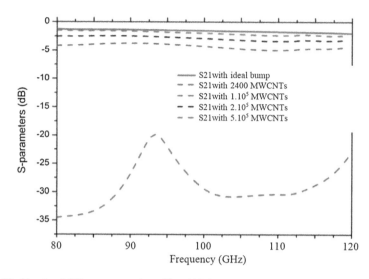

Fig. 8.22 Simulated S21-parameters from 80 to 120 GHz with several numbers of MWCNTs in a 20 × 20 μm bump (from 2400 to 500,000 MWCNTs)

number of CNTs in the bundles, we observe that with $N_{MWCNT} = 1.10^5$ and 2.10^5 (blue and black dashes), we have already enough CNTs to achieve insertion loss better than -4 dB and return loss below -10 dB. By increasing the number of CNTs to $N_{MWCNT} = 5.10^5$ (purple dash curve) we can even reach the same performance we obtained with ideal bumps (perfect metal). This represents a CNT density of 1.25×10^{15} CNT m^{-2}, which is a realistic density. In conclusion, an increase of the CNT density is essential in order to achieve a correct level of performance for submillimeter application. Thus, CNTs become credible candidates to outperform metal interconnections for flip chip technology. A fabrication of the test structure is currently under progress.

Acknowledgements The authors want to acknowledge all the contributors to this work and in particular Dr. Christophe Brun and Dr. Yap Chin Chong as main contributors during their Ph.D. thesis, Dr. Tan Chong Wei, Dr. Lu Congxiang, Dr. Chow Wai Leong, and Dunlin Tan for their help and valuable advice.

References

1. Naeemi A, Meindl JD (2008) Performance modeling for single- and multiwall carbon nanotubes as signal and power interconnects in gigascale systems. IEEE Trans Electron Devices 55:2574–2582
2. Banerjee K, Hong L, Srivastava N (2008) Current status and future perspectives of carbon nanotube interconnects. In: Proceedings of the 8th international conference on nanotechnology, NANO '08, IEEE, Arlington, TX, 18–21 August 2008, pp 432–436

3. Saraswat P (2002) http://www.stanford.edu/class/ee311/NOTES/Interconnect Scaling.pdf
4. Hong L, Chuan X, Srivastava N, Banerjee K (2009) Carbon nanomaterials for next-generation interconnects and passives: physics, status, and prospects. IEEE Trans Electron Devices 56:1799–1821
5. Awano Y, Sato S, Nihei M, Sakai T, Ohno Y, Mizutani T (2010) Carbon nanotubes for VLSI: interconnect and transistor applications. Proc IEEE 98:2015–2031
6. Zhengchun L, Lijie C, Kar S, Ajayan PM, Jian-Qiang L (2009) Fabrication and electrical characterization of densified carbon nanotube micropillars for IC interconnection. IEEE Trans Nanotechnol 8:196–203
7. Nihei M, Kawabata A, Sato M, Nozue T, Hyakushima T, Kondo D, Ohfuti M, Sato S, Yuji A (2010) Carbon nanotube interconnect technologies for future LSIs. In: Swart JW (ed) Solid state circuits technologies. InTech, Rijeka
8. Jiang D, Wang T, Chen S, Ye L, Liu J (2013) Paper-mediated controlled densification and low temperature transfer of carbon nanotube forests for electronic interconnect application. Microelectron Eng 103:177–180
9. Ting J-H, Chiu C-C, Huang F-Y (2009) Carbon nanotube array vias for interconnect applications. J Vac Sci Technol B Microelectron Nanometer Struct 27:1086–1092
10. Wen W, Krishnan S, Ke L, Xuhui S, Wu R, Yamada T et al (2009) Extracting resistances of carbon nanostructures in vias. In: IEEE international conference on microelectronic test structures, ICMTS 2009, IEEE, Oxnard, CA, 30 March 2009–2 April 2009, pp 27–30
11. Xuhui S, Ke L, Wu R, Wilhite P, Yang CY (2010) Contact resistances of carbon nanotubes grown under various conditions. In: Proceedings of the 2010 IEEE nanotechnology materials and devices conference, IEEE, Monterey, CA, 12–15 October 2010, pp 332–333
12. Sangsub S, Youngmin K, Jimin M, Heeseok L, Youngwoo K, Kwang-Seok S (2009) A millimeter-wave system-on-package technology using a thin-film substrate with a flip-chip interconnection. IEEE Trans Adv Packag 32:101–108
13. Hsu LH, Oh CW, Wu WC, Chang EY, Zirath H, Wang CT et al (2012) Design, fabrication, and reliability of low-cost flip-chip-on-board package for commercial applications up to 50 GHz. IEEE Trans Compon Packag Manuf Technol 2:402–409
14. Heinrich W (2005) The flip-chip approach for millimeter wave packaging. IEEE Microw Mag 6:36–45
15. Jae-Woong N, Kai C, Suh JO, Tu KN (2007) Electromigration study in flip chip solder joints. In: Proceedings of the 57th electronic components and technology conference, ECTC '07, IEEE, Reno, NV, 29 May 2007–1 June 2007, pp 1450–1455
16. Chhowalla M, Teo KBK, Ducati C, Rupesinghe NL, Amaratunga GAJ, Ferrari AC et al (2001) Growth process conditions of vertically aligned carbon nanotubes using plasma enhanced chemical vapor deposition. J Appl Phys 90:5308–5317
17. Tummala R, Wong CP, Markondeya Raj P (2009) Nanopackaging research at Georgia Tech. IEEE Nanotechnol Mag 3:20–25
18. Hermann S, Pahl B, Ecke R, Schulz SE, Gessner T (2010) Carbon nanotubes for nanoscale low temperature flip chip connections. Microelectron Eng 87:438–442
19. Jentzsch A, Heinrich W (2001) Theory and measurements of flip-chip interconnects for frequencies up to 100 GHz. IEEE Trans Microw Theory Tech 49:871–878
20. Iwai T, Shioya H, Kondo D, Hirose S, Kawabata A, Sato S et al (2005) Thermal and source bumps utilizing carbon nanotubes for flip-chip high power amplifiers. In: IEEE international electron devices meeting, 2005 IEDM technical digest, IEEE, Washington, DC, 5–5 December 2005, pp 257–260
21. Soga I, Kondo D, Yamaguchi Y, Iwai T, Mizukoshi M, Awano Y et al (2008) Carbon nanotube bumps for LSI interconnect. In: Proceedings of the 58th electronic components and technology conference, ECTC 2008, IEEE, Lake Buena Vista, FL, 27–30 May 2008, pp 1390–1394

22. Fan S, Chapline MG, Franklin NR, Tombler TW, Cassell AM, Dai H (1999) Self-oriented regular arrays of carbon nanotubes and their field emission properties. Science 283:512–514
23. Jun H, WonBong C (2008) Controlled growth and electrical characterization of bent single-walled carbon nanotubes. Nanotechnology 19:505601
24. Kumar A, Pushparaj VL, Kar S, Nalamasu O, Ajayan PM, Baskaran R (2006) Contact transfer of aligned carbon nanotube arrays onto conducting substrates. Appl Phys Lett 89: 163120–163123
25. Yung KP, Wei J, Tay BK (2009) Formation and assembly of carbon nanotube bumps for interconnection applications. Diam Relat Mater 18:1109–1113
26. Wang B, Liu X, Liu H, Wu D, Wang H, Jiang J et al (2003) Controllable preparation of patterns of aligned carbon nanotubes on metals and metal-coated silicon substrates. J Mater Chem 13:1124–1126
27. Tay BK, Wang ZF, Yung KP, Wei J (2008) Effects of under CNT metallization layers on carbon nanotubes growth. Mod Phys Lett 22:1827–1836
28. García-Céspedes J, Thomasson S, Teo KBK, Kinloch IA, Milne WI, Pascual E et al (2009) Efficient diffusion barrier layers for the catalytic growth of carbon nanotubes on copper substrates. Carbon 47:613–621
29. Bertrand N, Drevillon B, Gheorghiu A, Senemaud C, Martinu L, Klemberg-Sapieha JE (1998) Adhesion improvement of plasma-deposited silica thin films on stainless steel substrate studied by X-ray photoemission spectroscopy and in situ infrared ellipsometry. J Vac Sci Technol A 16:6–12
30. De Los Santos VL, Lee D, Seo J, Leon FL, Bustamante DA, Suzuki S et al (2009) Crystallization and surface morphology of Au/SiO2 thin films following furnace and flame annealing. Surf Sci 603:2978–2985
31. Wißmann P, Finzel H-U (2007) The effect of annealing on the electrical resistivity of thin gold films. In: Electrical resistivity of thin metal films, vol 223. Springer, Heidelberg, pp 35–52
32. Basa D (2010) Plasma treatment studies of MIS devices. Cent Eur J Phys 8:400–407
33. von Arnim VL, Fessmann J, Psotta L (1999) Plasma treatment of thin gold surfaces for wire bond applications. Surf Coat Technol 116–119:517–523
34. Burke PJ (2002) Luttinger liquid theory as a model of the gigahertz electrical properties of carbon nanotubes. IEEE Trans Nanotechnol 1:129–144
35. Minghui S, Zhiyong X, Yang C, Yuan L, Chan PCH (2011) Inductance properties of in situ-grown horizontally aligned carbon nanotubes. IEEE Trans Electron Devices 58:229–235
36. Yang C, Zhiyong X, Philip CHC (2010) Horizontally aligned carbon nanotube bundles for interconnect application: diameter-dependent contact resistance and mean free path. Nanotechnology 21:235705
37. Kociak M, Suenaga K et al (2002) Linking chiral indices and transport properties of double-walled carbon nanotubes, vol 89. American Physical Society, Ridge, NY
38. White CT, Mintmire JW (2004) Fundamental properties of single-wall carbon nanotubes. J Phys Chem B 109:52–65
39. Yap CC, Tan D, Brun C, Teo EHT, Wei J, Baillargeat D et al (2011) Characterization of novel CNT to CNT joining interconnections implemented for 1st level flip chip packaging. Presented at the electronics packaging technology conference, Singapore
40. Susi T, Kaskela A, Zhu Z, Ayala P, Arenal R, Tian Y et al (2011) Nitrogen-doped single-walled carbon nanotube thin films exhibiting anomalous sheet resistances. Chem Mater 23:2201–2208
41. Kim J-B, Kong S-J, Lee S-Y, Kim J-H, Lee H-R, Kim C-D et al (2012) Characteristics of nitrogen-doped carbon nanotubes synthesized by using PECVD and thermal CVD. J Korean Phys Soc 60:1124–1128
42. Yaglioglu O, Hart AJ, Martens R, Slocum AH (2006) Method of characterizing electrical contact properties of carbon nanotube coated surfaces. Rev Sci Instrum 77:095105–095103

43. Yoon Y-G, Mazzoni MSC, Choi HJ, Ihm J, Louie SG (2001) Structural deformation and intertube conductance of crossed carbon nanotube junctions. Phys Rev Lett 86:688–691
44. Fuhrer MS, Nygård J, Shih L, Forero M, Yoon Y-G, Mazzoni MSC et al (2000) Crossed nanotube junctions. Science 288:494–497
45. Nah J-W, Chen K (2007) Electromigration study in flip chip solder joints, Conference: Electronic Components and Technology Conference, 2007. ECTC '07. 1450–1455

Chapter 9
Electrothermal Modeling of Carbon Nanotube-Based TSVs

Wen-Yan Yin, Wen-Sheng Zhao, and Wenchao Chen

9.1 Introduction

Three-dimensional integrated circuits (3-D ICs) have attracted much interest in the past one decade because they can achieve technically breakthrough and significant enhancement in their electrical performance in comparison with that of conventional two-dimensional ICs. As the key technique for realization of ultra-high density integration and miniaturized packaging of 3-D ICs, through-silicon via (TSV) provides vertical electrical connection between different functional chips through multilayered silicon dies [1–3]. Moreover, it offers an opportunity for heterogeneously flexible integration, which is a powerful and effective solution called as "More-than-Moore" technology. Till now, there are numerous papers published on modeling, characterizing, and fabricating many different TSV structures for development of various 3-D ICs [4–13]. The most common TSV filling materials currently being used are copper (Cu), tungsten (W), and even doped poly-silicon. However, there are still some reliability and thermal management issues to be further studied in the realization of TSV-based 3-D ICs.

It is well known that single- and multi-walled carbon nanotubes (SWCNTs and MWCNTs) possess some extraordinary physical properties, including high current carrying capacity over 10^9 A/cm^2, large thermal conductivity of 1750–5800 W/m-K, and low coefficient of thermal expansion (CTE) of $\pm 0.4 \times 10^{-6}$/K, etc.

W.-Y. Yin (✉) • W. Chen
College of Information Science and Electronic Engineering, Zhejiang University, Hangzhou 310058, China
e-mail: wyyin@zju.edu.cn; wenchaochen@zju.edu.cn

W.-S. Zhao
Key Lab of RF Circuits and Systems of Ministry of Education, Microelectronic CAD Center, Hangzhou Dianzi University, Hangzhou 310018, China
e-mail: wsh.zhao@gmail.com

© Springer International Publishing Switzerland 2017
A. Todri-Sanial et al. (eds.), *Carbon Nanotubes for Interconnects*,
DOI 10.1007/978-3-319-29746-0_9

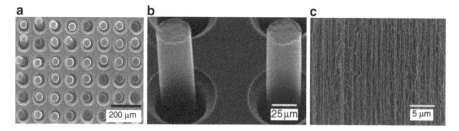

Fig. 9.1 Scanning electron microscope (SEM) images for some typical CNT-TSVs [18, 19]. (**a**) Top view; (**b**) 3-D view; and (**c**) high magnification SEM image

They can provide some novel alternative solutions for the realization of nanoscale interconnects with high electrothermal performance and reliability in comparison with their Cu counterpart [14]. In the past few years, significant progresses have been achieved in the development of SWCNT- and MWCNT-based interconnects [15–17]. It has been experimentally indicated that vertical integration of carbon nanotubes (CNTs) is much easier than that of horizontal case, and this implies that they are intrinsically suitable for design, fabrication, and realization of CNT-TSVs, as shown in Fig. 9.1a–c, respectively. For example, using thermal chemical vapor deposition (CVD) method, CNT-TSVs have been uniformly grown and realized as demonstrated in [18, 20, 21]. A CNT-Si structure has been fabricated with a post-growth planarization process employed recently. By using catalysts prepared with dip-coating, an aspect ratio (height to diameter ratio) of CNT-TSVs of 5–10 has been realized more recently. However, the applications of CNT-TSVs grown in such bottom-up approach are limited by high growth temperature of ∼700 °C and low volume fraction. In order to solve these problems, a novel approach has been proposed in [19] by densifying and transferring the as-grown CNT forests into the deep reactive ion etched (DRIE) holes, but there is a technical challenge in such post-growth transfer approach to form a small-diameter CNT-TSV due to unavoidable misalignment error of ∼2 μm. At low temperature below 550 °C, one novel bottom-up approach for growing CNT bundles has been achieved more recently, and it can be utilized for building sub-5 μm diameter CNT-TSVs with high aspect ratio [22].

In the development of any CNT-TSVs, it is necessary to at first develop an effective modeling method for accurately and fast capturing their electrical and thermal characteristics operating at different frequencies and temperatures. Similarly to W or Cu TSVs, currents in CNT-TSVs at high frequencies are pushed to its surface due to skin effect. The eddy current loss, induced by time-varying magnetic field and metal-oxide-semiconductor (MOS) effect, should be taken into account as the CNT-TSVs are embedded into lossy silicon substrate. However, the proximity effect, i.e., current crowding due to the impact of magnetic fields generated by the other TSVs in the vicinity, can be neglected as the pitch is usually larger than three times diameter of the TSVs. Besides these classical electromagnetic parameters, some characteristic parameters resulting from quantum effect in CNTs should also be considered and treated in an appropriate way. For example, the kinetic inductance of CNT must be involved for modeling CNT-TSVs, in particular at high frequencies.

This chapter will be focused on electrothermal modeling of CNT-TSVs. We start from some basic temperature-dependent properties of SWCNT and MWCNT, in particular including their thermal conductivity and specific heat. A set of closed-form equations is presented for fast calculating effective complex conductivity of both SWCNT and MWCNT bundles, which is indeed necessary for capturing their high-frequency behavior. The temperature effect is also taken into account for analyzing the electrical performances of CNT-TSVs. A pair of CNT-TSVs are modeled and studied in Sect. 9.3, and their S-parameters are compared with their Cu counterparts. In Sect. 9.4, the coupling noise among various CNT-TSVs arrays are investigated. Section 9.5 provides one novel 3-D carbon-based interconnect scheme for the next generation 3-D ICs by combining horizontal graphene interconnects and vertical CNT-TSVs. Finally, some conclusions are drawn in Sect. 9.6.

9.2   Temperature-Dependent Thermal Conductivity and Specific Heat of CNTs

In order to capture electrothermal effects in CNT-TSVs, electrical conductivity, thermal conductivity, and specific heat capacity of CNTs need to be at first known [23–32]. As already verified by theoretical and experimental studies, thermal conductivity and specific heat capacity of both SWCNT and MWCNT have strong dependence on temperature.

Thermal conductivity of single metallic SWCNT has two components. One is electron involved and the other is phonon involved. The phonon-involved component plays dominant role at low temperature. Following the Callaway's model [23], the phonon-involved temperature-dependent thermal conductivity for a pure SWCNT can be written as [24]

$$\kappa = \frac{k_B}{2\pi^2 v_g} \left(\frac{k_B T}{\hbar}\right)^3 \int_0^{\frac{\theta_D}{T}} \frac{\tau_R x^4 e^x}{(e^x - 1)^2} dx \tag{9.1a}$$

$$\tau_R = \left(Dx^4 + ET^2 + (v_g/L)\right)^{-1} \tag{9.1b}$$

$$D = \left(V_0 \Gamma_m / 4\pi v_g^3\right)(k_B T/\hbar)^4 \tag{9.1c}$$

$$E = (32/27)\,\gamma^4 \left(k_B/M v_g^2\right)^2 \omega_B \tag{9.1d}$$

where $x = \hbar\omega/k_B T$, and θ_D is the Debye temperature. The parameter k_B is the Boltzmann's constant, v_g is the phonon group velocity, $\hbar$ is the reduced Planck's constant. γ, ω_B, M, V_0, and Γ_m are the Gruneisen parameter, phonon branch frequency at the zone boundary, mass of the carbon atoms, volume per atom, and form factor, respectively [24]. When $T \to 0$, (9.1) can be simplified as [24]

$$\kappa_{Low} = A(T)\frac{4\pi^4}{15} \tag{9.2a}$$

$$A(T) = \left(k_B/2\pi^2 v_g\right)\left(k_B T/\hbar\right)^3\left(ET^2 + \left(v_g/L\right)\right)^{-1} \tag{9.2b}$$

While as $T \to \infty$, one can get [24]

$$\kappa_{High} = \frac{k_B}{2\pi^2 v_g}\left(\frac{k_B T}{\hbar}\right)^3\left[4\sqrt{2}(aD^3)^{\frac{1}{4}}\right]^{-1} B(T, \Gamma_m) \tag{9.3a}$$

$$B(T, \Gamma_m) = \ln\left[\left(\sqrt{D}(\theta_D/T)^2 + \sqrt{a} - \sqrt{2}(aD)^{1/4} \times (\theta_D/T)\right)/ \right. \\ \left.\left(\sqrt{D}(\theta_D/T)^2 + \sqrt{a} + \sqrt{2}(aD)^{1/4} \times (\theta_D/T)\right)\right] \\ + 2\tan^{-1}\left[\sqrt{2}\theta_D(D/a)^{1/4}/T\left(1 - (D/1)^{1/2}(\theta_D/T)^2\right)\right] \tag{9.3b}$$

in which $a = ET^2 + v_g/L$.

If the SWCNT is intrinsic, (9.3) can be simplified as [24]

$$\kappa_{High} = \frac{1}{3}A(T)\left(\frac{\theta_D}{T}\right)^3 \tag{9.4}$$

By combining (9.2) and (9.3), the thermal conductivity can be expressed as

$$\kappa^{-1} = \kappa_{Low}{}^{-1} + \kappa_{High}{}^{-1}. \tag{9.5}$$

The above equations illustrate temperature-dependent property of thermal conductivity of a single SWCNT. However, there are often surrounding materials around SWCNTs in their real application or experiment. Phonon interactions between SWCNT and surrounding materials may affect its thermal conductivity. Thermal conductivity of a metallic SWCNT on an SiO_2 substrate is investigated and measured, and an empirical equation is given by [26]

$$\kappa(L, T) = \left[3.7 \times 10^{-7}T + 9.7 \times 10^{-10}T^2 + 9.3\left(1 + 0.5/L\right)T^{-2}\right]^{-1} \tag{9.6}$$

where L is the length of metallic SWCNT in micrometer. Equation (9.6) has an approximately $1/T$ dependence above room temperature and a steeper $1/T^2$ component at temperatures approaching the burning point [26].

Some measurements also show that the thermal conductivity of SWCNT not only has strong non-linear temperature dependence, but also highly depends on its diameter, which is not taken into consideration in above closed-form equations. The measured thermal conductivities of SWCNTs are shown in Fig. 9.2a [27].

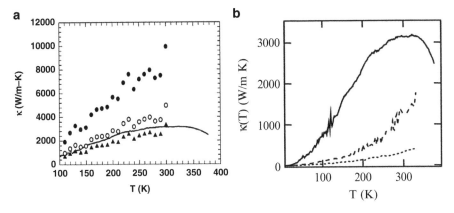

Fig. 9.2 Measured thermal conductivity of SWCNT and MWCNT. (**a**) Thermal conductivity of an SWCNT with diameter 1 nm (*filled circles*), 2 nm (*open circles*), and 3 nm (*filled triangles*) [27], the line is the measured result of an individual MWCNT from [28] and (**b**) thermal conductivity of an individual MWCNT with diameter 14 nm (*solid line*), small MWCNT bundle with diameter 80 nm (*broken line*), and large MWCNT bundle with diameter 200 nm (*dotted line*) [28]

Thermal conductivities of MWCNTs also have high non-linear dependence on temperature as in Fig. 9.2b, in which thermal conductivity of MWCNTs and MWCNT bundles is measured [28].

Similarly, specific heat capacity of SWCNT is highly dependent on temperature. By neglecting electronic contribution, its specific heat capacity can be determined by phonons. The specific heat capacity for each component of phonon dispersion relation can be described by [29–32]

$$C_{Vi} = \int k_B \left(\frac{\hbar\omega}{k_B T} \right)^2 \frac{e^{\hbar\omega/k_B T} g_i(\omega)}{\left(e^{\hbar\omega/k_B T} - 1 \right)^2} d\omega \tag{9.7}$$

where $g_i(\omega)$ is phonon density of states and ω is the phonon angular frequency. The six components are listed as follows [29]:

$$C_{V1} = C_1 T \int_0^{\theta_{\max}/T} \frac{x^2 e^x}{(e^x - 1)^2} dx \tag{9.8}$$

$$C_{V2} = C_2 T \int_0^{\theta_{\max}/T} \frac{x^2 e^x \left[\left(\frac{x^2}{x_0^2} - 1 + \sigma^2 \right)^2 + \sigma^2 (1 - \sigma^2) \right]}{(e^x - 1)^2 \left(\frac{x^2}{x_0^2} - 1 + \sigma^2 \right) \sqrt{ \left(\frac{x^2}{x_0^2} - 1 \right) \left(\frac{x^2}{x_0^2} - 1 + \sigma^2 \right) }} dx \tag{9.9}$$

$$C_{V3} = C_3 T \int_0^{\theta_{3\max}/T} \frac{x^2 e^x \left[\left(1 - \frac{x^2}{x_0^2} - \sigma^2 \right)^2 + \sigma^2 (1 - \sigma^2) \right]}{(e^x - 1)^2 \left(1 - \frac{x^2}{x_0^2} - \sigma^2 \right) \sqrt{ \left(1 - \frac{x^2}{x_0^2} \right) \left(1 - \frac{x^2}{x_0^2} - \sigma^2 \right) }} dx \tag{9.10}$$

$$C_{V4} = C_4 T^2 \int_{\theta_{4\,max}/T}^{\theta_{max}/T} \sum_{m=1}^{25} \frac{x^3 e^x}{(e^x-1)^2 \sqrt{\frac{x^2}{x_0^2} - \frac{1-\sigma}{2}m^2}} dx \qquad (9.11)$$

$$C_{V5} = C_5 T \int_0^{\theta_{max}/T} \sum_{m=1}^{16} \frac{x^2 e^x \left[\left(\frac{x^2}{x_0^2}-1\right)^2 + m^2\right]}{(e^x-1)^2 \left(\frac{x^2}{x_0^2}-1\right) \sqrt{\left(\frac{x^2}{x_0^2}-1\right)\left(\frac{x^2}{x_0^2}-1-m^2\right)}} dx \quad (9.12)$$

$$C_{V6} = C_6 T \int_0^{\theta_{6\,max}/T} \sum_{m=1}^{11} \frac{x^2 e^x \left[\left(1-\frac{x^2}{x_0^2}\right)^2 + m^2\right]}{(e^x-1)^2 \left(1-\frac{x^2}{x_0^2}\right) \sqrt{\left(1-\frac{x^2}{x_0^2}\right)\left(1+m^2-\frac{x^2}{x_0^2}\right)}} dx \quad (9.13)$$

where $x = \hbar\omega/k_B T$, $C_1 = 0.2468$ (J/Kg/K^2), $C_2 = C_3 = 0.16137$ (J/Kg/K^2), $C_4 = 6.5602 \times 10^{-5}$ (J/Kg/K^2), $C_5 = C_6 = 0.02468$ (J/Kg/K^2), $\theta_{max} = 16.3459\theta_0$ (K), $\theta_{3max} = 0.9855\theta_0$ (K), $\theta_{4max} = 160.84$ (K), $\theta_{6max} = 61.49$ (K), $\theta_0 = \hbar\omega/k_B$ = 245.988 (K), and $x_0 = \theta_0/T(K)$.

By performing summation of (9.8)–(9.13), the specific heat capacity of an SWCNT can be calculated [29], and its measured value as a function of temperature is shown and compared with the others in Fig. 9.3a.

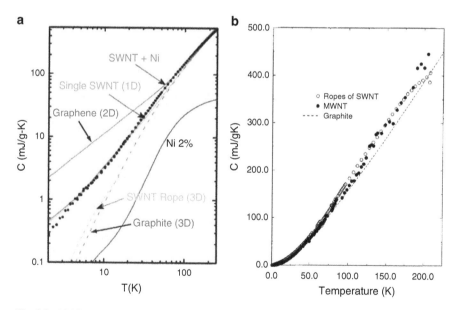

Fig. 9.3 (a) Measured specific heat capacity of SWCNT with calculations for 2-D graphene (*solid blue*), 3-D graphite (*dashed blue*), isolated tubes (*solid green*), and strongly coupled ropes (*dashed green*) [30]. (b) Measured specific heat for ropes of SWCNT and MWCNT, calculated graphite specific heat capacity [32]

The specific heat capacity of MWCNT has been investigated and measured as shown in [31, 32], respectively. It has linear dependence on temperature over the range of several hundreds K, which is similar to that of SWCNT. On the other hand, the measured specific heat capacity is shown and compared with that of SWCNT and graphite in Fig. 9.3b.

With these temperature-dependent parameters determined, electrothermal effects in CNT-TSVs can be further examined by solving heat conduction equation numerically until self-consistency is achieved with temperature-dependent property of each parameter considered for each loop. Moreover, as the structures of TSVs get complicated, two- as well as three-dimensional numerical model has to be developed.

9.3   Electrical Properties of CNT-TSVs

The schematic and cross-sectional views of a CNT-TSV are plotted in Fig. 9.4a, b, respectively. The radii of TSV and oxide layer are denoted by r_{via} and r_{ox}, respectively, and the oxide thickness is given by t_{ox}. As the silicon substrate is grounded, a depletion layer denoted by its radius r_{dep} appears when the applied voltage V exceeds the flatband one. Such CNT-TSV consists of N_{CNT} identical CNTs each with diameter D_{CNT}, and the spacing between two adjacent CNTs is given by δ. These CNTs can be SWCNT and MWCNT. An SWCNT can conceptually be viewed as one graphene sheet rolled into a hollow cylinder, while MWCNT appears like a coaxial assembly of cylinders of SWCNTs. According to some experimental results, the diameter ratio of innermost and outermost shells of an MWCNT can be 1/2 approximately, and the interval between two adjacent shells is $\delta' = 0.34$ nm, i.e., van der Waal's gap.

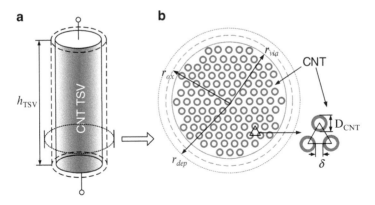

Fig. 9.4 (a) Schematic and (b) cross-sectional view of a CNT-TSV

For an ideally packed CNT bundle, δ is also equal to 0.34 nm, and the number of CNTs in Fig. 9.4b can be calculated by

$$N_{\text{CNT}} \approx \frac{2\pi r_{via}^2}{\sqrt{3}(D_{\text{CNT}} + \delta)^2} \tag{9.14}$$

As stated earlier, the kinetic inductance of CNTs must be taken into account for accurately evaluating the high-frequency behavior of CNT-TSV. Its effective complex conductivity can be represented as [33]

$$\sigma_{\text{eff}} = Fm \times h_{\text{TSV}} \left[\frac{\sqrt{3}}{2} (D_{\text{CNT}} + \delta)^2 Z_{\text{CNT}} \right]^{-1} \tag{9.15}$$

where Z_{CNT} is the intrinsic self-impedance of an isolated CNT, and Fm is the fraction of metallic CNTs in the bundle. As an MWCNT is always metallic, Fm in (9.15) is equal to 1, and for an SWCNT it depends on its chirality. It is noted that high $Fm = 0.91$ has been reported in [34], which should be suitable for the application of SWCNT bundle interconnect. In addition, it should be indicated that (9.15) is independent of the TSV radius, and is thereby suitable for other CNT-based interconnects as well.

The intrinsic self-impedance of a CNT shell with diameter D can be calculated by

$$\begin{aligned} Z_{\text{shell}}(D, T) &= R_{mc} + R_Q + R_S + j\omega L_K \\ &= \frac{R_{mc}}{N_{ch}(T)} + \frac{h}{2q^2} \frac{1}{N_{ch}(T)} \left(1 + \frac{h_{\text{TSV}}}{\lambda_{\text{eff}}(T)} + j\omega \frac{h_{\text{TSV}}}{2v_F} \right) \end{aligned} \tag{9.16}$$

where R_{mc} is the imperfect contact resistance, and it highly depends on fabrication process. R_Q, R_S, and L_K are the quantum contact resistance, scattering-induced resistance, and kinetic inductance, respectively. N_{ch} is the number of conducting channels, h is the Planck constant, q is the electron charge, and v_F ($= 8 \times 10^5$ m/s) is the Fermi velocity. According to the Matthiessen's rule, the effective mean free path (MFP) λ_{eff} can be calculated by [35]

$$\lambda_{\text{eff}}(T) = \left(\frac{1}{\lambda_{\text{ac}}(T)} + \frac{1}{\lambda_{\text{op,abs}}(T)} + \frac{1}{\lambda_{\text{op,ems}}^{\text{fld}}(T)} + \frac{1}{\lambda_{\text{op,ems}}^{\text{abs}}(T)} \right)^{-1} \tag{9.17}$$

where

$$\lambda_{\text{ac}}(T) = 400.46 \times 10^3 \frac{D}{T} \tag{9.18}$$

$$\lambda_{\text{op,abs}}(T) \approx \frac{\lambda_{\text{op}}}{f_{\text{op}}(T)} \tag{9.19}$$

$$\lambda_{op,ems}^{fld}(T) = \frac{\hbar\omega_{op}}{qV/h_{TSV}} + \frac{f_{op}(300) + 1}{f_{op}(T) + 1}\lambda_{op} \tag{9.20}$$

$$\lambda_{op,ems}^{abs}(T) = \lambda_{op,abs} + \frac{f_{op}(300) + 1}{f_{op}(T) + 1}\lambda_{op} \tag{9.21}$$

The average length of optical phonon emission by high-energy electrons is $\lambda_{op} = 56.4D$, the phonon energy $\hbar\omega_{op}$ is 0.18 eV approximately, and the Bose–Einstein distribution function is given by

$$f_{op}(T) = \frac{1}{e^{\hbar\omega_{op}/k_BT} - 1} \tag{9.22}$$

where k_B is the Boltzmann's constant, and T is temperature.

For a metallic SWCNT, N_{ch} is equal to 2, and the intrinsic self-impedance can be directly calculated. For an MWCNT, the number of shells can be obtained by

$$N_{shell} = 1 + Inter\left(\frac{D_{CNT}}{4\delta'}\right) \tag{9.23}$$

where "Inter(•)" indicates that only the integer part is taken into account. Therefore, the diameter of ith-shell, counted outside-in from the outermost shell, is given by

$$D_i = D_{CNT} - 2\delta' \times (i - 1), \qquad 1 \leq i \leq N_{shell} \tag{9.24}$$

It is noted that large diameter of CNT shell can be conductive despite of its chirality. The intrinsic self-impedance of an MWCNT can be calculated by

$$Z_{MWCNT} = \left(\sum_{shell} Z_{shell}^{-1}\right)^{-1} \tag{9.25}$$

Z_{shell} and λ_{eff} of the ith-shell of an MWCNT calculated by (9.24) and (9.25), respectively. Based on the Fermi–Dirac distribution function, N_{ch} of the ith-shell is determined by

$$N_{ch} = \sum_{subband} \frac{1}{1 + e^{|E_i-E_F|/k_BT}} \tag{9.26}$$

where E_i is the highest (or the lowest) energy of the subbands below (or above) the Fermi level E_F. For convenience, the parameter N_{ch} of MWCNT can be expressed by

$$N_{ch}(T) = \begin{cases} 2/3, & D_i < D_T/T \\ aD_iT + b, & D_i \geq D_T/T \end{cases} \tag{9.27}$$

where $a = 2.04 \times 10^{-4}$ nm^{-1} K^{-1}, $b = 0.425$, and $D_T = 1300$ nm K.

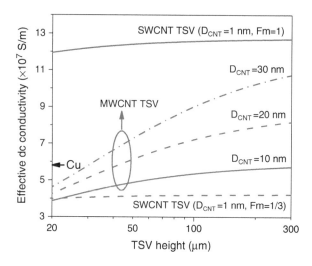

Fig. 9.5 Effective DC conductivity of the CNT-TSV versus its height at 300 K

The effective DC conductivity $\sigma_{\text{dc,eff}}$ of a CNT-TSV versus its height is plotted in Fig. 9.5 at $T = 300$ K. For an SWCNT-TSV, its conductivity increases with Fm, and it is equal to or larger than that of Cu when Fm exceeds 1/2. Noting that the TSV height is several to 100 times larger than the effective MFP of SWCNT, which is around several micrometers. Therefore, the effective dc conductivity of an SWCNT-TSV can be approximated as

$$\sigma_{\text{dc,eff}} \approx \frac{2Fm}{\sqrt{3}R_Q} \frac{\lambda_{\text{eff}}}{(D_{\text{CNT}} + \delta)^2} \qquad (9.28)$$

It is observed that $\sigma_{\text{dc,eff}}$ decreases with CNT diameter and it is independent of TSV height. In contrast, the conductivity of MWCNT-TSV is influenced by the CNT diameter as well as TSV height. This is because CNT shell with large diameter possesses long MFP, and its resistance is sensitive to the variations of diameter and length. Therefore, as shown in Fig. 9.5, the conductivity of an MWCNT-TSV increases with CNT diameter and TSV height. According to their dimensions, CNT-TSVs can be divided into wafer-to-wafer, wafer-to-chip, and chip-to-chip cases. It can be seen from Fig. 9.5 that MWCNT-TSVs are more suitable for the interposer application, and their performance could be degraded with the scaling trends of TSV height. However, SWCNT-TSVs can still be used extensively for 3-D ICs, and in particular for the sub-micrometer applications [34–36].

As thermal management is one of the most critical issues in the application of 3-D ICs, the temperature effects on effective conductivity of a CNT-TSV should be investigated. Here, the TSV height is 30 μm, and all the other geometrical parameters are the same as in Fig. 9.5. In order to make some comparisons,

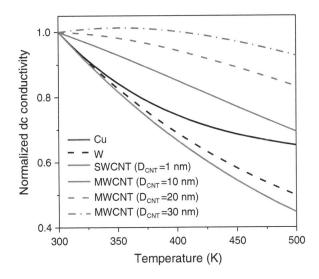

Fig. 9.6 Normalized DC conductivity of the CNT-TSV versus temperature

the temperature-dependent conductivities of bulk Cu and W are also considered and given by

$$\sigma(T) = \sum_{n=0}^{4} c_n T^n \qquad (9.29)$$

where the fitting coefficient c_n can be referred in [37]. As shown in Fig. 9.6, SWCNT conductivity decreases linearly with temperature, and its slope is not affected by the variation of Fm and SWCNT diameter. However, the change in MWCNT diameter has significant impact on the conductivity variation at different temperatures. When the MWCNT diameter is 30 nm, the conductivity is almost unaffected, in spite of the variation of temperature.

Figure 9.7a–d shows the effective complex conductivities of SWCNT- and MWCNT-TSVs at different temperatures. With the increase of frequency, the real part of SWCNT conductivity is kept almost unchanged, while the imaginary part decreases linearly. At room temperature, the real part is equal to 4.06×10^7, 8.13×10^7, and 12.19×10^7 S/m for an SWCNT-TSV with $Fm = 1/3$, $2/3$, and 1, respectively. For the MWCNT conductivity, its real part decreases significantly with frequency. It is worth noting that the imaginary to real ratio decreases linearly with frequency for both SWCNT and MWCNT conductivities. As frequency is increased to 100 GHz, it is about -0.47 for the SWCNT-TSV despite of its value of Fm, and it is equal to -2.79, -4.47, and -5.54 for the MWCNT-TSV, with $D_{CNT} = 10$, 20, and 30 nm assumed, respectively. As the temperature rises, the effective complex conductivity of SWCNT-TSV is decreased, and it has no significant effect for the MWCNT case.

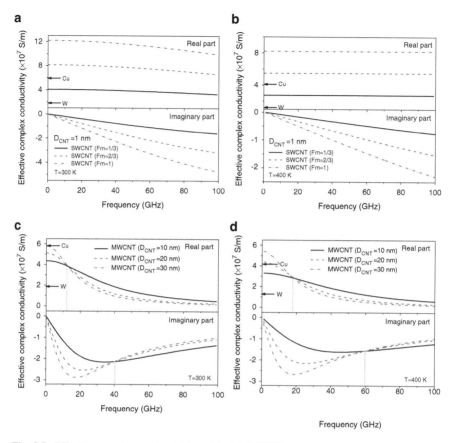

Fig. 9.7 Effective complex conductivities of (**a, b**) SWCNT- and (**c, d**) MWCNT-TSVs versus frequency at different temperatures

Further, the current density distribution over the CNT-TSV is studied. Assuming that the CNT-TSV is homogeneous over its cross-section and $\sigma \gg \omega\varepsilon$ in the frequency-domain, the current density along its radial direction can be described by [38]

$$\frac{d^2J}{dr^2} + \frac{1}{r}\frac{dJ}{dr} + k^2J = 0 \qquad (9.30)$$

where $k = \sqrt{-j\omega\mu\sigma\text{eff}(\omega,\ T)}$. By solving (9.30), the normalized current density is obtained by, i.e.,

$$J_{norm} = \frac{J(r)}{J(r_{via})} = \frac{\mathrm{J}_0(kr)}{\mathrm{J}_0(kr_{via})} \qquad (9.31)$$

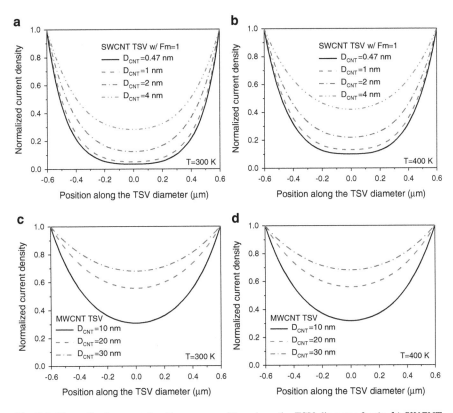

Fig. 9.8 Normalized current density versus position along the TSV diameter for (**a, b**) SWCNT- and (**c, d**) MWCNT-TSVs at different temperatures

where J_0 is the zero-order Bessel function of the first kind. Figure 9.8a–d shows the normalized current density distributions of both SWCNT- and MWCNT-TSVs at different temperatures. Here, the operating frequency is up to 100 GHz, and the TSV radius and height are 0.5 and 30 µm, respectively. With the increase of CNT diameter, the effective conductivity at high frequency decreases, thereby suppressing the skin effect in CNT-TSVs. In addition, an MWCNT-TSV has much smaller skin depth than that of the conventional metals even with the same electrical conductivity assumed [39]. As the temperature rises, the skin effect in an SWCNT-TSV is suppressed slightly, but it is unaffected for the MWCNT case.

9.4 Electrothermal Modeling of a Pair of CNT-TSVs

Figure 9.9a shows the geometry of a pair of CNT-TSVs with both bottom and top metal landings. It is noted that the coupling effects from metal contacts, redistribution layer (RDL), or back-end-of-line (BEOL) to the TSVs are not

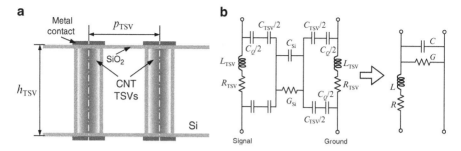

Fig. 9.9 (a) Schematic of a pair of CNT-TSVs, and (b) its equivalent circuit model, which can be simplified as a transmission line one

addressed here, together with feasibility of these connections and their impacts on the electrical performance excluded. Figure 9.9b shows the equivalent lumped-element circuit model of a pair of CNT-TSVs, which is similar to that of the Cu TSV case [40], but quantum effects have to be taken into account here.

For the p-type silicon, the hole mobility can be calculated by

$$\mu_p(T) = \left[\frac{\mu_{\max} - \mu_{\min}}{1 + (N_a/N_{ref})^\alpha} + \mu_{\min} \right] \times \left(\frac{T}{300} \right)^{-3/2} \tag{9.32}$$

where $\mu_{\max} = 495\,\text{cm}^2/\text{V s}$, $\mu_{\min} = 47.7\,\text{cm}^2/\text{V s}$, $\alpha = 0.76$, $N_{ref} = 6.3 \times 1016\,\text{cm}^{-3}$, and N_a is the concentration of substrate dopant impurity. Therefore, the temperature-dependent silicon conductivity can be described by

$$\sigma_{Si} = qN_a\mu_p(T) \qquad [\text{S/cm}] \tag{9.33}$$

The elements $G_{Si}(T)$ and $C_{Si}(T)$ represent silicon conductance and capacitance, i.e.,

$$G_{Si}(T) = \frac{\pi\sigma_{Si}(T)h_{TSV}}{\text{arccosh}\left(p_{TSV}/2r_{dep}(T)\right)} \tag{9.34}$$

$$C_{Si}(T) = \varepsilon_{Si}(T)G_{Si}(T)/\sigma_{Si}(T) \tag{9.35}$$

As introduced earlier, the MOS capacitance should be considered, and it can be obtained by solving 1-D Poisson equation in the cylindrical coordinate system [40], i.e.,

$$\frac{1}{r}\frac{d}{dr}\left(r\frac{d\varphi(r)}{dr}\right) = \frac{-q\left(p(r) - n(r) - N_a + N_d\right)}{\varepsilon_{Si}}, \qquad r \geq r_{ox} \tag{9.36}$$

where $\varphi(r)$ represents the electrostatic potential as a function of radius. N_d is the density of ionized donors, and it can be assumed to be 0 for the p-type silicon. The hole and electron concentrations can be described by

$$p(r) = N_a e^{-q\varphi(r)/k_B T} \tag{9.37}$$

$$n(r) = \frac{n_i^2}{N_a} e^{q\varphi(r)/k_B T} \tag{9.38}$$

where the intrinsic carrier concentration is [41]

$$n_i = 9.38 \times 10^{19} \left(\frac{T}{300}\right)^2 e^{-6884/T} \tag{9.39}$$

At room temperature, both electrons and hole charge densities can be neglected, and the initial maximum depletion radius can be calculated by solving

$$r_{ox}^2 - 2r_{dep}^2 \ln(r_{ox}) + r_{dep}^2 \left[2 \ln\left(r_{dep}\right) - 1\right] = \frac{4\varepsilon_{Si}\psi_s}{qN_a} \tag{9.40}$$

where the surface potential $\psi_s = 2k_B T/q \ln(N_a/n_i)$. The potential can be derived as

$$\varphi(r) = \frac{qN_a r^2}{4\varepsilon_{Si}} - \frac{qN_a r_{dep}^2}{2\varepsilon_{Si}} \ln(r) + \frac{qN_a r_{dep}^2}{4\varepsilon_{Si}} \left[2 \ln\left(r_{dep}\right) - 1\right] \tag{9.41}$$

By using these equations, the total hole and electron charge densities can be calculated by integrating the electron-hole charges over the depletion region. The new maximum depletion radius can be obtained by solving

$$r_{ox}^2 - 2r_{dep}^2 \ln(r_{ox}) + r_{dep}^2 \left[2 \ln\left(r_{dep}\right) - 1\right] = \frac{4\varepsilon_{Si}\psi_s}{q(N_a + p - n)} \tag{9.42}$$

To carry out the iteration described above until the maximum depletion radius reaches a steady value, and the TSV capacitance can be determined by

$$C_{TSV} = \left(\frac{1}{C_{ox}} + \frac{1}{C_{dep}}\right)^{-1} = 2\pi h_{TSV} \left(\frac{1}{\varepsilon_{ox}} \ln\left(\frac{r_{ox}}{r_{via}}\right) + \frac{1}{\varepsilon_{Si}} \ln\left(\frac{r_{dep}}{r_{ox}}\right)\right)^{-1} \tag{9.43}$$

Table 9.1 shows the comparison between the modeled and measured depletion capacitances. It is found that the increase in temperature increases the electron concentration in silicon, thereby leading to the reduction in depletion thickness and increase in TSV capacitance.

Table 9.1 Comparison between modeling and measured results

Temperature (°C)	TSV capacitance (fF) Model	Meas. [40]	Error (%) Model	Δ TSV capacitance (%) Meas. [40]
25	60.496	60.499	0.005	
50	60.811	61.698	1.438	0.52	1.982
75	62.236	62.961	1.152	2.88	4.069
100	63.886	65.187	1.996	5.604	7.749
125	65.799	67.12	1.969	8.766	10.944
150	68.057	–	–	12.498	–
175	70.793	–	–	17.021	–
200	74.236	–	–	22.712	–
225	78.828	–	–	30.303	–
250	85.647	–	–	41.575	–
275	99.098	–	–	63.809	–
300	124.523	–	–	105.837	–

In Fig. 9.9b, the quantum capacitance of CNT-TSV is calculated by [42]

$$C_Q = N_{\text{CNT}} h_{\text{TSV}} \sum_{\text{shell}} c_Q \approx N_{\text{CNT}} h_{\text{TSV}} \frac{4q^2}{h\upsilon_F} \sum_{\text{shell}} N_{ch} \qquad (9.44)$$

The shell number is equal to 1 for the SWCNT-TSV. In the real application, the quantum capacitance can be neglected, as it is much larger than the MOS capacitance.

The admittance between two adjacent CNT-TSVs is given by

$$Y = (G + j\omega C) h_{\text{TSV}} = \left(\frac{2}{j\omega C_Q} + \frac{2}{j\omega C_{ox}} + \frac{1}{G_{\text{Si}} + j\omega C_{\text{Si}}} \right)^{-1} \qquad (9.45)$$

Figure 9.10a shows the effective admittance between two adjacent TSVs, which can be made of Cu or CNTs. Here, the TSV radius is 2.5 μm, the oxide thickness is 0.5 μm, the pitch is 15 μm, and the doping concentration is 1.25×10^{15} cm^{-3}. It is found that the conductance increases and capacitance decreases when the frequency is higher than 1 GHz. Both of them are sensitive to the temperature change. At high frequencies, such as several tens of gigahertz, the effective conductance and capacitance tends to be a constant value, respectively, which can be written as

$$G \approx \frac{1}{h_{\text{TSV}}} \frac{C_{ox}^2}{(C_{ox} + 2C_{\text{Si}})^2} G_{\text{Si}} \qquad (9.46)$$

$$C \approx \frac{1}{h_{\text{TSV}}} \frac{C_{ox}}{2 + C_{ox}/C_{\text{Si}}} \qquad (9.47)$$

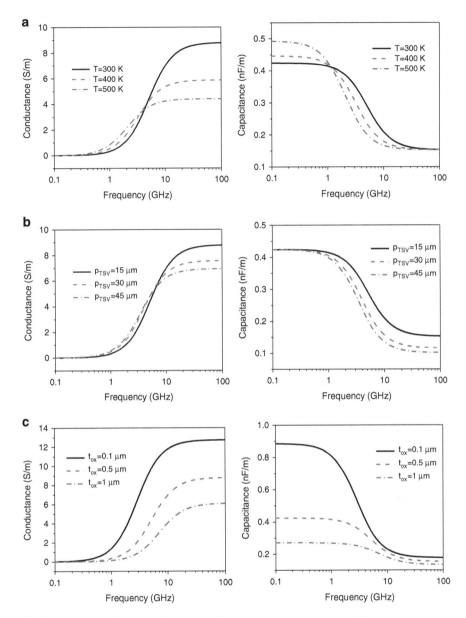

Fig. 9.10 Parallel admittance of a pair of TSVs versus frequency at (**a**) different temperatures, (**b**) different TSV pitches, and (**c**) different oxide thicknesses

Noting that the silicon capacitance C_{Si} decreases slightly as temperature rises, and therefore, the effective capacitance at high frequencies remains the same. The conductance decreases significantly due to the reduction in silicon substrate loss. The impacts of TSV pitch and oxide thickness on the effective admittance of TSV pair are also investigated, as shown in Fig. 9.10b, c, respectively.

The effective impedance of a pair of TSVs can be determined by [33]

$$Z = (R + jL) h_{\text{TSV}} = (2Z_{via} + R_{sub} + L_{outer}) h_{\text{TSV}} \tag{9.48}$$

where Z_{via}, R_{sub}, and L_{outer} are calculated by

$$Z_{via} = \frac{(1-j) \times J_0 \left((1-j) \, r_{via}/\delta_{\text{TSV}}\right)}{2\pi r_{via}\sigma_{\text{eff}}\delta_{\text{TSV}} \times J_1 \left((1-j) \, r_{via}/\delta_{\text{TSV}}\right)} \tag{9.49}$$

$$R_{sub} \approx \frac{\omega\mu}{2} \cdot \text{Re}\left[H_0^{(2)}\left(\frac{(1-j) \, r_{dep}}{\delta_{\text{Si}}}\right)\right] - \frac{\omega\mu}{2} \cdot \text{Re}\left[H_0^{(2)}\left(\frac{(1-j) \, p_{\text{TSV}}}{\delta_{\text{Si}}}\right)\right] \tag{9.50}$$

$$L_{outer} \approx \frac{\mu}{\pi} \text{arccosh}\left(\frac{p_{\text{TSV}}}{2r_{via}}\right) \tag{9.51}$$

where p_{TSV} is the pitch between two TSVs, $J_1(\bullet)$ is the first-order Bessel function of the first kind, $H(2) 0(\bullet)$ is the first-order Hankel function of the second kind, and the damping parameters δ_{TSV} and δ_{Si} are given by

$$\delta_{\text{TSV}} = \sqrt{2/\omega\mu\sigma_{\text{eff}}} \tag{9.52a}$$

$$\delta_{\text{Si}} = \sqrt{2/\omega\mu \left(\sigma_{\text{Si}} + j\omega\varepsilon_{\text{Si}}\right)} \tag{9.52b}$$

As the pitch of TSVs is much larger than each radius, proximity effect may be very weak and can be ignored. Figure 9.11a, b shows the effective resistance and inductance of a pair of TSVs, respectively. Here, the TSV height is 30 μm, the diameters of SWCNT and MWCNT are chosen to be 1 and 20 nm, respectively, and all the other parameters are the same as in Fig. 9.10. It is found that the SWCNT-TSV with $Fm = 1$ shows smaller resistance than that of Cu case. As $Fm = 1/3$,

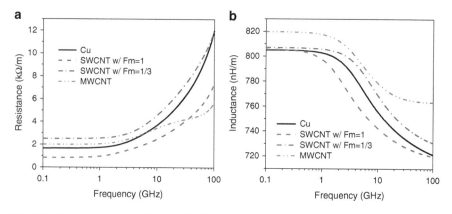

Fig. 9.11 Series impedance of a pair of TSVs versus frequency

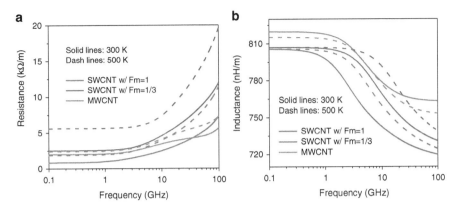

Fig. 9.12 Series impedance of a pair of CNT-based TSVs versus frequency at different temperatures

the resistance of SWCNT-TSV is greater than that of Cu case, but the difference between them decreases with frequency increasing. As shown in Fig. 9.11b, skin effect in the MWCNT-TSV can be suppressed significantly. Therefore, such MWCNT-TSV shows much smaller resistance than its Cu counterpart at high frequencies. The temperature effects on the series impedance of CNT-TSVs are plotted in Fig. 9.12a, b, respectively.

Based on the equivalent circuit model given in Fig. 9.9b, the S-parameters at different frequencies and temperatures can be calculated by

$$[S] = \frac{1}{A + B/Z_0 + CZ_0 + D} \begin{bmatrix} A + B/Z_0 - CZ_0 - D & AD - BC \\ 2 & -A + B/Z_0 - CZ_0 + D \end{bmatrix} \quad (9.53)$$

with $Z_0 = 50\ \Omega$, and

$$[ABCD] = \begin{bmatrix} \cosh(\lambda h_{\mathrm{TSV}}) & Z_c \sinh(\lambda h_{\mathrm{TSV}}) \\ \sinh(\lambda h_{\mathrm{TSV}})/Z_c & \cosh(\lambda h_{\mathrm{TSV}}) \end{bmatrix} \quad (9.54)$$

where the propagation constant and characteristic impedance are determined by

$$\gamma = \sqrt{(R + j\omega L)(G + j\omega C)} \quad (9.55)$$

$$Z_c = \sqrt{\frac{R + j\omega L}{G + j\omega C}} \quad (9.56)$$

Physically, a pair of TSVs with higher electrical conductivity has less conductive loss and better transmission characteristics. Therefore, our attention is focused

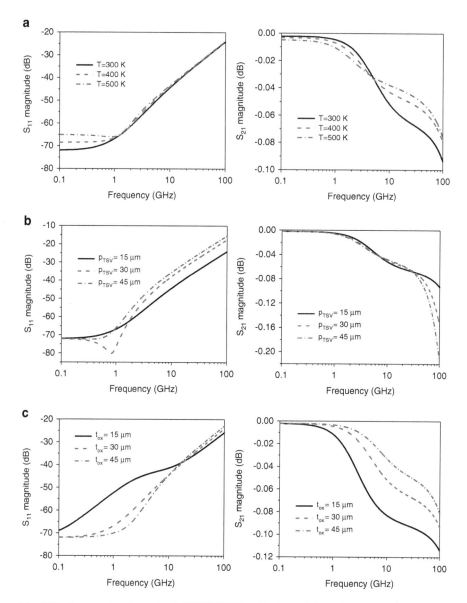

Fig. 9.13 *S*-parameter of a pair of SWCNT-TSVs with *Fm* = 1 for different (**a**) temperatures, (**b**) TSV pitches, and (**c**) oxide thicknesses

on a pair of SWCNT-TSVs with *Fm* = 1 as shown below. Figure 9.13a shows the magnitudes of backward and forward transmission coefficients of a pair of SWCNT-TSVs at different temperatures. At low frequencies, its performance is degraded with temperature increasing, which is due to the increase in conductive loss. However, silicon substrate loss turns to be dominant as the frequency exceeds a certain value. Therefore, the electrical performance of a pair of SWCNT-TSVs is sensitive to the variation of temperature (see Fig. 9.10a). Further, the impact of TSV pitch on the transmission characteristics is plotted in Fig. 9.13b. It is indicated that the effect of silicon loss becomes more significant with TSV pitch increasing, and the performance is degraded. Figure 9.13c shows the comparison in magnitudes of backward and forward transmission coefficients for different oxide thicknesses, and it is observed that larger oxide thickness leads to better electrical performance.

9.5 Crosstalk Effects in CNT-TSVs

Screening effect should be considered in the characterization of electrothermal performance of high density CNT-TSVs array. Here, 3-D transmission line method (TLM) is adopted for modeling the SWCNT-TSVs array. The equivalent circuit models of each unit cell are presented in Fig. 9.14a, b, respectively, with more details given as follows.

The unit cell resistances are determined by

$$R_{unit} = Rh_{unit}/2 \tag{9.57a}$$

$$R_{sub1} = l_{sub}/(2\sigma_{Si}w_{sub}h_{sub}) \tag{9.57b}$$

$$R_{sub2} = h_{sub}/(2\sigma_{Si}l_{sub}h_{sub}) \tag{9.57c}$$

$$R_{sub3} = h_{sub}/(2\sigma_{Si}w_{sub}l_{sub}) \tag{9.57d}$$

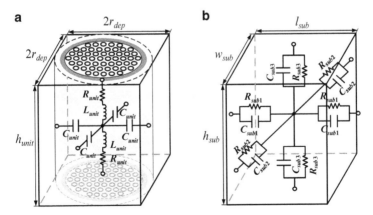

Fig. 9.14 Equivalent circuit models of (**a**) SWCNT-TSV and (**b**) silicon substrate unit cells

And the unit cell capacitances are calculated by

$$C_{unit} = h_{unit}/4 \left(C_Q^{-1} + C_{TSV}^{-1} \right) h_{TSV} \tag{9.58a}$$

$$C_{sub1} = 2\varepsilon_{Si} w_{sub} h_{sub}/l_{sub} \tag{9.58b}$$

$$C_{sub2} = 2\varepsilon_{Si} l_{sub} h_{sub}/w_{sub} \tag{9.58c}$$

$$C_{sub3} = 2\varepsilon_{Si} w_{sub} l_{sub}/h_{sub} \tag{9.58d}$$

The unit cell inductance is $L_{unit} = Lh_{unit}/2$. All these parameters depend on the configuration of SWCNT-TSVs array, with its resistance and inductance obtained by (9.57). Modeling of mutual inductance for such TSV array can be referred in [43].

As discussed above, electric field can be screened due to the presence of other TSVs. Under such circumstances, following the similar procedure presented in [33], substrate coupling can be characterized through Y_1 and Y_2 as shown in Fig. 9.15.

$$Y_{Si}(p_1) = Y_1 + \frac{Y_2}{2} \tag{9.59}$$

$$Y_{Si}\left(\frac{\sqrt{5}p_1}{2}\right) = Y_2 + \left(\frac{1}{Y_1} + \frac{1}{Y_2}\right)^{-1} \tag{9.60}$$

where $Y_{Si}(p_1)$ and $Y_{Si}\left(\sqrt{5p1}/2\right)$ are obtained by $G_{Si} + j\omega C_{Si}$ in (9.59) and (9.60). Therefore,

Fig. 9.15 Equivalent silicon admittance in the 3-SWCNT-TSV array

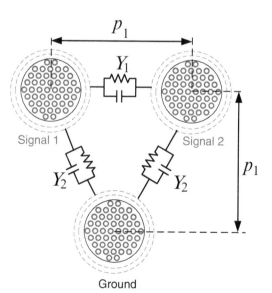

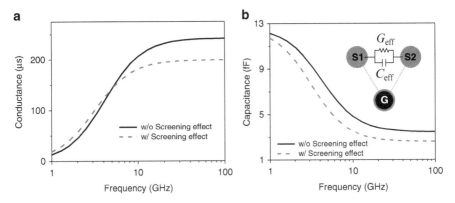

Fig. 9.16 (**a**) Effective conductance and (**b**) effective capacitance between the signal Vias 1 and 2, with and without screening effect considered, respectively

$$Y_1 = \frac{Y_{Si}(p_1) \cdot \left[2Y_{Si}(p_1) - Y_{Si}\left(\sqrt{5}p_1/2\right)\right]}{2Y_{Si}(p_1) - Y_{Si}\left(\sqrt{5}p_1/2\right)/2} \tag{9.61}$$

$$Y_2 = \frac{Y_{Si}(p_1) \cdot Y_{Si}\left(\sqrt{5}p_1/2\right)}{2Y_{Si}(p_1) - Y_{Si}\left(\sqrt{5}p_1/2\right)/2} \tag{9.62}$$

Numerically, Fig. 9.16a, b shows the effective conductance (G_{eff}) and capacitance (C_{eff}) between signal Vias 1 and 2, with and without screening effect considered, respectively. Here, the pitch between two adjacent TSVs is chosen as 30 μm, and the other parameters are the same as in Figs. 9.5, 9.17, and 9.18 that show the values of effective conductance and capacitance between the signal Vias 1 and 2 for different TSV pitches and silicon conductivities with screening effect taken into account. These numerical results are helpful for understanding mutual couplings in the SWCNT-TSVs array.

Based on the modeling method of unit cells as used in Fig. 9.14a, b, respectively, SWCNT-TSVs array is further meshed and simulated as below. A clock signal is injected into the input port Via 1, and the coupling noise on the signal Via 2 is captured and analyzed. It is found that the calculated peak-to-peak voltage amplitude V_{pk-pk} of the coupling noise is about 14.66 mV at 1 GHz, while at 10 GHz, it is increased to about 59.55 mV. V_{pk-pk} is also sensitive to the variation of rising and falling time (see Fig. 9.19a, b). At high frequency, mutual coupling is enhanced and dominated by capacitive coupling. Coupling noise can also be introduced by field-implantation, interlayer dielectrics/intermediate dielectrics (ILD/IMD), and bumps on the top and bottom of TSVs. These effects are not examined in detail here. Figure 9.20a, b shows the coupling noises among a 4-SWCNT-TSVs array of row-repeated and distributed, and all of them have the same geometrical parameters

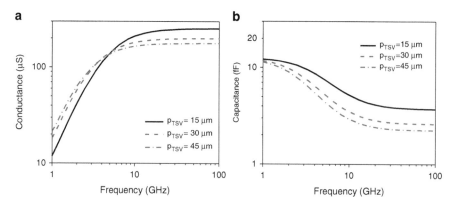

Fig. 9.17 (**a**) Effective conductance and (**b**) effective capacitance between signal Vias 1 and 2 for different pitches

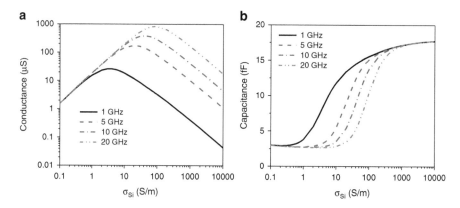

Fig. 9.18 (**a**) Effective conductance and (**b**) effective capacitance between signal Vias 1 and 2 at different operating frequencies

and material properties. It is observed that the peak-to-peak voltage of coupling noise is lower than that of the row-repeated. At frequencies of 1 and 10 GHz, $\left\{V_{pk-pk}^{(I)},\ V_{pk-pk}^{(II)}\right\} = \{11.85,\ 7.60\}$ and $\{51.08,\ 23.98\}$ mV, respectively. Therefore, a high density SWCNT-TSVs array can be designed in a distributed way, in particular operating at high frequencies.

9.6 3-D Carbon-Based Heterogeneous Interconnects

As 2-D graphene has the similar physical properties as CNT, it is much suitable for the realization of horizontal interconnects with remarkable performance [42, 44]. Recently, a 3-D carbon-based heterogeneous interconnect scheme, consisting of a

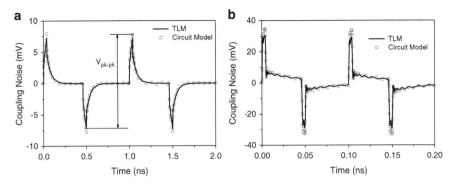

Fig. 9.19 Coupling noise in the 3-SWCNT-TSV array obtained by TLM and equivalent circuit model. (**a**) 1 GHz and (**b**) 10 GHz

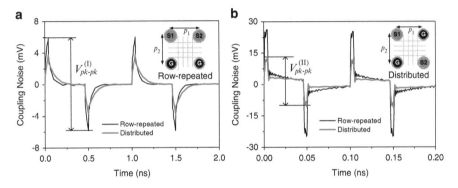

Fig. 9.20 Coupling noise in the 4-SWCNT-TSVs array. (**a**) 1 GHz and (**b**) 10 GHz

couple of horizontal multilayer graphene (MLG) interconnects and vertical CNT-TSVs, is proposed and investigated theoretically in [45], as plotted in Fig. 9.21a–c, respectively.

In Fig. 9.21a–c, there are MLG, bumps (see Fig. 9.22a), CNT-TSVs, silicon oxide isolation, and IMD layers. Similar to the modeling of CNT-TSVs performed above, a couple of MLG interconnects is at first studied as shown in Fig. 9.22b, c. The coupling capacitance between two MLG interconnects is denoted by C_c [4].

$$C_c = C_{air} + C_{pass} + C_{diel} \qquad (9.63)$$

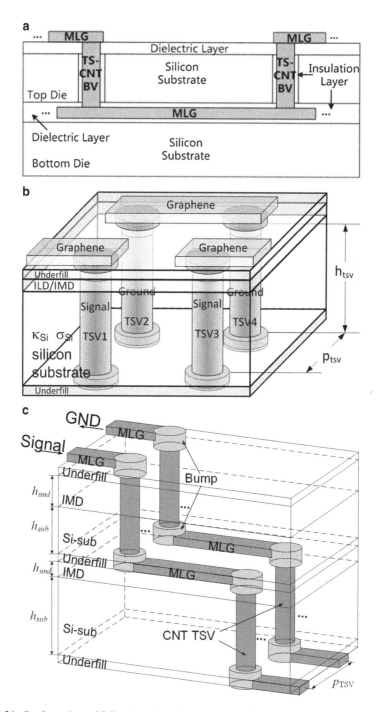

Fig. 9.21 Configurations of 3-D carbon-based heterogeneous interconnects

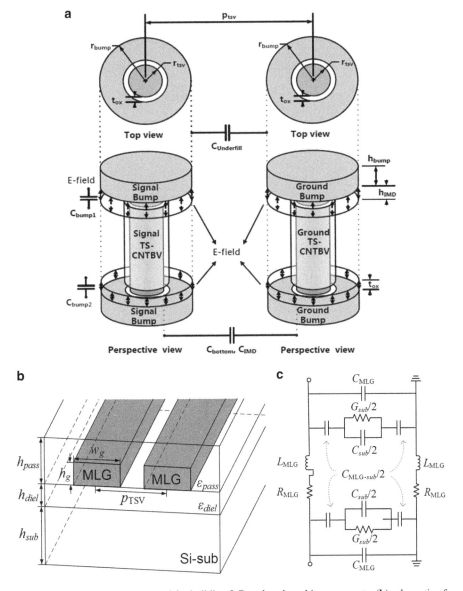

Fig. 9.22 (**a**) Metallic bumps used for building 3-D carbon-based interconnects; (**b**) schematic of a couple of MLG interconnects, and (**c**) its equivalent circuit model

The air, passivation layer, and dielectric capacitances can be obtained by

$$C_{air} = \varepsilon_{air} K' (k_0) / K (k_0) \tag{9.64}$$

$$C_{pass} = \left(\varepsilon_{pass} - \varepsilon_{air} \right) K' (k_1) / K (k_1) \tag{9.65}$$

$$C_{diel} = \left(\varepsilon_{diel} - \varepsilon_{pass}\right) K'\left(k_2\right) / K\left(k_2\right) \tag{9.66}$$

and

$$k_0 = \sqrt{1 - \left(w_g/p_{TSV}\right)^2} \tag{9.67a}$$

$$k_1 = \sqrt{1 - \frac{\sinh^2\left(\pi w_g/2h_{pass}\right)}{\sinh^2\left(\pi p_{TSV}/2h_{pass}\right)}} \tag{9.67b}$$

$$k_2 = \sqrt{1 - \frac{\sinh^2\left(\pi w_g/2h_{diel}\right)}{\sinh^2\left(\pi p_{TSV}/2h_{diel}\right)}} \tag{9.67c}$$

where $K(\bullet)$ is the complete elliptical integral of the first kind. w_g and h_g are the MLG width and thickness, respectively. h_{pass} is the thickness of passivation layer. ε_{air}, ε_{pass}, and ε_{diel} are the permittivities of air, passivation layer, and dielectrics, respectively. The parasitic capacitance between two MLG interconnects and substrate can be calculated by [46]

$$C_{MLG-sub} = C_b + 2C_f = \varepsilon_0 \varepsilon_{rox} \left[\frac{w_g}{h_{diel}} + \frac{K\left(k_{[VP]}\right)}{K'\left(k_{[Vp]}\right)} \right] \tag{9.68a}$$

and

$$k_{[VP]} = \sqrt{1 - \left(\frac{h_{diel}}{h_{diel} + w_g}\right)^2} \tag{9.68b}$$

where h_{diel} is the thickness of dielectric. The effect of silicon substrate can be modeled by its conductance and capacitance connected in a shunt way, and they are determined by

$$G_{sub}(T) = \sigma_{eff}(T)w_g/h_{eff}(f) \tag{9.69}$$

$$C_{sub}(T) = \varepsilon_{eff}(T)G_{sub}(T)/\sigma_{eff}(T) \tag{9.70}$$

where $\varepsilon_{eff}(T)$, $\sigma_{eff}(T)$, and $h_{eff}(T)$ are the effective permittivity, conductivity, and penetration depth of electric field in silicon, i.e.,

$$\varepsilon_{eff}(T) = \frac{\varepsilon_{Si}(f) + 1}{2} + \frac{\varepsilon_{Si}(f) - 1}{2\sqrt{1 + 10h_{sub}/w_g}} \tag{9.71a}$$

$$\sigma_{eff}(T) = \frac{\sigma_{Si}}{2} + \frac{\sigma_{Si}}{2\sqrt{1 + 10h_{sub}/w_g}} \tag{9.71b}$$

$$h_{eff}(T) = \frac{w_g}{2\pi \ln\left(8h_{sub}/w_g + w_g/4h_{sub}\right)} \tag{9.71c}$$

where h_{sub} is the thickness of silicon substrate. According to [47], the distribution function of carrier in AsF5-doped MLG interconnect can be described by

$$f = f_0 + f_n\left(\mathbf{v}, \mathbf{r}\right), \; f_0 = \left[1 + \exp\left(\frac{E - E_F}{k_B T}\right)\right]^{-1}. \tag{9.72}$$

where f_0 and f_n are the equilibrium and non-equilibrium distribution functions, respectively. $\mathbf{v}$ is the velocity vector, and $\mathbf{r}$ is the space vector. Based on the linear dispersion relation of graphene, one 3-D Boltzmann equation can be derived and turned into

$$v_x \partial f_n/\partial x + (1 + j\omega\tau)f_n/\tau = ev_y \varepsilon(x)\partial f/\partial E \tag{9.73}$$

where two velocity components v_x and v_y can be referred in [47]. Therefore, the general solution to (9.73) can be derived as

$$f_n\left(\omega\right) = \exp\left(-\frac{1 + j\omega\tau}{\tau v_x}x\right) \times \left[F(v) + \frac{e}{v_x}\frac{\partial f_0}{\partial E}\int_0^x v_y \varepsilon(x) \exp\left(\frac{1 + j\omega\tau}{\tau v_x}t\right)dt\right] \tag{9.74}$$

and it can be further solved with a set of boundary conditions applied. Further, an improved partial-element equivalent-circuit (PEEC) method is adopted for characterizing the frequency- and temperature-dependent impedance of MLGs. At first, such MLG is divided into many filaments, and we define

$$G_{jk} = \begin{cases} L_S, & \text{if } j = k \\ M_{jk}, & \text{otherwise} \end{cases} \tag{9.75}$$

where L_s and M_{ij} are the self- and mutual inductances (in μH/m) of single filament and between the filaments, respectively. The voltage drop across the jth-filament can be expressed as

$$A_j\left(\omega\right) = a_f \sum_k G_{jk}J_k\left(\omega\right) \tag{9.76}$$

where a_f is the cross-sectional area of single filament, J_k represents the current density at the center of kth-filament. The voltage drop across the jth-filament is $V_j(\omega) = [\varepsilon_j + j\omega Aj(\omega)]l_0$. As the average voltage drop V_{avg} is calculated, one new value of electric field distribution is obtained. The iteration is continued until the convergence is obtained for both real and imaginary parts of the electric field. Finally, the impedance of single MLG interconnect is calculated by [47]

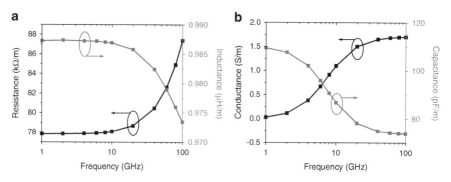

Fig. 9.23 (**a**) Resistance/inductance, and (**b**) conductance/capacitance per unit length of a couple of MLG interconnects as a function of frequency at room temperature

$$Z(\omega) = V_{avg}(\omega) / a_f \sum_k J_k(\omega) \qquad (9.77)$$

The circuit model in Fig. 9.22b can be further simplified into a transmission line one. The extracted resistance, inductance, conductance, and capacitance per unit length of a couple of MLG interconnects are plotted in Fig. 9.23a, b, respectively. Here, both MLG width and thickness are 1 μm, the pitch between two MLGs is 4 μm, the thickness of silicon substrate is 10 μm, and the silicon conductivity is 10 S/m.

As shown in Fig. 9.22a, metallic bumps must be introduced during the integration of a 3-D carbon-based interconnect. The contact resistances between metallic bumps and MLG interconnects can be extracted according to the circuit model proposed in [48]. The contact resistance at the interface between CNT-TSV and bump is fully determined by ballistic and diffusive regimes of electrons, i.e., $R_c = R_{nc} + R_o$, where R_{nc} and R_o are the non- and fully transparent contact resistances, respectively. R_o is equal to R_q, and as the MFP of electron in CNT is larger than the saturated one, electron transport at interface cannot be affected by scattering, and $R_{nc} \ll R_q$. Therefore, the contact resistance between bump and CNT is equal to R_q/N_{CNT} approximately. Our simulated results show that the contact resistances between metal bumps and carbon-based interconnects have negligible impact on the electrical performance of such 3-D carbon-based heterogeneous interconnect as shown in Fig. 9.21.

Finally, the S-parameters of such 3-D carbon-based heterogeneous interconnect are plotted in Fig. 9.24. The TSV radius and height are given by 0.5 and 10 μm, respectively, and all the other parameters are the same as those in Fig. 9.21. It is observed that it could provide superior performance over its Cu counterpart. In order to further decrease its insertion loss and improve its forward transmission performance at high frequencies, we can introduce some patterned ground shields (PGS) beneath MLG interconnects so as to suppress silicon loss [46].

On the other hand, it should be indicated that although MLG can provide superior performance over its Cu counterpart, state-of-the-art graphene samples

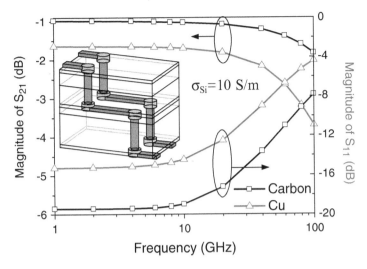

Fig. 9.24 S_{21}- and S_{11}-parameters of the 3-D carbon-based heterogeneous interconnect as a function of frequency

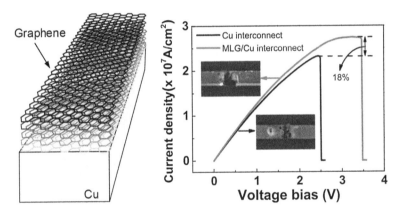

Fig. 9.25 Schematic of Cu-graphene heterogeneous interconnect [51], and its carrying current capability before its electrothermal breakdown [52]

cannot satisfy the requirements due to the limitation on fabrication quality in large scale. Fortunately, a promising alternative, i.e., Cu-graphene heterogeneous interconnect, has been proposed by combining Cu with graphene [49–54]. Such novel interconnect can be easily realized by transferring MLG to the surface of Cu, and subsequently patterning them using photolithography, as shown in Fig. 9.25. The reliability can be improved by 18 % by transferring MLG. Yeh et al. have fabricated such Cu-graphene heterogeneous interconnects by using electron cyclotron resonance chemical vapor deposition (ECR-CVD) at 400 °C, which is compatible with the CMOS technology. It is demonstrated that total current density

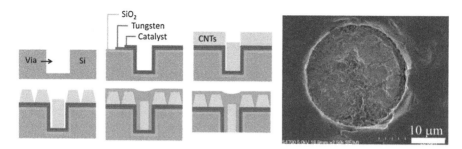

Fig. 9.26 Process flow for fabrication of TSVs filled with Cu–CNT composite, and its SEM image [57]

carried by such heterogeneous interconnect can be larger than that of Cu counterpart. Further, thermal properties of such structure have been studied more recently.

Similarly, a composite of Cu and CNT has also been developed for interconnect as well as TSV application [55–57]. Feng and Burkett [57] have used such composite to realize a TSV, as shown in Fig. 9.26. Compared to the pure CNT-TSV, such TSV made of both CNT and Cu possesses comparable conductivity to the pure Cu TSV. It is expected that an intermediate step by combining the Cu-graphene structure and Cu–CNT composite will be very helpful for developing the new generation 3-D carbon-based heterogeneous interconnects in the near future.

9.7   Conclusion

With continuous downscaling of feature dimensions, TSV reliability becomes one of the main concerns in the development of high density 3-D ICs. CNT should be a promising candidate as filling conductive material for the TSV realization. In this chapter, electrothermal modeling and performance analysis of CNT-TSVs have been performed. The effective complex conductivity of a CNT-TSV has been investigated for accurately evaluating impacts of CNT kinetic inductance and other distributed parameters. Further, coupling noises among various CNT-TSVs arrays have been studied by using an extended TLM. Finally, some novel 3-D carbon-based heterogeneous interconnect schemes, which combine MLG interconnect and vertical CNT-TSV, have been presented and their performance has been evaluated.

Acknowledgments This study was supported by the National Natural Science Foundation of China under Grants 61171037, 61431014, 61504033, and 61504121.

References

1. Motoyoshi M (2009) Through-silicon via (TSV). Proc IEEE 97(1):43–48
2. Katti G, Stucchi M, De Meyer K, Dehaene W (2010) Electrical modeling and characterization of through silicon via for three-dimensional ICs. IEEE Trans Electron Devices 57(1):256–261
3. Van der Plas G, Limaye P, Loi I, Mercha A, Oprins H, Torregiani C, Thijs S, Linten D, Stucchi M, Katti G, Velenis D, Cherman V, Vandevelde B, Simons V, De Wolf I, Labie R, Perry D, Bronckers S, Minas N,. Cupac M, Ruythooren W, Van Lomen J, Phommahaxay A, de Potter de ten Broeck M, Opdebeeck A, Rakowski M, De Wachter B, Dehan M, Nelis M, Agarwal R, Pullini A, Angiolini F, Benini L, Dehaene W, Travaly Y, Beyne E, Marchal P (2011) Design issues and considerations for low-cost 3-D TSV IC technology. IEEE J Solid State Circuits 46(1):293–307
4. Kim J, Park JS, Cho J, Song E, Cho J, Kim H, Song T, Lee J, Lee H, Park K, Yang S, Suh M, Byun K, Kim J (2011) High-frequency scalable electrical model and analysis of a through silicon via (TSV). IEEE Trans Compon Packag Manuf Technol 1(2):181–195
5. Pak JS, Kim J, Cho J, Kim K, Song T, Ahn S, Lee J, Lee H, Park K, Kim JH (2011) PDN impedance modeling and analysis of 3D TSV IC by using proposed P/G TSV array model based on separated P/G TSV and chip-PDN models. IEEE Trans Compon Packag Manuf Technol 1(2):208–219
6. Saccetto D, Zervas M, Temiz Y, De Micheli G, Leblebici Y (2012) Resistive programmable through-silicon vias for reconfigurable 3-D fabrics. IEEE Trans Nanotechnol 11(11):8–11
7. Kim K, Hwang C, Koo K, Cho J, Kim H, Kim JH, Lee J, Lee HD, Park KW, Pak JS (2012) Modeling and analysis of a power distribution network in TSV-based 3-D memory IC including P/G TSVs, on-chip decoupling capacitors, and silicon substrate effects. IEEE Trans Compon Packag Manuf Technol 2(12):2057–2070
8. Kim J, Cho J, Kim JH, Yook JM, Kim JC, Lee J, Park K, Pak JS (2014) High-frequency scalable modeling and analysis of a differential signal through-silicon via. IEEE Trans Compon Packag Manuf Technol 4(4):697–707
9. Lee DU, Kim KW, Kim KW, Lee KS, Byeon SJ, Kim JH, Lee J, Chun JH (2015) A 1.2 V 8 Gb 8-channel 128 GB/s high-bandwidth memory (HBM) stacked DRAM with effective I/O test circuits. IEEE J Solid State Circuits 50(1):191–203
10. Zhang X, Lin JK, Wickramanayaka S, Zhang S, Weerasekera R, Dutta R, Chang KF, Chui JJ, Li HY, Ho DSW, Ding L, Katti G, Bhattacharya S, Kwong DL (2015) Heterogeneous 2.5D integration on through silicon interposer. Appl Phys Rev 2:021308
11. Zhao WS, Yin WY, Wang XP, Xu XL (2011) Frequency- and temperature-dependent modeling of coaxial through-silicon via for 3-D ICs. IEEE Trans Electron Devices 58(10):3358–3368
12. Liang F, Wang G, Zhao D, Wang BZ (2013) Wideband impedance model for coaxial through-silicon vias in 3-D integration. IEEE Trans Electron Devices 60(8):2498–2504
13. Gambino JP, Adderly SA, Knickerbocker JU (2015) An overview of through-silicon-via technology and manufacturing challenges. Microelectron Eng 135:73–106
14. Zhao WS, Yin WY (2012) Carbon-based interconnects for RF nanoelectronics. In: Webster J (ed) Wiley encyclopedia of electrical and electronics engineering. Wiley, 1–20
15. Li H, Yin WY, Banerjee K, Mao JF (2008) Circuit modeling and performance analysis of multi-walled carbon nanotube interconnects. IEEE Trans Electron Devices 55(6):1328–1337
16. Chen WC, Yin WY, Jia L, Liu QH (2009) Electrothermal characterization of single-walled carbon nanotube (SWCNT) interconnect arrays. IEEE Trans Nanotechnol 8(6):718–728
17. Liang F, Wang G, Ding W (2011) Estimation of time delay and repeater insertion in multiwall carbon nanotube interconnects. IEEE Trans Electron Devices 58(8):2712–2720
18. Xie R, Zhang C, van der Veen MH, Arstila K, Hantschel T, Chen B, Zhong G, Robertson J (2013) Carbon nanotube growth for through silicon via application. Nanotechnology 24(12):125603
19. Wang T, Chen S, Jiang D, Fu Y, Jeppson K, Ye L, Liu J (2012) Through-silicon vias filled with densified and transferred carbon nanotube forests. IEEE Electron Device Lett 33(3):420–422

20. Xu T, Wang Z, Miao J, Chen X, Tan CM (2007) Aligned carbon nanotubes for through-wafer interconnects. Appl Phys Lett 91(4):042108
21. Wang T, Jeppson K, Olofsson N, Campbell EEB, Liu J (2009) Through silicon vias filled with planarized carbon nanotube bundles. Nanotechnology 20(48):485203
22. Ghosh K, Verma YK, Tan CS (2014) Implementation of carbon nanotube bundles in sub-5 micron diameter through-silicon-via structures for three-dimensionally stacked integrated circuits. Mater Today Commun 2:16–25
23. Callaway J (1959) Model for lattice thermal conductivity at low temperatures. Phys Rev 113(4):1046–1051
24. Bhttacharya S, Almaraj R, Mahapatra S (2011) Physics-based thermal conductivity model for metallic single-walled carbon nanotube interconnects. IEEE Electron Device Lett 32(2): 203–205
25. Verma R, Bhattacharya S, Mahapatra S (2011) Analytical solution of Joule-heating equation for metallic single-walled carbon nanotube interconnects. IEEE Trans Electron Dev 58(11): 3991–3996
26. Pop E, Mann DA, Goodson KE, Dai HJ (2007) Electrical and thermal transport in metallic single-wall carbon nanotubes on insulating substrates. J Appl Phys 101(9):093710
27. Yu C, Shi L, Yao Z, Li D, Majumdar A (2005) Thermal conductance and thermopower of an individual single-wall carbon nanotube. Nano Lett 9(9):1842–1846
28. Kim P, Shi L, Majumdar A, McEuen PL (2001) Thermal transport measurements of individual multiwalled nanotubes. Phys Rev Lett 87(21):215502
29. Zhang S, Xia M, Zhao S, Xu T, Zhang E (2003) Specific heat of single-walled carbon nanotubes. Phys Rev B 68:075415
30. Hone J, Batlogg B, Benes Z, Johnson AT, Fischer JE (2000) Quantized phonon spectrum of single-wall carbon nanotubes. Science 289:1730
31. Yi W, Lu L, Dian-lin Z, Pan ZW, Xie SS (1999) Linear specific heat of carbon nanotubes. Phys Rev B 59(14):9015
32. Mizel A, Benedict LX, Cohen ML, Louie SG, Zettl A, Budraa NK, Beyermann WP (1999) Analysis of the low-temperature specific heat of multi-walled carbon nanotubes and carbon nanotube ropes. Phys Rev B 60(5):3264
33. Xu C, Li H, Suaya R, Banerjee K (2010) Compact AC modeling and performance analysis of through-silicon vias in 3-D ICs. IEEE Trans Electron Devices 57(12): 3405–3417
34. Harutyunyan AR, Chen G, Paronyan TM, Pigos EM, Kuznetsov OA, Hewaparakrama K, Kim SM, Zakharov D, Stach EA, Sumanasekera GU (2009) Preferential growth of single-walled carbon nanotubes with metallic conductivity. Science 326(5949):116–120
35. Li H, Srivastava N, Mao JF, Yin WY, Banerjee K (2011) Carbon nanotube vias: does ballistic electron-phonon transport imply improved performance and reliability? IEEE Trans Electron Devices 58(8):2689–2701
36. Zhao WS, Sun L, Yin WY, Guo YX (2014) Electrothermal modelling and characterisation of submicron through-silicon carbon nanotube bundle vias for three-dimensional ICs. Micro Nano Lett 9(2):123–126
37. Wang XP, Yin WY, He S (2010) Multiphysics characterization of transient electrothermomechanical responses of through-silicon vias applied with a periodic voltage pulse. IEEE Trans Electron Devices 57(6):1382–1389
38. D'Amore M, Sarto MS, D'Aloia AG (2010) Skin-effect modeling of carbon nanotube bundles: the high-frequency effective impedance. In: 2010 IEEE international symposium on electromagnetic compatibility (EMC), IEEE, Fort Lauderdale, 25–30 July 2010, FL, pp 847–852
39. Chiariello AG, Maffucci A, Miano G (2012) Electrical modeling of carbon nanotube vias. IEEE Trans Electromagn Compat 54(1):158–166
40. Katti G, Stucchi M, Velenis D, De Meyer K, Dehaene W (2011) Temperature-dependent modeling and characterization of through-silicon via capacitance. IEEE Electron Device Lett 32(4):563–565

41. Zhao WS, Wang XP, Yin WY (2011) Electrothermal effects in high density through silicon via (TSV) arrays. Prog Electromagn Res 115:223–242
42. Chiariello AG, Maffucci A, Miano G (2013) Circuit models of carbon-based interconnects for nanopackaging. IEEE Trans Compon Packag Manuf Technol 3(11):1926–1937
43. Zhao WS, Yin WY, Guo YX (2012) Electromagnetic compatibility-oriented study on through silicon single-walled carbon nanotube bundle via (TS-SWCNTBV) arrays. IEEE Trans Electromagn Compat 54(1):149–157
44. Zhao WS, Yin WY (2014) Comparative study on multilayer graphene nanoribbon (MLGNR) interconnects. IEEE Trans Electromagn Compat 56(3):638–645
45. Liu YF, Zhao WS, Yong Z, Fang Y, Yin WY (2014) Electrical modeling of three-dimensional carbon-based heterogeneous interconnects. IEEE Trans Nanotechnol 13(3):488–495
46. Yin WY, Zhao WS, Webster J, (2013) Modeling and characterization of on-chip interconnects. In: Wiley encyclopedia of electrical and electronics engineering. Wiley 1–18
47. Sarkar D, Xu C, Li H, Banerjee K (2011) High-frequency behavior of graphene-based interconnects—part I: impedance modeling. IEEE Trans Electron Devices 58(3):843–852
48. Khatami Y, Li H, Xu C, Banerjee K (2012) Metal-to-multilayer graphene contact—part II: analysis of contact resistance. IEEE Trans Electron Devices 59(9):2453–2460
49. Mehta R, Chugh S, Chen Z (2015) Enhanced electrical and thermal conduction in graphene-encapsulated copper nanowires. Nano Lett 15(3):2024–2030
50. Zhao WS, Wang DW, Wang G, Yin WY (2015) Electrical modeling of on-chip Cu-graphene heterogeneous interconnects. IEEE Electron Device Lett 36(1):74–76
51. Zhao WS, Zhang R, Fang Y, Yin WY, Wang G, Kang K (2016) High-frequency modeling of Cu-graphene heterogeneous interconnects. Int J Numer Modell, 29(2): 157–165
52. Kang CG, Lim SK, Lee S, Lee SK, Cho C, Lee YG, Hwang HJ, Kim Y, Choi HJ, Choe SH, Ham MH, Lee BH (2013) Effects of multi-layer graphene capping on Cu interconnects. Nanotechnology 24:115707
53. Yeh CH, Medina H, Lu CC, Huang KP, Liu Z, Suenaga K, Chiu PW (2014) Scalable graphite/copper bishell composite for high-performance interconnects. ACS Nano 8(1): 275–282
54. Goli P, Ning H, Li X, Lu CY, Novoselov KS, Balandin AA (2014) Thermal properties of graphene-copper-graphene heterogeneous films. Nano Lett 14:1497–1503
55. Subramaniam C, Yamada T, Kobashi K, Sekiguchi A, Futaba DN, Yumura M, Hata K (2013) One hundred fold increase in current carrying capacity in a carbon nanotube-copper composite. Nat Commun 4:2202
56. Jordan MB, Feng Y, Burkett SL (2015) Development of seed layer for electrodeposition of copper on carbon nanotube bundles. J Vac Sci Technol B 33:021202
57. Feng Y, Burkett SL (2015) Fabrication and electrical performance of through silicon via interconnects filled with a copper/carbon nanotube composite. J Vac Sci Technol B 33:022004

Chapter 10
Exploring Carbon Nanotubes for 3D Power Delivery Networks

Aida Todri-Sanial

10.1 Introduction

Carbon nanotubes (CNTs) are a class of nanomaterials with unique mechanical, thermal, and electrical properties [11]. CNTs can be classified into two types: single-wall (SWCNTs) and multi-wall (MWCNTs). SWCNTs are rolled graphitic sheets with diameters on the order of 1 nm. MWCNTs consist of several rolled graphitic sheets nested inside each other and can have diameters as large as 100 nm. Depending on their chirality, the CNTs can be metallic or semiconductors. Metallic CNTs (m-CNTs) are ballistic conductors, which show promise for use as interconnects in nanoelectronics. On the other hand, semiconducting CNTs (s-CNTs) have a diameter-depended band-gap and do not have surface states that need passivation, thus can be used to make devices such as diodes and transistors [11, 12, 14, 15].

CNTs are cylindrical carbon molecules formed by one-atom thick sheets of carbon or graphene [5, 7–12, 15]. CNTs, both SWCNT and MWCNT are being investigated for a variety of nanoelectronics applications because of their unique properties [5, 7, 8]. Their extraordinary large electron mean free paths and resistance to electromigration make them potential candidates for interconnects in large scale systems. During the past decade, most of the research is focused on CNT growth, synthesis, modeling and simulation, and characterizing contact interfaces [2]. Detailed simulation for signal interconnects has been performed by Khan and Hassoun [7], Jirahara et al. [5], Leonard and Talin [8], Li et al. [9–12], and Naeemi et al. [15] and shown that CNTs have lower parasitics than Cu metal lines, however, the contact resistance between CNT-to-CNT and CNT-to-metal is large and can be detrimental for timing issues. Additionally, researchers are looking into

A. Todri-Sanial (✉)
CNRS-LIRMM/University of Montpellier, Montpellier, France
e-mail: aida.todri@lirmm.fr

© Springer International Publishing Switzerland 2017
A. Todri-Sanial et al. (eds.), *Carbon Nanotubes for Interconnects*,
DOI 10.1007/978-3-319-29746-0_10

different CNT growth techniques that are compatible with CMOS process and lab measurements indicate the potential of integrating CNTs on-chip [13, 16, 17].

One essential and most interesting application of the nanotubes in microelectronics is as interconnects using the ballistic (without scattering) transport of electrons and the extremely high thermal conductivity along the tube axis [2]. Electronic transport in SWCNTs and MWCNTs can go over long nanotube lengths, $1\,\mu m$, enabling CNTs to carry very high currents (i.e., $>10^9\,A/cm^2$) with essentially no heating due to nearly 1D electronic structure.

In literature, the comparison of copper and CNTs has been limited to signal interconnects. Investigation of CNTs for power and clock delivery would also have a significant importance. It would reveal whether or not CNTs can potentially replace both signal and power/ground copper wires. Additionally, clock and power networks are most vulnerable to electromigration, it is therefore critical to know whether or not CNTs improve their reliability.

There are many works in literature that investigate CNT interconnects. The first group of works focuses on modeling aspects of CNT interconnects [2, 11, 12, 16]. The second group of works focuses on performance comparison of CNT interconnects versus copper (Cu) interconnects [12, 14, 20, 22]. Almost all these works have considered the application of CNT interconnects for signaling and few works focus on power delivery [7, 15]. Complementary to these efforts, in this chapter we investigate the application of horizontally aligned CNTs for power delivery network (i.e., both 2D and 3D ICs) while exploiting their unique electrical and thermal properties.

The rest of this chapter is organized as follows. Section 10.2 describes the modeling techniques that we utilize in this work. In Sect. 10.3, we explore 2D power delivery networks with CNTs. In Sect. 10.4, we present the analysis of CNT Through-Silicon-Vias (TSVs) for 3D power delivery networks. Section 10.5 concludes this chapter.

10.2 Modeling of CNTs

There are many papers in literature that focus on CNT modeling and understanding its transport properties [2, 11, 12, 14, 15]. In this section, we provide a brief description of CNT modeling that we utilize in this work. A generalized model for CNT interconnects is depicted in Fig. 10.1. In Fig. 10.1a the model of an individual MWCNT is shown with parasitics represent both dc conductance and high-frequency impedance, i.e., inductance and capacitance effects. Multiple shells of a MWCNT are presented by the individual parasitics of each shell. Such model can also be applicable to SWCNTs where only a single shell is represented.

Each shell has a lumped ballistic resistance (R_i), and lumped contact resistance (R_c) due to imperfect metal-nanotube contacts. These contacts are typically constructed of Gold, Palladium, or Rhodium [11]. The nanotubes have also a distributed ohmic resistance (R_o), which is dependent on length, l_b, and mean free path of

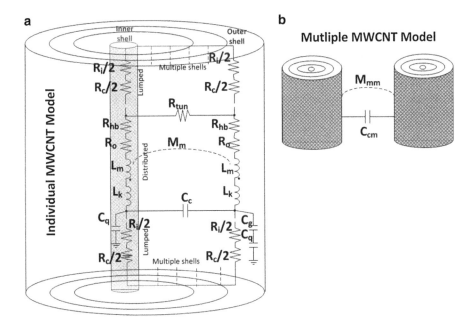

Fig. 10.1 (**a**) Circuit model of an individual MWCNT and (**b**) multiple MWCNTs. This is general enough model to be applicable to MWCNTs of different diameters and shell numbers. It can also be applicable to SWCNTs where the model of a single shell can be utilized

acoustic phonon scattering (λ_{ap}). Overall CNT resistance depends also on the applied bias voltage, $R_{hb} = V_{bias}/I_o$, where I_o is the maximum saturation current (I_o values 15–30 µA [16]). Between shells in MWCNTs, there is also an intershell tunneling resistance (R_{tun}). As the applied bias voltage to each shell is the same, the impact of R_{tun} is relatively small. All the aforementioned ballistic, ohmic and contact resistance depend on the number of 1-D conducting channels, N_c. For metallic SWCNTs the number of conducting channels is always $N_c = 2$ due to lattice degeneracy [2]. Whereas, for semiconducting SWCNTs and small diameter semiconducting MWCNTs, $N_c = 0$. For any conducting shells in an MWCNT, the intrinsic resistance, $R_i = R_q/N_c$, where R_q is the quanta conductance for a 1D conduction channel ($R_q = 12.9\,\Omega$)[2]. Also, contact resistance is $R_c = 2R_{co}/N_c$ where R_{co} is the nominal contact resistance [2]. Ohmic resistance is derived as $R_o = R_q L/N_c d_s C_\lambda$, where L is the length of the MWCNT, d_s is the diameter of the shell, and C_λ is the acoustic phonon scattering mean free path (λ_{ap}). Thus, the total resistance of an individual nanotube, (R_t), can be obtained by computing resistance of each shell, $R_{shell} = R_i + R_c + R_{hb} + R_o$:

$$G_t = \sum_{i=0}^{N} \frac{1}{R_{shell_i}} = \sum_{i=0}^{N} \frac{1}{\frac{R_f}{N_c} + \frac{R_q L}{C_\lambda (d_{in} + iS_a)N_c}} \tag{10.1}$$

where $R_f = R_i + R_c + R_{hb}$, d_{in} is the diameter of the inner shell, S_a is the space between shells where typically 0.34 nm is shell thickness and 0.34 nm is shell-to-shell spacing, and N is the number of shells in MWCNT as $N = (d_{out} - d_{in})/S_a$ where d_{out} is the diameter of the outer shell. In a bundle of SWCNTs or MWCNTs, the total resistance can be derived as, $R_b = R_t/n_b$, where n_b is the number of bundles. For example, a metal track with width, w, and height, h, the number of horizontally aligned nanotube bundles can be expressed as in [16]:

$$n_b = P_m(n_h n_w - \lfloor n_h/2 \rfloor) \tag{10.2}$$

where P_m is the probability that a nanotube is metallic and usually $P_m = 0.3$ [2], n_h is the number of nanotubes in vertical direction as $n_h = \lfloor h/d_{out} \rfloor$ and n_w is the number of nanotubes in horizontal direction as $n_w = \lfloor w/d_{out} \rfloor$.

The capacitance of nanotubes consists of both quantum, C_q, and electrostatic capacitances, C_e, that can impact power supply noise on power tracks. Additionally, there is coupling capacitance between: (1) conducting shells in an individual MWCNT, C_c, and (2) individual MWCNTs depending on the proximity between them, C_{cm}. Using Luttinger liquid theory [2], quantum capacitance can be derived as $4e^2/h_p v_f \approx 193$ aF/μm per conducting channel, where h_p is Planck's constant, e is charge of single electron, and v_F is Fermi velocity in graphene. Therefore, for each shell, quantum capacitance is as $C_q = \frac{4e^2}{h_p v_F} N_c L$ and the total quantum capacitance of a CNT bundle is as:

$$C_{q_t} = n_b \sum_{i=1}^{N} C_{qi} \tag{10.3}$$

Electrostatic coupling depends on the geometry of the CNT and also the bundle density (i.e., number of bundles, n_b). It is shown in [20] that CNT bundles have slightly smaller electrostatic capacitance compared to Cu interconnects with same dimensions. Capacitance of CNT bundles would decrease slowly with increase of bundle density [2]. However, for a MWCNT, these capacitances cannot be assumed equal due to the fringing coupling effects between shells. The electrostatic capacitance of an MWCNT which is equivalent to ground capacitance from the outer shell to the ground plane, distance y, can be obtained as:

$$C_{e_t} = \frac{2\pi\varepsilon}{ln(y/d_{out})} \tag{10.4}$$

The shell-to-shell coupling capacitance is as in [14, 20]:

$$C_c = \frac{2\pi\varepsilon}{ln(d_{out}/d_{in})} \tag{10.5}$$

and coupling capacitance C_{cm} between two CNT bundles with space, s can be expressed as:

$$C_{cm} = \frac{2\pi\varepsilon}{s/d_{out}} \tag{10.6}$$

As for inductance, CNTs have both kinetic and magnetic inductance that impact power supply noise and high-frequency effects on power tracks. Again, based on the Luttinger liquid theory, the kinetic inductance per conducting shell can be theoretically expressed as $L_k = h_p L/4e^2 v_F N_c$ or $\approx 8\,\text{nH}/\mu\text{m}$ per conducting shell. Thus, the total kinetic inductance for all shells in a CNT bundle is derived as:

$$L_{k_t} = \frac{1}{n_b \sum_{i=1}^{N} \frac{1}{L_{k_i}}} \tag{10.7}$$

where L_{k_i} is the kinetic inductance of each shell i. Magnetic inductance L_m and mutual inductance M_m are also of importance as they can have an impact on dynamic voltage drop behavior. For each shell $L_m = \frac{\mu}{2\pi}ln(y/d)$ and for a CNT bundle is derived as:

$$L_{m_t} = \frac{1}{n_b \sum_{i=1}^{N} \frac{1}{L_{m_i}}} \tag{10.8}$$

Scalable mutual inductance model between any two shells i and $i+1$ with space distance, S_a was presented in [11, 12, 20] and can be estimated as:

$$M_{m_i} = \frac{\mu_o l}{\pi}ln(S_a/(d_{i+1} - d_i)) \tag{10.9}$$

and mutual inductance M_{mm}, between two CNT bundles with space, s can be similarly expressed as:

$$M_{mm} = \frac{\mu_o l}{\pi}ln(s/d_{out}) \tag{10.10}$$

Resistance, capacitance, and inductance models for MWCNTs are further utilized to study the dynamic voltage drop behavior on power delivery networks and TSVs.

10.3 CNTs for 2D Power Delivery Network

In this section, we explore CNTs as global tracks for on-chip power distribution.

Typically power distribution networks are hierarchical in nature and can be classified as local, intermediate, and global power delivery networks. In this work, we focus solely on global power delivery networks as already pointed out from [15, 20] that compared to Cu interconnect CNTs would be beneficial at the global interconnects.

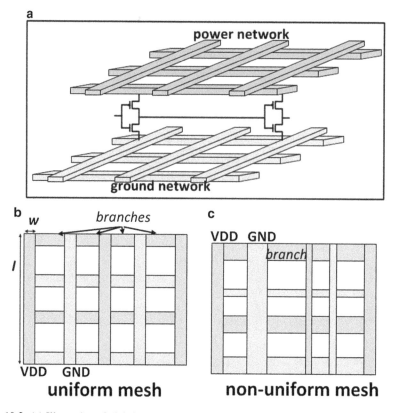

Fig. 10.2 (**a**) Illustration of global power delivery network for a single tier, and description of **b**) uniform and **c**) non-uniform power delivery networks. The meshes have the same area and regular structure but some tracks have different widths, thus varying the branches lengths

Additionally, power delivery networks are structured as meshes with power tracks running in parallel with each other and vias are inserted on their perpendicular intersections as shown in Fig. 10.2a. Also depending on the circuit current demand, these meshes can be designed as uniform (i.e., all branches are equal lengths) and non-uniform grids (i.e., branches of different lengths) while still being regular meshes as shown in Fig. 10.2b.

Power branches as shown in Fig. 10.2a, b are the simplest composing element of the power delivery network, which is the segment between two intersecting metal tracks that can be composed of horizontally aligned CNTs. To check the power integrity of the network, we check the voltage drop along the branch and derive it as:

$$V_{branch} = I_{branch}R_{branch} + L_{branch}\frac{dI_{branch}}{dt} + R_{branch}C_{branch}\frac{dV_{branch}}{dt} \quad (10.11)$$

where R_{branch}, C_{branch}, $and L_{branch}$ represent the parasitic of the branch segment that can be either a copper metal layer or CNTs (i.e., SWCNTs, MWCNTs) and I_{branch} is the current flow on the branch. Regardless of the mesh (i.e., uniform or non-uniform), V_{branch} can be computed for each branch and serves as a quality metric for the power delivery network.

10.3.1 Branch Analysis with CNTs

To predict the voltage drop on a power delivery network, we analyze a single branch implemented with MWCNT bundles. The branch width 100 nm and height 100 nm are fixed which are typical values for global power grid branches on advanced scaled technologies, i.e., 28 nm or 32 nm, whereas the branch length and diameter of MWCNT bundles are varied. Figure 10.3 shows the resistivity of MWCNT bundles computed with Eqs. (10.1) and (10.2). Bundle lengths are varied from 1 to 100 μm and MWCNT outer diameter varying from 1 to 100 nm. Figure 10.3 shows that for a fixed length, the smallest possible diameter, *dout*, provides the lowest resistivity.

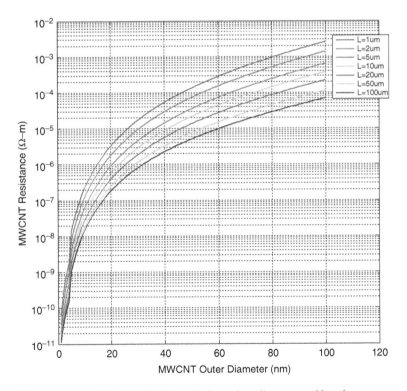

Fig. 10.3 Branch resistivity of MWCNT bundle for various diameters and lengths

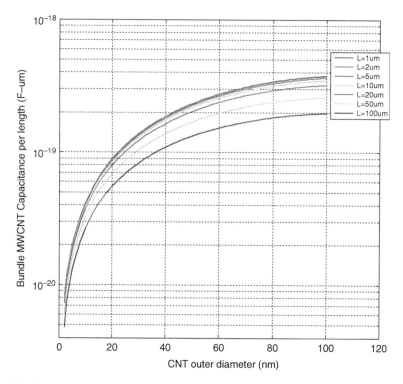

Fig. 10.4 Branch capacitance (quantum and electrostatic capacitance) of MWCNT bundle for various diameters and lengths

Whereas, for a fixed diameter, the largest possible length, L, provides the smallest resistivity. Therefore, selecting optimal diameter in terms of resistance is depended on MWCNT bundle length.

Similarly, we analyze the capacitance (both quantum and electrostatic) of MWCNT bundle as a function of bundle length and diameter as shown in Fig. 10.4. We utilize Eqs. (10.3)–(10.6) to derive branch capacitance and we observe that for branches of fixed bundle length, small diameter bundles provide smaller capacitance. Whereas, for branches with fixed diameter, large bundle lengths provide the smallest capacitance. Both quantum and electrostatic capacitances are equally important for deriving branch capacitance. In Fig. 10.5, the kinetic and magnetic inductance values are plotted for MWCNT bundles of various diameters and lengths. Utilizing Eqs. (10.7)–(10.10), kinetic inductance was derived while assuming 1D structure of MWCNT bundle. As length of MWCNT bundles is larger than the mean free path (λ), kinetic inductance has small impact [16]. Whereas, magnetic inductances have larger values which decrease with the diameter of MWCNT bundles. For short length MWCNT bundles, magnetic inductance is somewhat similar for bundles of different diameters. However, for large length bundles, small magnetic inductance values are obtained at larger diameters. The amount of voltage

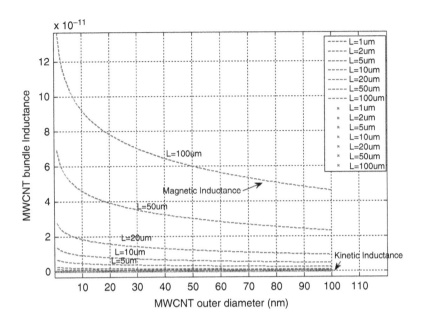

Fig. 10.5 Branch inductance (kinetic and magnetic inductance) of MWCNT bundle for various diameters and lengths

drop contributed independently from resistance, capacitance, and inductance are shown in Fig. 10.6. To compute voltage drop on the branch, we make assumptions that current flowing on the branch, $I_{branch} = 1\,\mu A$, $dI_{branch} = 1\,\mu A$, $dt = 1\,ns$ (or 1 GHz switching frequency), and $dV_{branch}=0.1\,V$. Please note that these values are simply chosen to quantify branch voltage drop when $1\,\mu A$ current is flowing on the branch for varying MWCNT bundle lengths and diameters. We notice that resistance impact (IR) on branch voltage drop increases with the MWCNT bundle diameter and decreases with the bundle length. A similar but less dominant effect is obtained from capacitance ($RCdV_{branch}/dt$) impact on branch voltage drop. Whereas, inductive effects (LdI_{branch}/dt) are very minimal at this frequency. In Fig. 10.7, the branch voltage drop contour plots are shown with respect to MWCNT bundle length and outer diameter. Please note that the branch length is represented by the MWCNT bundle length. Voltage drop is computed as in Eq. (10.11). Each contour represents the amount of branch voltage drop which can vary from 2 to 960 mV for large diameters and lengths. It is important to note that contours can indicate a region for which MWCNT bundle lengths and diameters would result in allowable branch voltage drop. For example, for 0.1 V allowable voltage drop, the area to the left of thick contour represents the MWCNT bundle widths and lengths that can be used to construct a power grid branch. Thus, power grid branches can be implemented by either long and small diameter bundles or short and large diameter bundles. Hence, it is important to investigate the topology of the CNT-based power delivery network (i.e., sizing and location of power tracks which create

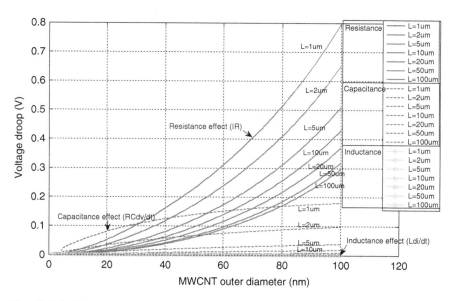

Fig. 10.6 Individual impact of parasitic resistance, inductance, and capacitance to voltage drop on a power grid branch

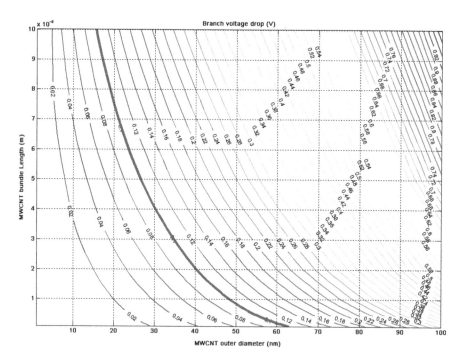

Fig. 10.7 Contour plot of voltage drop on a power grid *branch* as a function of MWCNT bundle length (1–100 μm) and diameter (1–100 nm)

branches) and placement of CNT tubes that would lead to minimum power supply noise generation. The problem of optimal power delivery sizing and placement is the focus of on-going research.

10.4 CNTs for 3D Power Delivery Network

Three-dimensional integration technology provides the opportunity to implement multi-layer circuits for higher density, heterogeneity, and small footprint. Utilization of TSVs as interconnects allows for shorter connections with improved delays and increased bandwidths. As wire width continue to shrink, copper interconnects in high-performance systems will suffer from significant increase in resistivity due to surface roughness and grain boundary scattering and from electromigration problems due to the low current densities supported by copper conductors. Hence, despite the advantages of 3D integration, copper (Cu)-based interconnects, i.e., Cu TSVs will hinder performance and reliability of interconnects, thus motivating the need for alternative interconnect materials for future process technologies.

Typically, 3D power distribution networks are hierarchical in nature and can be classified as local, intermediate, and global 3D power delivery networks. In this work, we focus solely on global power delivery networks as already pointed out from [15, 20] that compared to Cu interconnect CNTs would be beneficial at the global interconnects.

In Fig. 10.8, 3D power delivery network for two-tiers is shown connected using TSVs for power and ground. For this work, we assume that TSVs connect global to global interconnects. Depending on the stacking configuration such as face-to-face or face-to-back and processing approach such as via-first, -middle, or -last, TSVs can vary in pitch. Here, we study high-density TSVs that can vary from 1 to 5 nm in diameter and 5 to 100 μm in thickness, thus reaching density up to 10,000 TSVs/mm^2. Regardless of the structure, the power networks should deliver voltage with minimum voltage drop. Power branches as shown in Fig. 10.2a, b are the simplest composing element of the power delivery network, which is the segment between two intersecting metal tracks that can be composed of horizontally aligned CNTs. Whereas, TSVs can be implemented using vertically grown CNTs, which make them suitable for TSV process. In comparison with copper (Cu) or tungsten (W) filled TSV, CNT-based TSVs provide a long mean free path, a large thermal conductivity, and a high current-carrying capacity. CNTs can provide competitive solution for high-density vertical interconnects.

To check the power integrity of the network, we check the voltage drop from the power supply pin to the voltage node of the switching circuit. For example, for a two-tier stack, such current path would include the voltage drop on the branches of the top tier (i.e., Tier 2), the voltage drop on TSVs between top and bottom tiers, and the voltage drop on the branches of the bottom tier (i.e., Tier 1). This is also illustrated in Fig. 10.9.

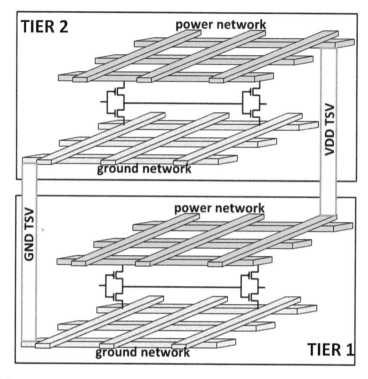

Fig. 10.8 Illustration of 3D power delivery network for two tiers connected via Through-Silicon-Vias (TSVs)

Fig. 10.9 Current path from power pin to switching circuit passing through the shortest path on branches of Tier 2, Tier 1, and TSVs

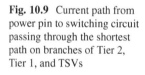

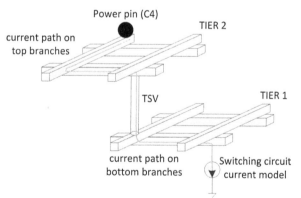

To compute the worst-case voltage drop that the switching circuits will experience when CNT interconnects and TSV are utilized, we first compute the voltage on top branches (i.e., Tier 2) with respect to different dimensions of CNT interconnect lengths and diameters. Voltage drop on a single branch can be computed based on the RLC parasitics of the CNT bundle interconnect as:

$$V_{topbranch} = I_{topbranch}R_{topbranch} + L_{topbranch}\frac{dI_{topbranch}}{dt} + \\ R_{topbranch}C_{topbranch}\frac{dV_{topbranch}}{dt} \quad (10.12)$$

where $R_{topbranch}$, $C_{topbranch}$, $and L_{topbranch}$ represent the parasitic of the top branch segment that are in Tier 2 and are CNT bundles (i.e., either SWCNTs or MWCNTs) and $I_{topbranch}$ is the current flow on the branch representing the current demand of the switching circuit. Please note that the voltage drop can also be computed as a summation of individual branches voltage drop to obtain the total voltage drop on top branches.

Similarly, the voltage drop on bottom branches (i.e., Tier 1) can be computed as:

$$V_{botbranch} = I_{botbranch}R_{botbranch} + L_{botbranch}\frac{dI_{botbranch}}{dt} + \\ R_{botbranch}C_{botbranch}\frac{dV_{botbranch}}{dt} \quad (10.13)$$

where $R_{botbranch}$, $C_{botbranch}$, $and L_{botbranch}$ represent the parasitic of the bottom branch segment that are in Tier 1 and are CNT bundles (i.e., either SWCNTs or MWCNTs) and $I_{botbranch}$ is the current flow on the branch representing the current demand of the switching circuit. Similarly, if the current path flows through several branches, then the voltage drop can also be represented as the summation of voltage drops from each branch.

Cross-section of a power CNT TSV is shown in Fig. 10.10. The metallic CNT bundles can be densely packed to create many parallel connections and reduce the overall resistance of the TSV. Metallic nanotubes are distributed in the sparsely bundle with probability 1/3 since approximately one-third of tubes are metallic [18]. It has been shown that nanotubes with diameter less than 0.5 nm tend to be metallic regardless of the chirality due to the steep angle of curvature in the graphene sheet [5, 9]. The resistance of an individual nanotube depends on the applied bias voltage. If low bias voltage (i.e., $V_b \leq 0.1$) is applied, then the resistance is the summation of lumped intrinsic ($R_i \approx 6.5$ kΩ) and contact resistance (R_c) resistances and distributed per unit length ohmic resistance (R_o).

$$R_{low} = R_i + R_c \quad \text{if} \quad l_b \leq \lambda_{ap} \quad (10.14)$$

$$R_{low} = R_o + R_c \quad \text{if} \quad l_b > \lambda_{ap} \quad (10.15)$$

Fig. 10.10 Cross-sectional view of a CNT TSVs made of several individual carbon nanotubes

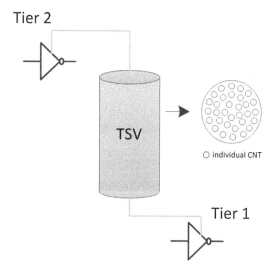

where l_b is the length of the nanotube, and λ_{ap} is the mean free path for acoustic phonon scattering. For high bias voltages (i.e., $V_{bias} \geq 0.1$), the resistance of an individual tube depends on the applied bias voltage as:

$$R_{high} = R_{low} + \frac{V_b}{I_o} \tag{10.16}$$

where I_o is the maximum current flow through an individual nanotube, which is approximately 20–25 μA [6].

Thus, the resistance of an individual CNT TSV for power and ground lines, will have resistance as R_{high}. Contact resistance of CNT TSV models the increased resistance due to imperfect metal contacts. Recent studies have shown that R_c of an SWCNTs greatly increases when the diameter of the nanotube (d_t) is less than 1 nm [8, 21]. Ohmic resistivity of an SWCNT is defined as [16]:

$$\rho_t = \frac{h}{4e^2 C_\lambda d_t} \tag{10.17}$$

where C_λ is a mean free path-to-nanotube diameter proportionality constant defined as $C_\lambda = \lambda_{ap}/d_t$. The ohmic resistance is proportional to its diameter $1/d_t$, while for standard copper conductors, resistance has $1/d^2$ dependence. Thus, ohmic resistance of CNTs is proportional to surface area of the tube, whereas resistance of metallic conductor is proportional to its cross-sectional area of the conductor.

Thus, the overall TSV resistance made of SWCNT bundle is defined as the parallel combination of the individual SWCNT resistances as

$$R_{tsv} = \frac{R_o + R_c + \frac{V_b}{I_o}}{n_b} \tag{10.18}$$

where n_b is the number of bundles and it was defined in Eq. (10.2). Here, we compute the voltage drop on a TSV based on its resistance as:

$$V_{tsv} = R_{tsv}I_o \qquad (10.19)$$

Hence, the total voltage drop on a current path passing through the CNT-based branches on Tier 1, TSVs, and Tier 2 can be computed as:

$$V_{drop} = V_{topbranch} + V_{tsv} + V_{botbranch} \qquad (10.20)$$

10.4.1 TSV CNT Analysis

Here, we analyze the CNT TSV resistance when SWCNT bundles are used. Assuming there is no current redistribution due to magnetic inductance, the resistance of CNT TSVs (or SWCNT bundle) is defined as the parallel combination of individual SWCNT resistances. This is also described in Eq. (10.18). We compute TSV resistance while varying the diameter and length of individual SWCNTs. The experiments are performed on d_t=0.5–4 nm diameter range for lengths (or TSV heights) varying from 10 to 300 μm, where R_c=20 kΩ and I_o=25 μA. We assume supply voltage is V_{bias}=0.8 and driver resistance is 2.5 kΩ which corresponds to predictions of ITRS for 28 nm process technology. The CNT-based TSV resistance is displayed in Fig. 10.11.

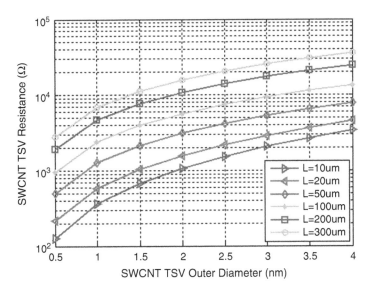

Fig. 10.11 Resistance of Through-Silicon-Via made of SWCNTs with varying diameters and lengths

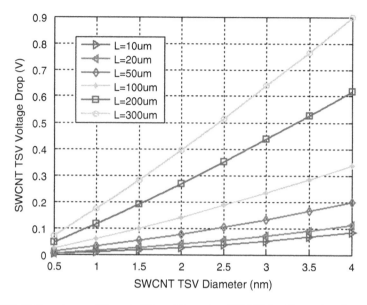

Fig. 10.12 Voltage drop on Through-Silicon-Via made of SWCNTs with varying diameters and lengths

As shown in the above figure, the TSV resistance can be categorized based on the diameter and length of the SWCNT bundles. For long SWCNT bundles, the TSV resistance increases with the increase in diameter size. SWCNT bundles are also at a disadvantage for short TSV lengths and large diameters. Whereas for short bundle lengths with small diameters, SWCNT TSV becomes more favorable. As SWCNT TSV resistance varies significantly due to both diameter and length, selecting the optimal TSV dimensions is not trivial as it can lead to large voltage drop. Figure 10.12 displays the voltage drop on SWCNT TSVs for various diameter and length sizes.

We observe that the TSV voltage drop varies significantly due to SWCNT length. For the same size diameter SWCNT (i.e., d_t=2 nm), voltage drop varies from 26.6 to 400 mV when length is 10 μm and 300 μm, respectively. We also note, that for short SWCNT lengths, the impact of diameter on voltage drop is minimal such as a variation of 3–85 mV for SWCNT length of 10 μm. However, for longer SWCNT lengths, the impact of SWCNT diameter becomes more dominant on voltage drop such as 5–620 mV for SWCNT lengths of 200 μm. Hence, for a given allowed voltage drop threshold, one can easily determine the lengths and diameter ranges of SWCNT TSVs to satisfy such voltage drop threshold. For example, for a TSV voltage drop threshold of 0.1 V, we can determine that TSV lengths of 10 and 20 μm can be utilized for any diameter size SWCNTs. However, for longer TSV lengths, the range of SWCNT diameter sizes becomes more limited in order to satisfy the 0.1 V voltage drop threshold.

10.4.2   Voltage Drop Analysis on a 3D PDN

Here, we analyze the voltage drop on a 3D power delivery network composed of two stacked tiers as depicted in Figs. 10.8 and 10.9. Assuming that there is a single switching circuit, we can identify the current path from the power pin to the circuit traversing branches on top tier, TSVs, and bottom tier. The same principle would also apply if several switching circuits were present. This is due to the superposition principle on linear systems.

As heterogeneous technologies and functionalities can be implemented in 3D, each tier can have its individual topology which might differ from the rest of the tiers. In this work, we assume that both Tier 1 and Tier 2 have uniform topologies. However, the applied analysis method and analytical formulas are general enough to be applied to any type of topology.

We consider different geometries for SWCNT branches for both Tier 1 and 2. The experiments are performed on CNT-based branches (i.e., either SWCNTs or MWCNTs) with lengths of 1–100 μm with inner wall diameter of 1 nm and outer diameter ranges from 1 to 20 nm. Whereas, SWCNT parameters with lengths of 5–100 μm and diameter ranges 1–4 nm are applied to SWCNT TSVs. Figure 10.13 displays the voltage drop on Tier 1 CNT-based branches with respect to branch

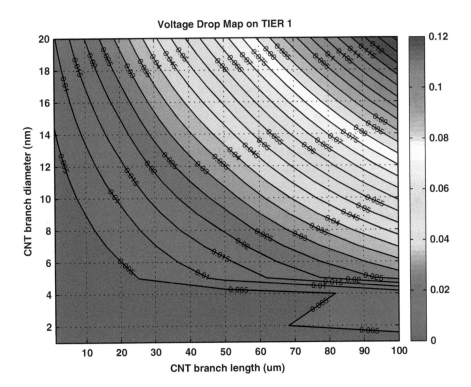

Fig. 10.13 Voltage drop on CNT branches located in Tier 1

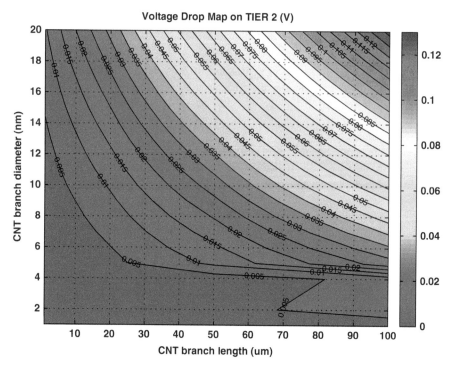

Fig. 10.14 Voltage drop on CNT branches located in Tier 2

length and diameter. Please note that voltage drop is computed based on Eq. (10.12).
As we are applying the same parameters for branches in Tier 2, the voltage drop is
similar as shown in Fig. 10.14.

We note a large variation in branch voltage drop due to both its length and
CNT bundle diameter size. As power grid branches can be implemented with either
SWCNTs or MWCNTs, diameter-depended voltage drop became more dominant
for large diameters and longer length branches. Similarly, one can determine the
optimal CNT-based branch lengths and diameter for a given voltage drop threshold
that is allowed on the power grid branches.

In Fig. 10.15, the TSV only voltage drop is plotted with respect to SWCNT
TSV length and diameter. In comparison to CNT-based branches, larger amount
of voltage drop can occur on SWCNT TSVs. This could be due to the impact of
contact resistances and geometry of the SWCNT TSV which are different from
CNT-based branches. We also plot the total voltage drop that occurs on Tier 1, TSV,
and Tier 2 branches for different geometries of branches and TSVs. Figure 10.16
shows different voltage drops maps when SWCNT TSVs of different sizes are used.
The x- and y-axis of each sub-figure represent the CNT-based branch diameter and
lengths. Each sub-figure represents different voltage drop with respect to SWCNT
TSV diameter and length also noted on top of each sub-figure. Such analysis helps to

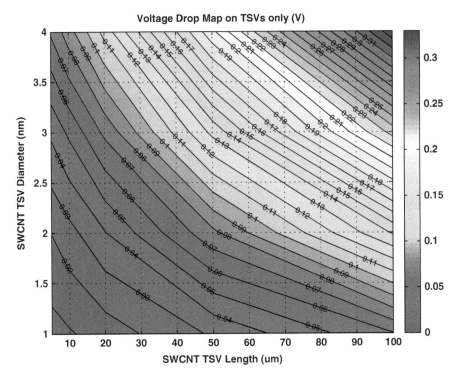

Fig. 10.15 Voltage drop on CNT TSVs for varying diameter and height dimensions

determine the TSV dimensions that would be suitable for a given topology on each tier while providing minimal voltage drop. As shown in each sub-figure, the overall voltage drop gets worse with the increase of TSV diameter and length. From these analyses, we are able to understand better the potential of CNT-based interconnects and TSVs for designing optimal 3D power delivery networks. Additionally, such early-on analyses are valuable for performing design space exploration of 3D PDNs with CNT interconnects.

10.5   Thermal Modeling for CNTs

In an integrated chip, the temperature may rise above $80\,°C$ which will impact the behavior of devices and parasitics of interconnects. As interconnect lengths are usually larger than the free mean path of CNTs, there will be electron–phonon scattering along the length of nanotube that would lead to self-heating and temperature rise along the nanotube interconnect. The temperature variations along a nanotube are also dependent on the defect density, alignment, and contact resistance. Large contact resistances create large potential barriers at the interfaces

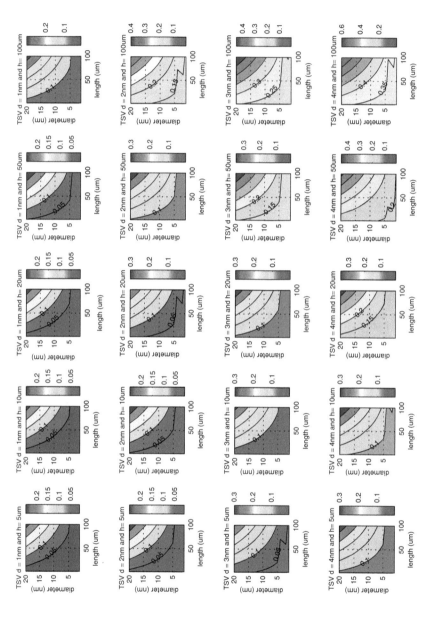

Fig. 10.16 Total voltage drop on 3D power delivery network considering the voltage drop on Tier 1, TSVs, and Tier 2. Each figure shows the total voltage drop as a function of branch CNT length and diameters (both Tier 1 and Tier 2 branches are uniform) and also as dimensions of TSV diameter and height

of nanotubes for electrons to tunnel through. Additionally, large defect densities cause electrons to localize and conduction happens through thermally activated electron hopping. As already mentioned in the previous sections, there are many contradicting and inconsistent reports on the thermal conductivity and temperature coefficient of resistance (TCR) for CNTs. TCR is the change in resistance for every 1-K of temperature rise. In this section, we will exploit the existing physical models for CNTs and express its properties as a function of temperature.

We investigate both SWCNTs and MWCNTs with various lengths and diameters. We start by deriving the thermal coefficient for resistance for SWCNTs as in [13]:

$$TCR_{swcnt} = \frac{(L/10^3 d_s)/T_o}{1 + (L/10^3 d_s)(T/T_o - 2)} \qquad (10.21)$$

where L and d_s are the length and shell diameter of the nanotube. T is the temperature and $T_o = 300$ K. For a single-wall nanotube with small diameter, the number of conduction channels N_c is independent of temperature for long length interconnects (i.e., significantly larger than mean free path at room temperature). Whereas, for large diameter nanotubes with increasing temperature, the number of conduction channels increase which consequently increases N_c. Similarly, nanotube conductance also depends on nanotube length and temperature. Neutral length L_N is the length at which resistance becomes independent of temperature. For lengths smaller than L_N, the nanotube resistance is mainly influenced by the number of conduction channels, and increasing temperature lowers resistance. Whereas, for lengths larger than L_N, the mean free path is more important and increasing temperature, increases resistance. Neutral length is derived as in [13]:

$$L_N = \frac{10^3 a T_o d_s^2}{b + 2a T_o d_s} \qquad (10.22)$$

where a is 2.04×10^{-4} nm^{-1} K^{-1} and b=0.425. Figure 10.17 shows the thermal coefficient of resistance for SWCNTs with different lengths and diameters. We observe that small diameter nanotubes have larger TCR than nanotubes with large diameter. For example, for nanotubes with length 1 μm and diameter 1 nm, the TCR=6.5 whereas for nanotubes with diameter 10 nm, the TCR=4. For short length and large diameter nanotubes, the TCR is relative small (i.e., <1), then TCR increases linearly with nanotube length. It is important to note, that TCR is a positive coefficient for SWCNTs with different lengths and diameters.

In Fig. 10.18, the ratio of resistances with respect to resistance at T=300 K is shown for nanotubes with diameter 1 nm and various lengths. We note that for short length nanotubes, the resistance ratio is small (i.e., <1.05) for temperature ranges of 300–350 K. There is a linear increase to the resistance ratio as the nanotube length increases (i.e., ratio=1.3 for L=9 μm). Hence, the nanotube resistance increases with temperature but at different rates depending on nanotube length. A similar plot is shown in Fig.10.19 where the nanotube diameter is set to 10 nm. We observe a similar trend as in Fig. 10.18, but the resistance ratios are smaller.

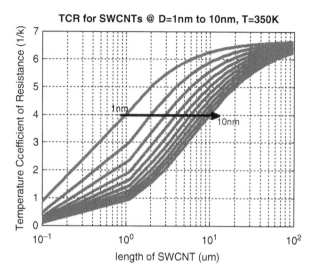

Fig. 10.17 Temperature coefficient of resistance for SWCNTs

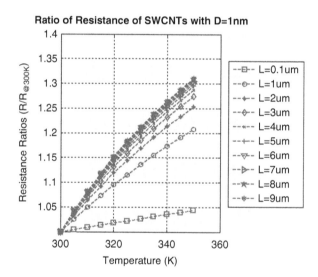

Fig. 10.18 Ratio of resistances due to temperature impact for SWCNTs with d=1 nm

This is an important observation for nanotube interconnects where diameter and length of nanotubes can be exploited as a knob for alleviating temperature impact on resistance. For example, the resistance ratio is minor when increasing nanotube diameter from 1 to 10 nm for nanotube lengths of 10 μm.

In Fig. 10.20, the resistance distribution is shown for SWCNTs with different lengths and temperature values for nanotubes of d=1 nm. As already mentioned above, for short length nanotubes, the temperature increase does not change

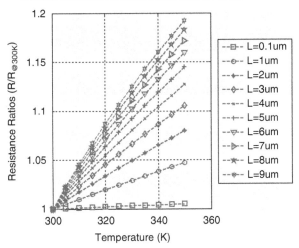

Fig. 10.19 Ratio of resistances due to temperature impact for SWCNTs with d=10 nm

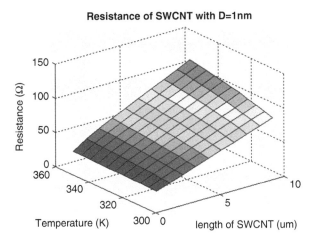

Fig. 10.20 Resistance distribution for SWCNTs with diameter=1 nm for various lengths and temperature values

the conduction channels, hence no change in resistance. Whereas, as the length increases, it also increases conduction channels, N_c, which further increase resistance. Additionally, as temperature increases, we observe an increase on resistance due to temperature impact on N_c.

To derive TCR for MWCNTs, it is important to understand the heat transport and distribution on individual shells and bundles. Once heat is introduced in the outer shell of MWCNT, the high thermal conductivity along the graphene layer transfers heat at a high flow rate in the circumferential direction, as well as along the tube.

Due to close proximity between shells, there is thermal coupling that enables heat flow between shells. Authors in [19] demonstrated that heat introduced at outer shells is evenly distributed to all shells within a short distance, $L \sim 50$ nm. The total heat flowing through the outer shell is always higher than the current in the inner shells. We compute the thermal coefficient of resistance based for each shell, which then can be used to compute the resistance of MWCNT with respect to temperature. As the outer shell conducts most of the heat, we derive its TCR as in [13]:

$$TCR_{mwcnt} = \frac{\delta R/\delta T}{R_{i_{th}}} \tag{10.23}$$

where

$$\frac{\delta R}{\delta T} = \left(\frac{R_q}{10^3 d_{out}}\right)\left(\frac{2ad_{out} + b/T_o}{(aTd_{out} + b)^2}\right)(L - L_N) \tag{10.24}$$

$$R_{i_{th}} = R_q + \left(\frac{R_q L}{\lambda_{ac}} + \frac{R_q L}{\lambda_{opabs}} + \frac{R_q L}{\lambda_{opems}^{fld}} + \frac{R_q L}{\lambda_{opems}^{abs}}\right)/N_c \tag{10.25}$$

where the mean free path for acoustic scattering is as [4] $\lambda_{ac} = 10^3 d_{out} T_1/T$ where $T_1 = 400$ K. Optical phonon scattering can occur if an electron obtains adequate energy (i.e., $\hbar w_{op} \approx 0.18$ eV), it can emit an optical phonon and get backscattered. The scattering length is (much shorter than acoustic scattering) computed as $\lambda_{op} = 56 d_{out}$ and measured with smaller coefficients (i.e., $\sim$15–20)[17]. The mean free path for absorbing an optical phonon is as $\lambda_{opabs} \approx \lambda_{op}/n_{phonons}$ where $n_{phonons} = 1/(e^{\hbar w/K_B T} - 1)$ is the number of phonons and K_B Boltzmann constant.

An electron can obtain sufficient energy for emitting an optical phonon either by getting accelerated long enough by electrical field or by absorbing an optical phonon. The scattering lengths can be calculated as:

$$\lambda_{opems}^{fld} = \hbar w_{op}/(qV_{bias}/L) + \lambda_{op} \tag{10.26}$$

and

$$\lambda_{opems}^{abs} = \lambda_{opabs} + \lambda_{op} \tag{10.27}$$

where V_{bias} is the applied voltage. It has been shown that increasing the diameter or temperature linearly increases the number of conduction channels in large shells [13]. The average number of channels in a shell can be estimated as:

$$N_c \approx \begin{cases} aTd_s + b, & d_{out} > d_T/T \\ 2/3, & d_{out} < d_T/T \end{cases} \tag{10.28}$$

where d_T is 1300 nm K, whose value is determined by the thermal energy of electrons.

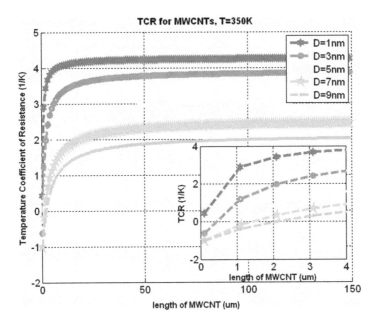

Fig. 10.21 Temperature coefficient of resistance for MWCNTs for longer lengths (i.e., up to 150 μm) and outer diameters (i.e., 1–10 nm). The *inset* shows the TCR for shorter lengths MWCNTs

Figure 10.21 shows the TCR for MWCNTs for various lengths and diameters. The inset figure shows the TCR for shorter length MWCNTs. It is important to note that the TCR for large diameter (i.e., $d_{out} > 1$ nm) is negative, where the increase in temperature leads to decrease in resistance. This behavior can be explained by Joule heating (or self-heating) effect and is consistent with negative TCRs obtained by Cassell et al. [3] and Naeemi and Meindl [13]. Theoretical analysis has also shown that negative TCRs for MWCNTs [10, 13]. Negative TCRs can be explained from the fact that there are more channels in MWCNTs contributing to conductance at higher temperatures as per Fermi-Dirac distribution. Larger number of channels lowers both scattering resistance and contact resistance. The negative TCR is opposing with other metals (i.e., copper) conductors, which presents an advantage for CNTs to be implemented as on-chip interconnect material. We also note that as length of nanotube increases, the TCR also increases till it saturates for lengths longer than 50 μm.

In Fig. 10.22, the TCR for large diameter MWCNTs is shown. The diameter varies from 10 to 50 nm for various nanotube lengths. The inset figure shows the TCR for short length MWCNTs. We note a negative TCR that for large diameter and short length MWCNTs. As length increases, TCR also increases and saturates for MWCNT lengths of 50 μm. In comparison with Fig. 10.21, large diameter nanotubes have negative TCRs and small increase in TCR is observed for longer length nanotubes.

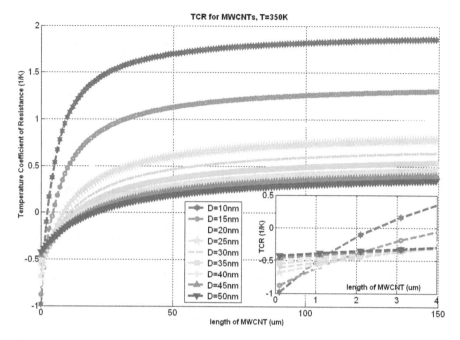

Fig. 10.22 Temperature coefficient of resistance for MWCNTs for longer lengths (i.e., up to 150 μm) and outer diameters (i.e., 10–50 nm). The *inset* shows the TCR for smaller lengths MWCNTs

From these experiments, we derive that temperature effects can be coped with for interconnects with MWCNTs where for short lengths even a decrease in resistance can be obtained. Figures 10.23 and 10.24 show the resistance distribution for MWCNTs with diameter 10 nm and 50 nm, respectively. The resistance of MWCNTs with diameter 10 nm is at least 1× order of magnitude larger from nanotubes with diameter d=50 nm for T=350 K. Such observation is significant for application of CNTs as signaling and power interconnects where large timing errors and voltage drops would be attained. Such large changes in resistance at high temperatures would impact the dimensions of MWCNTs that can be used for building reliable interconnects.

Another important metric is the resistance ratio with respect to resistance at T=300 K. Figures 10.25 and 10.26 show the resistance ratio for MWCNTs with diameter 10 nm and 50 nm, respectively. Depending on the diameter and temperature, the neutral length, L_N, is obtained and also shown the title of each figure. L_N represents the MWCNT length that is independent of temperature. For MWCNT with diameter d=10 nm, the neutral length is $L_N = 2.449$ μm. This can also be deduced from Fig. 10.25, where resistance ratio is less than 1 for MWCNT

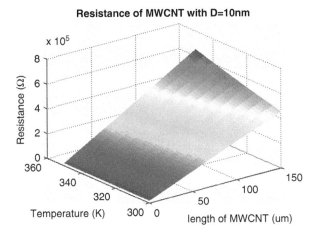

Fig. 10.23 Resistance distribution for MWCNTs with diameter=10 nm for various lengths and temperature values

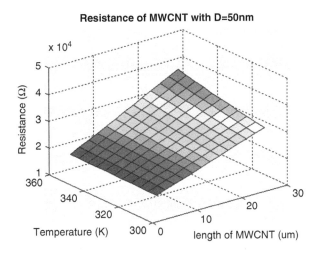

Fig. 10.24 Resistance distribution for MWCNTs with diameter=50 nm for various lengths and temperature values

lengths of 1 μm. For MWCNT lengths of 3 μm, the resistance ratios are almost constant to 1 (i.e., as it is close to L_N). The resistance ratio linearly increases with increase in MWCNT length and reaching maximum of 7 % increase (i.e., maximum resistance ratio 1.07 for 25 μm nanotube length).

Figure 10.26 shows resistance ratios for 50 nm diameter nanotubes where $L_N = 20.6$ μm. Such large L_N means that long MWCNT interconnects can be used without any dominant impact from temperature. This can also be noted from Fig. 10.26 where most of the resistance ratios are below 1. For example, MWCNTs with lengths L< L_N have lower resistance ratio and temperature have a negative

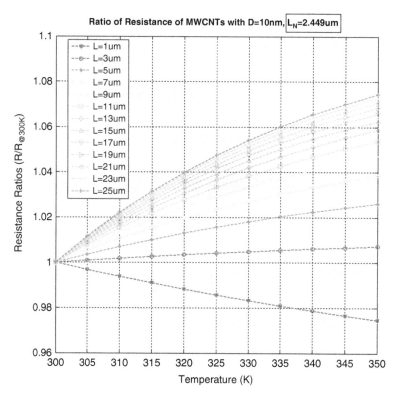

Fig. 10.25 Resistance ratio for MWCNTs with D=10 nm for various lengths. Please note that $L=2.449\,\mu m$

impact on the resistance. Such observations are important for selecting interconnect dimensions that immune to joule heating effects.

To predict the voltage drop on a power delivery network, we analyze a single branch implemented with MWCNTs. To compute voltage drop on the branch, we make assumptions that current flowing on the branch, $I_{branch} = 1\,\mu A$, $dI_{branch} = 1\,\mu A$, $dt = 1\,ns$ (or 1 GHz switching frequency), and dV_{branch}=0.1 V. Please note that these values are simply chosen to quantify branch voltage drop when $1\,\mu A$ current is flowing on the branch for varying MWCNT lengths and diameters. Figures 10.27 and 10.28 show the voltage drop for T=300 K and T=350 K, respectively. Overall, we observe minor voltage drop changes due to the impact of negative TCR on resistance. We deduce that selecting MWCNT interconnects length and diameter are essential for limiting the amount of voltage drop.

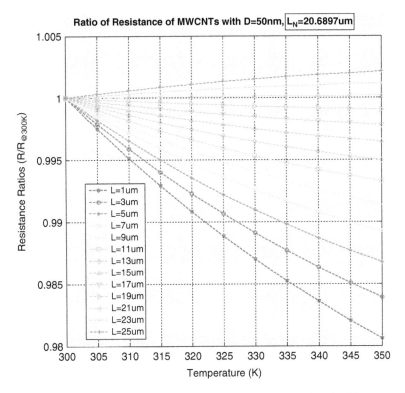

Fig. 10.26 Resistance ratio for MWCNTs with D=50 nm for various lengths. Please note that $L_N = 20.6897\,\mu\text{m}$

10.6   Conclusion

CNTs due their unique mechanical, thermal, and electrical properties are being investigated as promising candidate material for power delivery. The attractive mechanical, electrical, and thermal properties of CNTs offer great advantage for reliable and strong interconnects, and even more so for 3D integration. In this chapter, we performed a detailed design space exploration of horizontally and vertically aligned CNTs for implementing power delivery and TSVs. Analyses demonstrate that CNTs can be efficiently exploited for both 2D and 3D power delivery networks while resulting in minimal voltage drop and minimal impact due to temperature.

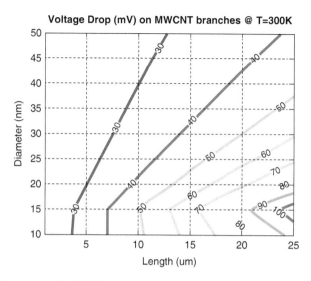

Fig. 10.27 Voltage drop for MWCNT interconnects with different lengths and diameters for T=300 K

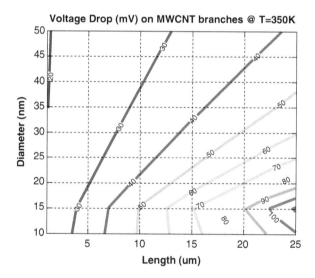

Fig. 10.28 Voltage drop for MWCNT interconnects with different lengths and diameters for T=350 K

References

1. Aliev AE, Lima MH, Silverman EM, Baughman RH (2010) Thermal conductivity of multi-walled carbon nanotube sheets: radiation losses and quenching of phonon modes. Nanotechnology. 21(3)
2. Burke PJ (2002) Luttinger liquid theory as a model of the gigahertz electrical properties of carbon nanotubes. IEEE Trans Nanotechnol 1(3):129–144
3. Cassell AM, Kreupl F, Li H, Liu W, Banerjee K (2013) Low-resistivity long-length horizontal carbon nanotube bundles for interconnect applications - part ii characterization. IEEE Trans Electron Devices 60:2870–2876
4. Jiang J, Saito R, Grüneis A, Chou SG, Ge Samsonidze G, Jorio A, Dresselhaus G, Dresselhaus MS (2005) Photoexcited electron relaxation processes in single-wall carbon nanotubes. Phys Rev B 71:045417
5. Jirahara K, Miyamoto Y, Ando Y, Qin L-C, Zhao X, Iijima S (2000) The smallest carbon nanotube. Nature 408(6808):50
6. Kane CL, Yao Z, Dekker C (2000) High-field electrical transport in single-wall carbon nanotubes. Phys Rev Lett 84(13):2941–2944
7. Khan NH, Hassoun S (2012) The feasibility of carbon nanotubes for power delivery in 3-d integrated circuits. In: 17th Asia and South Pacific design automation conference (ASP-DAC), pp 53–58
8. Leonard F, Talin AA(2006) Electrical contacts to nanotubes and nanowires: why size matters. Condens Matter. ArXiv e-prints. http://arxiv.org/abs/cond-mat/0602003
9. Li GD, Wang N, Tang ZK, Chen JS (2000) Single-walled carbon 4 a carbon nanotube arrays. Nature 408(6808):50–51
10. Li H, Srivastava N, Mao J-F, Yin W-Y, Banerjee K (2007) Carbon nanotube vias: a reality check. In: IEEE international electron devices meeting, 2007 (IEDM, 2007), pp 207–210
11. Li H, Xu C, Srivastava N, Banerjee K (2009) Carbon nanomaterials for next-generation interconnects and passives: physics, status, and prospects. IEEE Trans Electron Devices 56(9):1799–1821
12. Li H, Liu W, Cassell AM, Kreupl F, Banerjee K (2013) Low-resistivity long-length horizontal carbon nanotube bundles for interconnect applications 2014 - part ii - characterization. IEEE Trans Electron Devices 60(9):2870–2876
13. Naeemi A, Meindl JD (2007) Physical modeling of temperature coefficient of resistance for single- and multi-wall carbon nanotube interconnects. IEEE Electron Device Lett 28:135–138
14. Naeemi A, Sarvari R, Meindl JD (2005) Performance comparison between carbon nanotube and copper interconnects for gigascale integration (GSI). IEEE Electron Device Lett 26(2):84–86
15. Naeemi A, Huang G, Meindl JD (2007) Performance modeling for carbon nanotube interconnects in on-chip power distribution. In: Electronic components and technology conference, pp 420–428
16. Nieuwoudt A, Massoud Y (2006) Evaluating the impact of resistance in carbon nanotube bundles for VLSI interconnect using diameter-dependent modeling techniques. IEEE Trans Electron Devices 53(10):2460–2466
17. Pop E, Mann D, Reifenberg J, Goodson K, Dai H (2005) Electro-thermal transport in metallic single-wall carbon nanotubes for interconnect applications. In: IEEE international electron devices meeting (IEDM), Technical Digest, pp 253–256
18. Rinzler AC, Smalley RE, Wildoeer JWG, Venema LC, Dekkler C (1998) Electronic structure of atomically resolved carbon nanotubes. Nature 391(6662):59–61
19. Aliev AE, Lima MH, Silverman EM, Baughman RH (2010) Thermal conductivity of multi-walled carbon nanotube sheets: radiation losses and quenching of phonon modes. Nanotechnology, IOP Publishing, 21(3):035709

20. Srivastava N, Banerjee K (2005) Performance analysis of carbon nanotube interconnects for VLSI applications. In: Proceedings of the 2005 IEEE/ACM international conference on computer-aided design, pp 383–390
21. Tu R, Cao J, Wang Q, Kim W, Javey A (2005) Electrical contacts to carbon nanotubes down to 1 nm in diameter. Appl Phys Lett 87(17):173101
22. Zhu L, Sun Y, Xu J, Zhang Z, Hess DW, Wong CP (2005) Aligned carbon nanotubes for electrical interconnect and thermal management, vol 1, pp 44–50

Chapter 11
Carbon Nanotubes for Monolithic 3D ICs

Max Marcel Shulaker, Hai Wei, Subhasish Mitra, and H.-S. Philip Wong

The diversity of applications for carbon nanotubes (CNTs) is rather remarkable. This is in part due to their remarkable range of physical and electronic properties [1, 2]. In addition to metallic CNTs which may serve as interconnects that may complement conventional bulk metal wires [3, 4], semiconducting CNTs are the ideal transistor channel material for an ultimately-scaled high-performance and energy-efficient digital logic technology [2, 5]. Analysis for very-large-scale integrated (VLSI) systems (modeled using an entire IBM Power 7 processor) reveals CNT field-effect transistors (CNFETs) would provide an order of magnitude benefit in the energy-delay product (EDP, a measure of energy efficiency) over silicon CMOS [6–9].

Yet even with improved next-generation logic devices and interconnects, system-level performance will remain severally constrained by the growing memory-logic communication bottleneck [10, 11]. This is particularly true for future abundant-data[1] applications, which have the potential to revolutionize information technology with applications ranging from social networks, e-commerce, genomics, and multi-media analytics.

To overcome this bottleneck and enable next-generation abundant-data applications, revolutionary digital system architectures which provide massive bandwidth between computation and memory will be required. One such promising technological option for providing highly fine-grained integration of logic circuits with massive amounts of memory is monolithic three-dimensional (3D) integrated

---

[1]Abundant-data: massive amounts of highly unstructured data, with little or no locality, often streamed in terabytes.

M.M. Shulaker • H. Wei • S. Mitra • H.-S.P. Wong (✉)
Stanford University, Stanford, CA, USA
e-mail: hspwong@stanford.edu

© Springer International Publishing Switzerland 2017   315
A. Todri-Sanial et al. (eds.), *Carbon Nanotubes for Interconnects*,
DOI 10.1007/978-3-319-29746-0_11

circuits (ICs), which represents a significant departure from today's conventional computing systems [12, 13]. In this chapter, we elucidate how CNTs naturally enable not just next-generation interconnects and transistors, but also provide a path towards realizing monolithic 3D ICs.

11.1   Introduction to Monolithic 3D Integration

The traditional path for improving the energy efficiency and performance of digital systems has been silicon CMOS scaling (i.e., Dennard scaling) [14]. As Dennard scaling grew increasingly difficult (i.e., oxide thickness scaling limited by leakage current, supply voltage scaling limited by non-ideal subthreshold slope, etc.), alternative means to boosting transistor performance were investigated. This "equivalent scaling" path included strained silicon, high-k gate dielectric and metal gate, and advanced device geometries (e.g., FinFETs) [15]. Yet today, equivalent scaling is also hitting limits, and has been unable to maintain the historic year-on-year gains in transistor performance [16]. Alternative technologies and approaches are needed to continue improving computing performance. Three-dimensional integration, whereby circuits are stacked vertically over one another, is one such path [17]. Today's 2.5D[2] and 3D[3] integration is typically achieved through chip-stacking [18, 19]. To perform chip-stacking, each vertical circuit layer is fabricated as separate 2D circuits on different substrates. After fabricating the 2D circuits, the different layers are then physically stacked and bonded one on top of the other. Yet such 3D integration is severely limited, as it typically relies on Through Silicon Vias (TSVs) as vertical interconnects between different layers of the 3D IC. These TSVs occupy a large footprint (typical TSV dimensions are 5 μm in diameter with a 20 μm inter-TSV pitch), which in turn limit the density of vertical connections between vertical layers (for a thorough discussion of today's 3D integration, the interested reader can turn to numerous works: [20–23]) [24].

   An alternative technological option to today's conventional 2.5D and 3D integration is monolithic 3D integration. To perform monolithic 3D integration, each vertically stacked layer of the 3D IC is fabricated *directly over* the previously fabricated layers. While the reader might not immediately notice a substantial difference between monolithic 3D integration and chip-stacking, there is a key change. As each vertical layer in the monolithic 3D IC is fabricated over the same starting substrate, nano-scale Inter-Layer Vias (ILVs) can be used to connect between the vertical circuit layers (i.e., no TSVs are required). These ILVs can be fabricated with a traditional damascene process (similar to the global metal wiring

---

[2]2.5D integration refers to integration of several 2D chips or dies, which are mounted in a package in a single plane, for instance, over a silicon interposer.

[3]Conventional 3D integration refers to integration of several 2D chips or dies, which are mounted vertically over one-another, in different vertical planes.

in chips today), or can leverage a novel interconnect technology such as those discussed previously in this book. These nano-scale ILVs have the same pitch and dimensions as tight-pitched metal layer vias, and are therefore orders of magnitude smaller than TSVs [25, 26]. Given the ratio between TSV and ILV pitch, monolithic 3D integration enables massive increase in vertical integration, allowing for orders of magnitudes ($\sim1000\times$) denser vertical connections compared to today's TSV-based 3D ICs. When monolithic 3D integration is used to interleave layers of logic *and* memory, such massive connectivity between layers allows for true computation immersed in memory in an ultra-fine-grained manner, directly translating into an unprecedented logic-memory bandwidth, resulting in significant performance and energy efficiency gains [13].

11.1.1   Challenges for Monolithic 3D ICs

While monolithic 3D integration is a promising technological option, it is *extremely challenging* (if not impossible) to realize with today's silicon-based technologies [12]. Unlike chip-stacking, where the fabrication of the multiple vertical layers are decoupled (i.e., performed on their own substrate, and then stacked), the fabrication of the top vertical layers in a monolithic 3D IC can impact the bottom layers (as the upper layers are fabricated over all of the previously fabricated bottom layers). This imposes stringent limitations for the upper-layer processing. Specifically, the upper layers must be fabricated at a low temperature ($<400\,°C$), as to not damage or destroy the bottom layer transistors (i.e., impact precise dopant profiles or metal-semiconductor junctions) and metal interconnects (which can be destroyed or significantly diffuse at high processing temperatures) [27]. This is referred to as the processing thermal budget (PTB). Silicon CMOS fabrication involves processing temperatures which exceed $1000\,°C$ (for instance, for dopant activation annealing), and thus violates the PTB [28]. While techniques for fabricating silicon CMOS within the PTB have been investigated, these processes suffer for several important limitations: either resulting in sub-par quality transistors, has only been demonstrated for the minimum two layers of stacked transistors, or slightly exceeds the PTB ($>400\,°C$ but $<1000\,°C$) and therefore constrains the back-end-of-the-line (BEOL) processing (such as requiring high temperature and higher resistivity tungsten wires rather than copper) [25, 29–32].

11.1.2   Enabling Monolithic 3D: CNTs and Emerging Nanotechnologies

Unlike silicon CMOS, CNFETs can be fabricated at very low processing temperatures $<250\,°C$, well within the PTB [33]. Therefore, using CNFET logic rather than traditional silicon CMOS is the key to enabling monolithic 3D integration.

It is important to note that for monolithic 3D integration of logic *and* memory, the memory layers must also be fabricated within the PTB. Conventional trench or stacked-capacitor DRAM and Flash are therefore not suitable (moreover, the physical height of the device layers must be small enough to enable dense vias, as the aspect ratio of vertical interconnect wires is finite; stacked-capacitor DRAM and stacked control gate Flash are not suitable for monolithic 3D integration due to this limitation as well) [34, 35]. Therefore, emerging memories technologies, such as spin-transfer torque magnetic RAM (STT-MRAM), resistive RAM (RRAM), and conductive-bridging RAM (CBRAM) are amenable to be integrated as the upper layers of memory [36–38]. By capitalizing on emerging logic and memory technologies to realize monolithic 3D integration, architectural-level benefits are supplemented by benefits gained from the device-level [13]. Therefore, such an approach realizes greater gains than by focusing on improving devices or architectures alone.

Figure 11.1 shows an example monolithic 3D IC enabled by the device technologies mentioned above. The computing elements and memory access circuitry are built using layers of high-performance and energy-efficient CNFET logic. The memory layers are chosen to best match the properties of the memory technology to the function of the memory subsystem. For instance, in Fig. 11.1, STT-MRAM is used for caches (L2 and higher) and main memory, in order to utilize its fast access

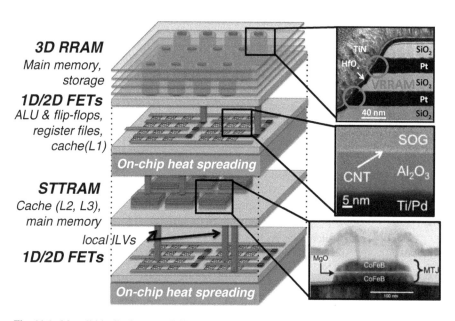

Fig. 11.1 Monolithically integrated 3D computing platform with CNFET-based logic circuits (for computing elements and memory access), STT-MRAM-based caches and main memory, and RRAM-based massive storage. The right-half portion of the figure includes the transmission electron microscopy (TEM) images of the different technologies. TEMs (*top*-to-*bottom*) from [37, 41, 42], schematic from [12]

time, energy retention, and endurance characteristics. RRAM (specifically 3D RRAM) is used for massive on-chip storage to minimize off-chip communication. The various layers of the 3D IC are connected with nano-scale ILVs, permitting massive connectivity between the vertical layers. Additionally, appropriate inter-layer cooling techniques must also be integrated [39, 40].

11.2   CNFETs for Monolithic 3D ICs

A guarded reader would raise the objection that while CNTs might have the right characteristics to enable monolithic 3D ICs (namely their low processing temperature and inherent energy efficiency), one must first be able to realize even traditional 2D CNFET digital systems before attempting 3D systems. Indeed, for many years (around a decade after the first CNFET experimental demonstrations [43, 44]), no digital systems could be fabricated using CNTs. Yet the status-quo of CNFETs has changed drastically in recent years. CNTs have morphed from a long-shot technology to the most advanced emerging nanotechnology for high-performance digital logic. Here we describe the key techniques needed to realize a CNFET digital system. We then discuss how these techniques are applied to realize monolithic 3D CNFET ICs.

11.2.1   CNTs as a Digital Logic Technology

A precursor to realizing monolithic 3D CNFET ICs is the ability to realize functional CNFET digital logic systems. While the challenges and techniques for realizing CNT interconnects have been discussed in length previously, here we give a brief discussion of the obstacles, along with the processing and design solutions, for VLSI CNFET digital logic.

Several significant imperfections inherent to CNTs prevented the demonstration of CNFET VLSI digital systems in the past [45]:

1. Mis-positioned CNTs: It is currently not possible to guarantee exact alignment and positioning of all CNTs on a wafer, especially for large-scale CNFET circuits. The resulting mis-positioned CNTs can cause stray conducting paths inside CNFET circuits, resulting in incorrect logic functionality [46].
2. Metallic CNTs: Due to imprecise control over CNT diameters (and therefore bandgap), some proportion of all grown CNTs are metallic (i.e., CNTs which have little or no bandgap). Contrary to the requirements for CNT interconnects (mainly high packing density of metallic CNTs), metallic CNTs for CNFET logic result in CNFETs with low I_{ON}/I_{OFF}, increased leakage power, and incorrect logic functionality, and are therefore unwanted [47].
3. Carbon nanotube-specific variations: In addition to process variations that exist for silicon FETs (e.g., variations in channel length, oxide thickness, and threshold

voltage), CNFETs are also subject to CNT-specific variations. CNT-specific variations range from variations in CNT type (see Metallic CNTs above), CNT alignment, CNT doping, and local CNT density (due to non-uniform spacing between CNTs resulting from the CNT growth or subsequent processing steps) (details can be found in [45, 48, 49]). These variations can lead to significantly reduced circuit yield, increased susceptibility to noise, and large variations in CNFET circuit delay.

All of these imperfections and variations must be overcome to realize not only functional but also high-performance VLSI CNFET circuits.

11.2.2   Overcoming CNT Obstacles

The set of combined processing and design solutions which overcome these imperfections and variability is referred to as the *imperfection-immune paradigm (IIP)* [45].

To overcome mis-positioned CNTs, highly aligned CNTs are grown on crystalline ST-cut quartz substrates (although this technique has also been shown on alternative substrate types, such as sapphire) [50, 51]. The CNTs grow preferentially along the crystalline plane of the substrate. Such growths can achieve >99.5 % CNT alignment, with CNTs growing >1 mm long (the length is a design decision, as it is determined by the spacing between the lithographically patterned catalyst on the substrate [50]). After growth, the CNTs are transferred to a traditional amorphous SiO_2 substrate through a low temperature (125 °C) transfer method, which maintains the alignment and density of the CNTs [50, 52]. The CNT transfer is also a key to enabling monolithic 3D CNFET ICs, and will be discussed in additional detail shortly.

Yet even 99.5 % CNT alignment is insufficient for VLSI CNFET digital systems, which require billions of CNTs. To render CNFET circuits immune to any remaining mis-positioned CNTs, a design technique, referred to as Mis-positioned CNT-immune Layout Design, is used [46]. Mis-positioned CNT-immune Layout Design relies on etched regions defined within the standard cell, which etches away a pre-defined area of circuit and any CNTs within this region (details in [46, 53]). The positioning of these etched regions ensures that if any mis-positioned CNTs could have caused incorrect logic functionality, part of that mis-positioned CNT must have passed through an etch region, and thus would have been removed from the circuit. Importantly, as the technique is implemented entirely within the standard cell, it does not require any die-specific customization (i.e., no imaging or other characterization is required), a requirement for VLSI process flows and VLSI design methodologies. It has been shown that Mis-positioned CNT-immune Layout Design has significantly smaller area, power, and speed impact compared to traditional redundancy-based defect and fault tolerance techniques, and can be applied to any arbitrary logic functions.

Several techniques exist for overcoming the presence of metallic CNTs. While CNT growth techniques exist for growing preferentially semiconducting CNTs, none achieve >99.9999 % semiconducting CNTs (the requirement for VLSI CNFET digital logic systems) [54]. Therefore, metallic CNTs must be removed post-growth. Several methods exist for removing metallic CNTs post-growth. Many variations of solution-based sorting, whereby CNTs are dispersed and separated by type in solution, have been reported, with varying degrees of success [55, 56]. However, even the most successful solution-based sorting methods have been unable to achieve 99.9999 % pure semiconducting CNT solutions. Moreover, solution-based sorting itself has inherent trade-offs; proponents tout its relative processing ease and economic feasibility, though these come at the expense of having much shorter CNTs (typical solution-based sorting techniques result in short <1 μm long CNTs) with much less alignment than the aligned CNT growths—both of which contribute to increased CNFET variability.

An alternative to solution-based sorting is single device electrical breakdown (SDB) [57]. SDB is performed by using the gate of the transistor to turn off all semiconducting CNTs, followed by applying a large source-drain bias. Under sufficient source-drain biasing, the metallic CNTs (which are not turned off by the gate by virtue of being metallic) heat to >600 °C due to Joule self-heating, at which temperature they are oxidized and therefore most literally "burnt" out of the circuit. Unlike solution-based sorting which achieves selectivity due to, for instance, slight changes in CNT diameter, electrical breakdown leverages the exponential changes in conductance between semiconducting and metallic CNTs. As a consequence, electrical breakdown has experimentally achieved >99.99 % metallic CNT removed with only <4 % inadvertent semiconducting CNT removal—sufficiently selective for VLSI applications (assuming a starting growth purity of 99 % semiconducting CNTs, which has been experimentally demonstrated) [58].

Yet SDB suffers from scalability challenges which limit its applicability to VLSI circuits: not every CNFET source and drain contact can be contacted for a VLSI-scale circuit. Remaining metallic CNT fragments would still cause CNFET source and drain shorts. To overcome this SDB obstacle, a combined processing and design technique, referred to as VLSI-compatible Metallic CNT Removal (VMR), is performed [47]. VMR relies on an additional fabrication step during circuit fabrication, and allows electrical breakdown to be performed at the entire chip scale. Any arbitrary logic function can be implemented with VMR, while incurring minimal impact at the system level with an area overhead of less than 1 %.

The steps for performing VMR are [47]:

1. Patterning inter-digitated electrodes at the minimum lithographic pitch. A subset of these electrodes become the CNFET source and drain contacts for the final CNFET circuit. The fact that the electrodes are patterned at the minimum lithographic pitch is essential; this guarantees that after electrical breakdown, no remaining metallic CNT fragments can short any CNFET source and drain contacts.

2. Electrical breakdown is performed once on the entire VMR structures. A global back-gate (highly doped wafer substrate which acts as the gate for all CNTs) is used to turn off all semiconducting CNTs, and the source-drain breakdown voltage is applied across the VMR structure. The metallic CNTs are not turned off by the gate, and therefore flow sufficient current to breakdown.
3. The final circuit is fabricated: unneeded sections of the VMR structure are removed (the remaining sections of the VMR structure act as the source and drain contacts for the CNFETs in the final circuit design), along with any mis-positioned CNTs (following mis-positioned CNT-immune design). The CNFET fabrication flow continues as normal (i.e., CNFET doping, top-gates, etc.).

It is important to note that significant work today is focusing on continuing to improve metallic CNT removal techniques. For instance, the reader should also note that the above techniques are not mutually exclusive and can be performed in tandem. For instance, VMR can be performed with starting solution-based sorted CNTs, combining the selectivity of both techniques. Recent work has also expanded upon the concept of electrical breakdown, allowing for the removal of the entire metallic CNTs [59, 60]; this can also be combined with VMR design to scale to VLSI CNFET systems. It has been experimentally demonstrated that the process for removing the entire metallic CNT achieves record selectivity of >99.99 % metallic CNT removal versus <1 % semiconducting CNT removal, can be applied to high CNT densities (i.e., CNTs with small inter-CNT spacing, required for high current drive CNFETs), and is applicable to any arbitrarily scaled technology node. Moreover, the area cost for the technique can be <1 % [60].

Variations represent the third major obstacles towards realizing VLSI CNFET systems. As noted previous, in addition to variations that exist for silicon FETs, CNFETs are also subject to CNT-specific variations. These variations translate to large variations in CNFET circuit delays, increased susceptibility to noise, and significantly reduced functional circuit yield [48, 61]. For example, for VLSI CNFET circuits today consisting of >100,000 logic gates, the functional yield can be *near zero* due to the probability of a CNFET having no semiconducting CNTs (resulting from local CNT density variations as well as the probabilistic presence of metallic CNTs and the probabilistic removal of semiconducting CNTs). One naïve method to counteract these variations is to upsize all CNFETs. Such naïve upsizing statistically averages variations, reducing their circuit-level impact. Yet upsizing all CNFETs in a circuit comes at a significant energy and area cost, diminishing the benefits of a CNT technology. Instead, an energy-efficient method of overcoming the circuit-level impact of CNT variations is to *jointly co-optimize* CNT process improvements with CNFET circuit design techniques (e.g., *selective* transistor upsizing techniques, and standard cell layouts for CNFET standard library cells). Using such techniques, the overall speed degradation of VLSI-scale CNFET circuits can be limited to less than 5 %, with less than 5 % energy cost, while simultaneously meeting circuit-level noise immunity and functional yield constraints (e.g., functional yield >99.999 %), which can be achieved for circuit modules with >100,000 logic gates [61].

By overcoming the above inherent CNT imperfections, the first CNFET digital systems have been experimentally demonstrated [58, 62–64]. To date, the most advanced CNFET digital system is a turing-complete microprocessor [33]. While this microprocessor is small by today's standard (implementing the single SUBNEG instruction), it remains a milestone for CNT technologies as it explicitly demonstrated the ability to fabricate any arbitrary digital logic system using CNFETs. To date of this publication, it remains the most complex nanoelectronic system yet realized, and the first (and only) system-level demonstration among promising nanotechnologies for high-performance and highly energy-efficient digital systems.

It is important to briefly note that to realize a CNT technology, multiple axes of the technology space must be advanced. We have focused on integration (realizing functional larger-scale CNFET systems). Yet other, equally important axes exist. For instance, both continued scaling of CNFET geometry, together with achieving high-performance CNFETs, are pivotal goals towards realizing this promising digital logic technology. To these ends, recent work has demonstrated scalability of CNFETs to advanced technology nodes as well as high-performance CNFETs. CNFETs with sub-10 nm channel lengths have been demonstrated [65]; 32 nm channel length-CNFET circuits have been demonstrated [62]; and high current drive CNFETs with high I_{ON}/I_{OFF} which compete with similarly scaled and similarly biased silicon MOSFETs in production in major semiconductor foundries have been demonstrated [64]. Controlled CNT doping, metal-CNT contact resistance, and hysteresis need to be further improved to realize a high-performance and energy-efficient CNFET digital VLSI technology, and recent work has shown progress in these areas [66–70]. For the interested reader, a detailed discussion on current and future work concludes this chapter.

11.2.3 Fabricating Monolithic 3D CNFET ICs

With a maturing CNT technology for traditional 2D circuits, the vision of a monolithic 3D CNFET technology becomes feasible. A process flow for such a monolithic 3D CNFET IC is shown in Fig. 11.2. Each layer of the monolithic 3D CNFET IC follows the same repeated processing steps. These are the same processing steps discussed for realizing a 2D CNFET circuit layer, with minor modifications:

1. Carbon nanotube growth: highly aligned CNTs are grown on a crystalline ST-cut quartz substrate. The CNT growth is a high-temperature step (>850 °C), well above the PTB. Therefore, the CNT growth cannot be performed on the same substrate as the monolithic 3D CNFET IC.

2. Carbon nanotube transfer: the CNTs are transferred through a low temperature (125 °C) layer transfer process from the CNT growth substrate onto the target substrate for the monolithic 3D CNFET IC. This layer transfer is a key step in the monolithic 3D fabrication flow, as it decouples the high temperature

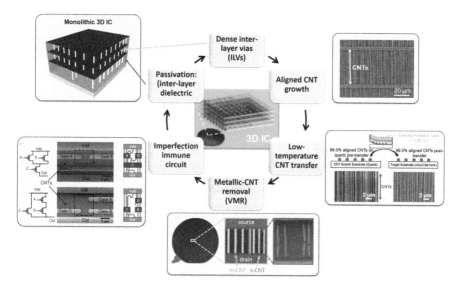

Fig. 11.2 CNFET-based monolithic 3D IC fabrication flow. Processing details are in [41, 73]

CNT growth which exceeds the PTB from the monolithic 3D substrate; that substrate never goes above the transfer temperature of <125 °C [41, 71]. The layer transfer maintains both the density and alignment of the original CNT growth. An additional benefit of the layer transfer is that is can also be performed multiple times onto the same target substrate, effectively combining multiple CNT growths together (thereby achieving higher CNT densities and in turn high CNFET drive currents). It is also important to note to a reader who might be concerned about the practicality of performing layer transfers that layer transfers are already common practice, as the majority of silicon-on-insulator (SOI) wafers in production are manufactured using similar layer transfer concepts.

3. Imperfection-immune paradigm: Mis-positioned CNT-immune Design is applied to each layer to render every CNFET circuit layer immune to mis-positioned CNTs. VMR is also performed on each circuit layer to removed >99.99 % of all metallic CNTs across the 3D stack. A slight modification is made to apply VMR to a monolithic 3D stack. VMR for 2D circuits uses a highly doped silicon substrate to act as a global back-gate to turn off semiconducting CNTs during electrical breakdown. However, the global back-gate has diminishing control over semiconducting CNTs on the upper layers of a monolithic 3D IC. Therefore, a modified version of VMR, 3D-VMR, uses the local bottom-gates of CNFETs on each layer to maintain strong gate control, resulting in effective metallic CNT removal on each layer of the monolithic 3D IC [41].

4. Passivation: A thin inter-layer dielectric (ILD) is deposited through low-temperature (<250 °C) chemical vapor deposition (CVD). The substrate is then

smoothed (through various means, such as back-etching or chemical-mechanical polishing (CMP)) to reduce topology from the bottom layer circuits [72].

5. Vertical connections: Conventional nano-scale ILVs are patterned, etched, and filled with metal to form connections between the vertical layers. Because this is monolithic 3D integration, these vias are extremely dense, and can be at the same pitch and dimensions as tight-pitched metal layer vias. Emerging interconnect technologies, such as those discussed previously in this book, should also be leveraged to further improve via density and performance.

6. The process (steps 1–5) is repeated for as many vertical layers of circuits are required.

11.3 Experimental Demonstrations

Monolithic 3D integration leveraging CNTs has extended far beyond concept. To date, several landmark experimental demonstrations have illustrated the feasibility and practicality of realizing monolithic 3D CNFET ICs.

11.3.1 Monolithic 3D CNFET ICs

Initial demonstrations of monolithic 3D CNFET ICs achieved three vertically stacked layers of CNFET logic, using the design and fabrication processes described previously [71]. In addition, this work used CNTs for both CNFETs and interconnects, demonstrating their multiple applications. The authors demonstrated fully complementary and cascadable logic, which could operate correctly with supply voltages scaled all the way down to 0.2 V [41]. These were achieved spanning multiple layers of the monolithic 3D IC, illustrating that vertical integration of CNFETs does not come at the cost of performance. To demonstrate the flexibility of monolithic 3D CNFET integration, both intra-layer 3D logic gates (logic gates implemented on different layers, cascaded to each other through conventional ILVs) and inter-layer 3D logic gates (logic gates consisting of CNFETs split between multiple vertical layers) were built. Transmission electron microscopy (TEM) images showing three vertically stacked layers of CNFET circuits, as well the measured circuit results, are shown in Fig. 11.3.

11.3.2 Hybrid CNFET-Silicon CMOS Monolithic 3D ICs

Beyond monolithic 3D integration of CNFETs, further work demonstrated monolithic 3D integration of CNFETs over a starting silicon CMOS substrate [74]. While the use of a starting silicon CMOS substrate is certainly not necessary

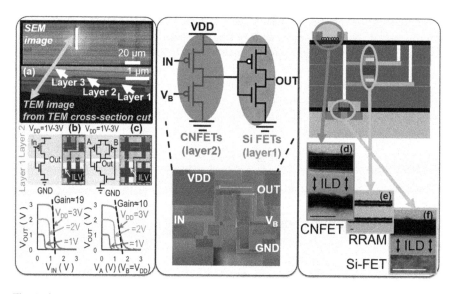

Fig. 11.3 Monolithic 3D experimental demonstrations. *Left*: (**a**) SEM and TEM of a monolithic 3D IC with three vertical layers of CNFET logic circuits. (**b**, **c**) Fully complementary inter-layer CNFET monolithic 3D logic gates. (*top*) Schematics and SEM images of inter-layer INV and NAND logic gates. Conventional nano-scale ILVs instead of TSVs are used to connect the pull-up and pull-down networks on two separate layers. (*bottom*) Menasured voltage transfer curves showing correct operation and gain ≥ 10 for VDD = 3 V all the way to VDD = 1 V. *Middle*: Fully complementary and cascading hybrid CNFET-silicon FET inter-layer logic circuit, demonstrating the monolithic 3D processing and CNFET fabrication is silicon CMOS compatible. *Right*: Monolithic 3D integration of logic and memory, with four vertically stacked layers: bottom layer of silicon CMOS, middle two layers of RRAM memory, and top layer of CNFET logic. *Bottom* images are TEM images of the cross-section of each layer of the monolithic 3D IC

(i.e., the bottom layer could be a layer of high-performance CNFET logic), it does demonstrate an important notion: that the entire monolithic 3D CNFET process flow is silicon CMOS compatible. Silicon CMOS compatibility, while not a technological requirement for a new technology, might very well be an attribute that lowers the barrier for entry for emerging nanotechnologies, due to the dominance of silicon CMOS. As shown in Fig. 11.3, a true hybrid CNFET-silicon CMOS monolithic 3D process can be achieved, with fine-grained monolithic 3D integration of CNFET logic over a silicon CMOS substrate. Fine-grained integration is achieved at the logic gate level, whereby individual fully complementary logic gates are composed of both CNFETs and silicon FETs stacked directly vertically over each other and interconnected using conventional ILVs; integration is also achieved at the circuit-level, with CNFET logic gates on the upper circuit layers cascaded to silicon CMOS logic gates underneath. While the reader might not envision a processor split between vertical layers and between two technologies, the notion of a highly scaled silicon processor on the bottom substrate, followed by periphery logic (such as memory controllers, voltage regulators, or low-power sensor interface circuits) implemented with CNFETs at a lagging technology node on the upper layers, holds significant promise as a ramp to adopting a beyond-silicon technology.

11.3.3 Monolithic 3D Integration: Logic + Memory

The works above demonstrate the natural ability of CNFETs to realize monolithic 3D ICs. However, while such fine-grained monolithic 3D integration of logic gates has been proposed in literature [75, 76], monolithic 3D integration restricted to logic gates alone has limited benefits. In contrast, the massive connectivity enabled by monolithic 3D integration of logic *and* memory (and memory interface circuits) can directly translate into unprecedented memory bandwidth, which in turn translates into significant improvements in performance and energy efficiency [13].

Such a chip—with logic and memory densely integrated in an interleaving monolithic 3D fashion—has already been experimentally demonstrated, with working circuit demonstrations [73]. A TEM of the monolithic 3D IC is shown in Fig. 11.3. There are four vertical layers: a bottom-layer of logic, a memory layer, another memory layer, and another logic layer. The four layers directly vertically overlap one another, and are connected using nano-scale ILVs. Similar to the hybrid CNFET-silicon CMOS demonstration discussed above, the bottom layer of logic was implemented with silicon CMOS, while the top layer of logic was implemented with CNFETs. The middle two layers of memory were implemented with RRAM. RRAM was chosen as it is a promising emerging technology for high-capacity storage and BEOL compatible nonvolatile memory. Moreover, the fabrication process for RRAM is low-temperature (<200 °C), and therefore stays within the PTB for monolithic 3D fabrication.

The flexibility of such a monolithic 3D IC was shown by demonstrating arbitrary connectivity between the vertical layers. For instance, *any* element fabricated on *any* of the vertical layers could be connected to form logic gates and circuits, without impacting the circuit's performance. Inherent in that statement is that the performance of both logic and memory elements themselves in the monolithic 3D IC is invariant to their vertical stacking order (i.e., the performance of the silicon FETs, RRAM, and CNFET logic show negligible change before and after the vertical layers are fabricated above them).

To demonstrate the monolithic 3D integration of logic and memory, a complete four-layer routing element of a fully programmable gate array (FPGA) was realized [73]. Based on these works, one can readily envision VLSI digital systems with arbitrary vertical stacking of logic and memory layers, truly embodying computation finely immersed in memory.

11.4 Outlook: Ongoing and Future Work

Recent experimental demonstrations have transformed CNTs from an intriguing emerging nanotechnology to a realistic beyond-silicon technology option. It should be clear that beyond realizing next-generation interconnects and transistors, CNTs can enable revolutionary monolithic 3D architectures, greatly expanding their impact. Despite their exciting (and now ever-more tangible) promise, significant obstacles still remain towards achieving their full potential. We leave the reader with the following major remaining challenges towards realizing a high-performance CNFET technology, while highlighting recent works which present promising paths for overcoming them.

Doping CNTs is necessary for complementary logic, which is essential for a highly energy-efficient technology. Arguably the single largest misconception surrounding CNT technologies is the inability to dope a CNT to be n- or p-type. The reader must remember that an intrinsic material, including a CNT, is simply that—*intrinsic*. CNFETs are commonly p-type today not due to intrinsic properties of the CNTs, but rather due to the commonplace CNFET fabrication techniques which have been developed and subsequently adopted by the broader community (for instance, most CNTs are back-gates and leave the CNTs exposed to air, resulting in p-type doping from the ambient oxygen [77], and most metal contacts to CNTs leverage stable metals such as gold, platinum, or palladium, which have high work-functions resulting in Schottky barriers more transparent to holes than electrons [78]). Several works have shown that by engineering the source and drain contact metal work-functions, both p-type (high work-function metals such as Pd and Pt) and n-type (low work-function metals such as Sc and Y) can be made, with similar performance and yield [79, 80]. Moreover, an alternative doping technique leveraging low work-function metal oxides as the gate dielectrics resulted in long-term air-stable CNFETs that exhibited n-type behavior with high I_{on}/I_{off} of 10^6 and reasonable inverse subthreshold slope of 95 mV/dec [67]. Most importantly, the technique was completely solid-state and wafer-scale, imperative for VLSI technologies. With a foundation of being able to dope either n- or p-type, the next step is to perform controllable doping, as threshold voltage control becomes increasingly important as supply voltage scales for smaller technology nodes. In addition to controlled doping for controlled threshold voltage, CNFET hysteresis must also be understood and then controlled. This will require careful engineering of the CNT-dielectric interface, which therefore should be viewed in the context of these dielectric-based doping techniques.

In addition to controlled CNT doping, metal-CNT contact resistance must be improved. Moreover, as devices continue to scale, contact resistance becomes an increasingly dominant concern for every technology, including silicon CMOS. Several promising techniques, ranging from interfacial layers of graphene between CNTs and metal to contacting the edge instead of the body of a CNT, have been proposed [66, 69]. While no solution today appears to be complete, there is a significant amount of work on contacts for low-dimensional materials which show significant promise to date of this book. Therefore, the reader can be rest assured that contacts are neither an ignored nor insurmountable obstacle.

Moreover, in addition to the above obstacles, realizing a CNT technology requires advancing multiple axes of the technology. Clearly, significant progress has been made in achieving increased levels of integration, and we have summarized several challenges towards achieving high-performance CNFET devices. After each challenge is overcome in turn, the solutions must then be integrated together. This in itself presents its own set of challenges. For instance, doping techniques might not be compatible with techniques for controlling hysteresis. Moreover, all of the different techniques must be seamlessly integrated and compatible with a final device structure that minimizes parasitic capacitances and parasitic resistances—important considerations for system performance. Yet the integration of many disjoint works into a final high-performance technology also gives rise to significant opportunities, which one must not overlook. For instance, when approaching metallic CNT removal, both increasingly selective semiconducting CNT growths can be combined with solution-based processing, which can be combined with electrical breakdown, to achieve a design point not realizable by any technique alone.

While the reader should not trivialize the above challenges, it would be equally naïve to discount the entire technology due to the existence of these challenges. Because despite these future challenges, CNTs still stand unique among all emerging nanotechnologies. No other emerging nanotechnology for high-performance digital logic has achieved the level of integration or performance as CNTs. And the novel applications of CNTs, specifically monolithic 3D integration, are morphing from ideas to reality in near real-time.

Acknowledgements This work was supported in part by STARnet SONIC, the National Science Foundation, the Stanford SystemX Alliance, and the Hertz Fellowship and Stanford Graduate Fellowship for Max Shulaker. We acknowledge Gage Hills, Tony Wu, Rebecca Park, Gregory Pitner, Luckshitha Suriyasena Liyanage, and Professor Eric Pop of Stanford University for fruitful discussions and collaborations. Works by previous generations of former students have laid the foundation for the work described in this chapter.

References

1. Javey A et al (2003) Ballistic carbon nanotube field-effect transistors. Nature 424(6949): 654–657
2. Appenzeller J (2008) Carbon nanotubes for high-performance electronics—progress and prospect. Proc IEEE 96(2):201–211
3. Naeemi A, Sarvari R, Meindl JD (2005) Performance comparison between carbon nanotube and copper interconnects for gigascale integration (GSI). Electron Device Lett IEEE 26(2): 84–86
4. Kreupl F et al (2002) Carbon nanotubes in interconnect applications. Microelectron Eng 64(1):399–408

5. Wong H-SP et al (2011) Carbon nanotube electronics—materials, devices, circuits, design, modeling, and performance projection. In: Electron devices meeting (IEDM), 2011 IEEE international, IEEE, Washington DC, USA
6. Deng J et al (2008) Carbon nanotube transistor compact model for circuit design and performance optimization. ACM J Emerg Technol Comput Syst 4(2):7
7. Chang L (2012) Short course. In: Electron devices meeting (IEDM), 2011 IEEE international, IEEE, Washington DC, USA
8. Wei L et al (2009) A non-iterative compact model for carbon nanotube FETs incorporating source exhaustion effects. In: Electron devices meeting (IEDM), 2009 IEEE international, IEEE, Baltimore, Maryland, USA
9. Lee CS et al (2015) A compact virtual-source model for carbon nanotube FETs in the sub-10-nm regime—part II: extrinsic elements, performance assessment, and design optimization. IEEE Trans Electron Devices 62(9):3070–3078
10. Stanley-Marbell P, Caparros Cabezas V, Luijten R (2011) Pinned to the walls: impact of packaging and application properties on the memory and power walls. In: Proceedings of the 17th IEEE/ACM international symposium on low-power electronics and design, IEEE, Fukuoka, Japan
11. Dally B (2011) Power, programmability, and granularity: the challenges of exascale computing. In: Test conference (ITC), 2011 IEEE international, IEEE, Anaheim, CA, USA
12. Shulaker MM et al (2015) Monolithic 3D integration: a path from concept to reality. In: Proceedings of the 2015 design, automation and test in Europe conference and exhibition, EDA Consortium, Grenoble, France
13. Ebrahimi M et al (2014) Monolithic 3D integration advances and challenges: from technology to system levels. In: SOI-3D-subthreshold microelectronics technology unified conference (S3S), 2014 IEEE, IEEE, San Francisco, CA, USA
14. Dennard RH et al (1974) Design of ion-implanted MOSFET's with very small physical dimensions. Solid State Circuits IEEE J 9(5):256–268
15. Haensch W et al (2006) Silicon CMOS devices beyond scaling. IBM J Res Dev 50(4.5): 339–361
16. Skotnicki T et al (2005) The end of CMOS scaling: toward the introduction of new materials and structural changes to improve MOSFET performance. Circuits Devices Mag IEEE 21(1):16–26
17. Banerjee K et al (2001) 3-D ICs: a novel chip design for improving deep-submicrometer interconnect performance and systems-on-chip integration. Proc IEEE 89(5):602–633
18. Garrou P, Koyanagi M, Ramm P (2014) Handbook of 3D integration: volume 3-3D process technology. Wiley, http://eu.wiley.com/WileyCDA/WileyTitle/productCd-3527332650, subjectCd-PH62.html
19. Lau JH (2011) TSV interposers: the most cost-effective integrator for 3D IC integration. Chip Scale Rev 15(5):23–27
20. Black B et al (2006) Die stacking (3D) microarchitecture. In: Microarchitecture, 2006 MICRO-39, 39th annual IEEE/ACM international symposium on, IEEE, Orlando, Florida, USA
21. Sakuma K et al (2008) 3D chip-stacking technology with through-silicon vias and low-volume lead-free interconnections. IBM J Res Dev 52(6):611–622
22. Ko C-T, Chen K-N (2010) Wafer-level bonding/stacking technology for 3D integration. Microelectron Reliab 50(4):481–488
23. Topol AW et al (2006) Three-dimensional integrated circuits. IBM J Res Dev 50(4.5):491–506
24. Xu Z, Lu J-Q (2013) Through-silicon-via fabrication technologies, passives extraction, and electrical modeling for 3-D integration/packaging. Semicond Manuf IEEE Trans 26(1):23–34
25. Batude P et al (2011) Advances, challenges and opportunities in 3D CMOS sequential integration. In: Electron devices meeting (IEDM), 2011 IEEE international, IEEE, Washington DC, USA
26. Panth S et al (2013) High-density integration of functional modules using monolithic 3D-IC technology. In: Design automation conference (ASP-DAC), 2013 18th Asia and South Pacific, IEEE, Yokohama, Japan

27. Wong S et al (2007) Monolithic 3D integrated circuits. In: VLSI technology, systems and applications, VLSI-TSA 2007, international symposium on, IEEE, Hsinchu, Taiwan
28. Yang F-L et al (2002) 35 nm CMOS FinFETs. In: VLSI technology, digest of technical papers, 2002 symposium on, IEEE, Honolulu, Hawaii
29. Tsai JC (1966) Integrated complementary MOS circuits. In: Electron devices meeting, 1966 international, IEEE, Washington DC, USA
30. Hamaguchi T et al (1985) Novel LSI/SOI wafer fabrication using device layer transfer technique. In: Electron devices meeting, 1985 international, IEEE, Washington DC, USA
31. Shen C-H et al (2013) Monolithic 3D chip integrated with 500 ns NVM, 3 ps logic circuits and SRAM. In: Electron devices meeting (IEDM), 2013 IEEE international, IEEE, Washington DC, USA
32. Yang C-C et al (2013) Record-high 121/62 μA/μm on-currents 3D stacked epi-like Si FETs with and without metal back gate. In: Electron devices meeting (IEDM), 2013 IEEE international, IEEE, Washington DC, USA
33. Shulaker MM et al (2013) Carbon nanotube computer. Nature 501(7468):526–530
34. Hubert A et al (2009) A stacked SONOS technology, up to 4 levels and 6 nm crystalline nanowires, with gate-all-around or independent gates (Φ-Flash), suitable for full 3D integration. In: Electron devices meeting (IEDM), 2009 IEEE international, IEEE, Washington DC, USA
35. Divakauni R et al (2003) SOI stacked DRAM logic. Google Patents
36. Huai Y (2008) Spin-transfer torque MRAM (STT-MRAM): challenges and prospects. AAPPS Bull 18(6):33–40
37. Wong H-SP et al (2012) Metal–oxide RRAM. Proc IEEE 100(6):1951–1970
38. Kund M et al (2005) Conductive bridging RAM (CBRAM): an emerging non-volatile memory technology scalable to sub 20 nm. In: IEEE international electron devices meeting, IEDM Technical Digest, Washington DC, USA
39. Fuensanta M et al (2013) Thermal properties of a novel nanoencapsulated phase change material for thermal energy storage. Thermochim Acta 565:95–101
40. Pop E, Varshney V, Roy AK (2012) Thermal properties of graphene: fundamentals and applications. MRS Bull 37(12):1273–1281
41. Hai W et al (2013) Monolithic three-dimensional integration of carbon nanotube FET complementary logic circuits. In: Electron devices meeting (IEDM), 2013 IEEE international, IEEE, Washington DC, USA
42. Smullen CW et al (2011) Relaxing non-volatility for fast and energy-efficient STT-RAM caches. In: High performance computer architecture (HPCA), 2011 IEEE 17th international symposium on, IEEE, San Antonio, Texas, USA
43. Martel RA et al (1998) Single-and multi-wall carbon nanotube field-effect transistors. Appl Phys Lett 73:2447
44. Tans SJ, Verschueren AR, Dekker C (1998) Room-temperature transistor based on a single carbon nanotube. Nature 393(6680):49–52
45. Zhang J et al (2012) Robust digital VLSI using carbon nanotubes. IEEE Trans Comput Aided Des Integr Circuits Syst 31(4):453–471
46. Patil N et al (2008) Design methods for misaligned and mispositioned carbon-nanotube immune circuits. Comput Aided Des Integr Circuits Syst IEEE Trans 27(10):1725–1736
47. Patil N et al (2009) VMR: VLSI-compatible metallic carbon nanotube removal for imperfection-immune cascaded multi-stage digital logic circuits using carbon nanotube FETs. In: Electron devices meeting (IEDM), 2009 IEEE international, IEEE, Washington DC, USA
48. Hills G et al (2013) Rapid exploration of processing and design guidelines to overcome carbon nanotube variations. In: Proceedings of the 50th annual design automation conference, ACM, Austin, Tx, USA
49. Zhang J et al (2011) Characterization and design of logic circuits in the presence of carbon nanotube density variations. Comput Aided Des Integr Circuits Syst IEEE Trans 30(8): 1103–1113

50. Patil N et al (2008) Integrated wafer-scale growth and transfer of directional carbon nanotubes and misaligned-carbon-nanotube-immune logic structures. In: VLSI Technology, 2008 symposium on, IEEE, Honolulu, Hawaii
51. Xiao J et al (2009) Alignment controlled growth of single-walled carbon nanotubes on quartz substrates. Nano Lett 9(12):4311–4319
52. Shulaker MM et al (2011) Linear increases in carbon nanotube density through multiple transfer technique. Nano Lett 11(5):1881–1886
53. Patil N et al (2011) Scalable carbon nanotube computational and storage circuits immune to metallic and mispositioned carbon nanotubes. Nanotechnol IEEE Trans 10(4):744–750
54. Ding L et al (2009) Selective growth of well-aligned semiconducting single-walled carbon nanotubes. Nano Lett 9(2):800–805
55. Liu J, Hersam MC (2010) Recent developments in carbon nanotube sorting and selective growth. MRS Bull 35:315–321
56. Arnold MS et al (2006) Sorting carbon nanotubes by electronic structure using density differentiation. Nat Nanotechnol 1(1):60–65
57. Collins PG, Arnold MS, Avouris P (2001) Engineering carbon nanotubes and nanotube circuits using electrical breakdown. Science 292(5517):706–709
58. Shulaker MM et al (2014) Sensor-to-digital interface built entirely with carbon nanotube FETs. *IEEE J Solid State Circuits* 49(1):190–201
59. Jin SH et al (2013) Using nanoscale thermocapillary flows to create arrays of purely semiconducting single-walled carbon nanotubes. Nat Nanotechnol 8(5):347–355
60. Shulaker MM, Hills G, Wu TF, Bao Z, Wong HSP, Mitra S (2015) Efficient metallic carbon nanotube removal for highly-scaled technologies. IEEE International Electron Devices Meeting (IEDM) 32–4
61. Hills G et al Rapid co-optimization of processing and circuit design to overcome carbon nanotube variations
62. Shulaker MM et al (2014) Carbon nanotube circuit integration up to sub-20 nm. *ACS Nano* 8(4):3434–3443
63. Shulaker M et al (2013) Experimental demonstration of a fully digital capacitive sensor interface built entirely using carbon-nanotube FETs. In: Solid-state circuits conference digest of technical papers (ISSCC), 2013 IEEE international, IEEE, San Francisco, CA, USA
64. Shulaker MM et al (2014) High-performance carbon nanotube field-effect transistors. In: Electron devices meeting (IEDM), 2014 IEEE international, IEEE, Washington DC, USA
65. Franklin AD et al (2012) Sub-10 nm carbon nanotube transistor. Nano Lett 12(2):758–762
66. Chai Y et al (2012) Low-resistance electrical contact to carbon nanotubes with graphitic interfacial layer. Electron Devices IEEE Trans 59(1):12–19
67. Suriyasena Liyanage L et al (2014) VLSI-compatible carbon nanotube doping technique with low work-function metal oxides. Nano Lett 14(4):1884–1890
68. Franklin AD, Chen Z (2010) Length scaling of carbon nanotube transistors. Nat Nanotechnol 5(12):858–862
69. Cao Q et al (2015) End-bonded contacts for carbon nanotube transistors with low, size-independent resistance. Science 350(6256):68–72
70. Park RS, Shulaker MM, Hills G, Suriyasena Liyanage L, Lee S, Tang A, Mitra S, Wong HSP (2016) Hysteresis in carbon nanotube transistors: measurement and analysis of trap density, energy level, and spatial distribution. ACS Nano 10(4):4599–4608. doi:10.1021/acsnano.6b00792, Publication Date (Web): March 22, 2016 (Article)
71. Wei H, et al (2009) Monolithic three-dimensional integrated circuits using carbon nanotube FETs and interconnects. In: Electron devices meeting (IEDM), 2009 IEEE international, IEEE, Springer-Verlag Berlin Heidelberg, ISBN 978-3-540-43181-7, Washington DC, USA
72. Oliver MR (2004) Chemical-mechanical planarization of semiconductor materials, vol. 69. Springer, Springer-Verlag Berlin Heidelberg, ISBN 978-3-540-43181-7
73. Shulaker MM et al (2014) Monolithic 3D integration of logic and memory: carbon nanotube FETs, resistive RAM, and silicon FETs. In: Electron devices meeting (IEDM), 2014 IEEE International, IEEE, Washington DC, USA

74. Shulaker MM et al (2014) Monolithic three-dimensional integration of carbon nanotube FETs with silicon CMOS. In: VLSI technology (VLSI-technology): Digest of technical papers, 2014 symposium on, IEEE, Honolulu, Hawaii
75. Bobba S et al (2011) CELONCEL: Effective design technique for 3-D monolithic integration targeting high performance integrated circuits. In: Proceedings of the 16th Asia and South Pacific design automation conference, IEEE, Yokohama, Japan
76. Lee Y-J, Lim SK (2013) Ultrahigh density logic designs using monolithic 3-D integration. Comput Aided Des Integr Circuits Syst IEEE Trans 32(12):1892–1905
77. Collins PG et al (2000) Extreme oxygen sensitivity of electronic properties of carbon nanotubes. Science 287(5459):1801–1804
78. Heinze S et al (2002) Carbon nanotubes as Schottky barrier transistors. Phys Rev Lett 89(10):106801
79. Shahrjerdi D et al (2013) High-performance air-stable n-type carbon nanotube transistors with erbium contacts. ACS Nano 7(9):8303–8308
80. Ding L et al (2009) Y-contacted high-performance n-type single-walled carbon nanotube field-effect transistors: scaling and comparison with Sc-contacted devices. Nano Lett 9(12): 4209–4214

CPSIA information can be obtained
at www.ICGtesting.com
Printed in the USA
LVHW06s1945290718
585294LV00002B/9/P